Masonry Design and Detailing

About the Author
Christine Beall, NCARB, CCS, is a member of ASTM Committees C-12 (Masonry Mortar), C-15 (Manufactured Masonry Units), and E-6 (Building Performance). She has more than 30 years' experience in the design, specification, construction, and inspection of residential, commercial, institutional, and industrial buildings. Christine is also the author of the McGraw-Hill publications, *Concrete and Masonry Databook* and *Thermal and Moisture Protection Manual.*

Masonry Design and Detailing

Christine Beall, NCARB, CCS

Sixth Edition

New York   Chicago   San Francisco
Lisbon   London   Madrid   Mexico City
Milan   New Delhi   San Juan
Seoul   Singapore   Sydney   Toronto

The McGraw·Hill Companies

Cataloging-in-Publication Data is on file with the Library of Congress.

McGraw-Hill books are available at special quantity discounts to use as premiums and sales promotions, or for use in corporate training programs. To contact a representative please e-mail us at bulksales@mcgraw-hill.com.

Masonry Design and Detailing, Sixth Edition

Copyright © 2012, 2004, 1997, 1993, 1987 by Christine Beall. All rights reserved. Printed in the United States of America. Except as permitted under the United States Copyright Act of 1976, no part of this publication may be reproduced or distributed in any form or by any means, or stored in a data base or retrieval system, without the prior written permission of the publisher.

1 2 3 4 5 6 7 8 9 0   DOC/DOC   1 9 8 7 6 5 4 3 2

ISBN 978-0-07-176639-5
MHID 0-07-176639-1

The pages within this book were printed on acid-free paper.

Sponsoring Editor	**Project Manager**	**Production Supervisor**
Judy Bass	Sheena Uprety, Cenveo Publisher Services	Richard C. Ruzycka
Acquisitions Coordinator	**Copy Editor**	**Composition**
Molly Wyand	Cenveo Publisher Services	Cenveo Publisher Services
Editorial Supervisor	**Proofreader**	**Art Director, Cover**
David E. Fogarty	Cenveo Publisher Services	Jeff Weeks

Information contained in this work has been obtained by The McGraw-Hill Companies, Inc. ("McGraw-Hill") from sources believed to be reliable. However, neither McGraw-Hill nor its authors guarantee the accuracy or completeness of any information published herein, and neither McGraw-Hill nor its authors shall be responsible for any errors, omissions, or damages arising out of use of this information. This work is published with the understanding that McGraw-Hill and its authors are supplying information but are not attempting to render engineering or other professional services. If such services are required, the assistance of an appropriate professional should be sought.

To Star, again

Contents

Preface xv

1 Introduction to Masonry 1

2 Brick, Concrete Masonry Units, and Stone 3
   2.1 Brick 3
       2.1.1 Types of Brick 3
       2.1.2 Brick Sizes and Shapes 7
       2.1.3 Brick Colors and Textures 11
       2.1.4 Brick Absorption 11
   2.2 Concrete Masonry Units 12
       2.2.1 Types of CMUs 13
       2.2.2 CMU Colors and Textures 15
       2.2.3 CMU Coring 18
       2.2.4 CMU Grading and Moisture Content 19
       2.2.5 CMU Sizes and Shapes 19
       2.2.6 CMU Strength 20
       2.2.7 CMU Absorption 21
   2.3 Natural Stone 22
       2.3.1 Granite 22
       2.3.2 Limestone 23
       2.3.3 Marble 24
       2.3.4 Sandstone 25
       2.3.5 Selecting Stone 25
   2.4 Cast Stone 27
   2.5 Adhered Manufactured Stone Masonry Veneer 31
   2.6 Glass Block 32

3 Mortar and Grout 35
   3.1 Mortar Ingredients 35
       3.1.1 Cementitious Materials 35
       3.1.2 Sand 37
       3.1.3 Mortar Admixtures 37
       3.1.4 Coloring Pigments 38
   3.2 Mortar Properties 39
       3.2.1 Workability 40
       3.2.2 Water Retention 40
       3.2.3 Bond Strength 41
       3.2.4 Compressive Strength 42
   3.3 Three Kinds of Mortar 43

		3.3.1	Cement-Lime Mortars	43
		3.3.2	Masonry Cement Mortars	43
		3.3.3	Mortar Cement Mortars	43
		3.3.4	Choosing Cement-Lime or Masonry Cement Mortar	44
	3.4	Mortar Types		44
		3.4.1	Type M Mortar	44
		3.4.2	Type S Mortar	44
		3.4.3	Type N Mortar	45
		3.4.4	Type O Mortar	45
		3.4.5	Type K Mortar	45
		3.4.6	Choosing the Right Mortar Type	45
		3.4.7	Proportion Versus Property Method of Specifying Mortar	46
	3.5	Masonry Grout		50
		3.5.1	Grout Materials	50
		3.5.2	Grout Admixtures	51
		3.5.3	Grout Properties	52
		3.5.4	Grout Types and Proportions	52
		3.5.5	Self-Consolidating Grout	54
4	**Masonry Accessories**			**55**
	4.1	Metals and Corrosion		55
	4.2	Joint Reinforcement		58
	4.3	Connectors		60
		4.3.1	Ties	62
		4.3.2	Anchors	64
	4.4	Reinforcing Bars and Bar Positioners		66
	4.5	Flashing Materials		68
	4.6	Weeps		70
	4.7	Drainage Accessories		72
	4.8	Flashing and Drainage Systems		73
5	**Masonry Properties**			**75**
	5.1	Durability and Freeze-Thaw Resistance		75
	5.2	Fire-Resistance Characteristics		76
		5.2.1	Fire-Resistance Ratings	77
		5.2.2	UL Ratings	80
		5.2.3	Steel Fireproofing	80
		5.2.4	Compartmentation	80
	5.3	Thermal Properties		81
		5.3.1	Heat Capacity	82
		5.3.2	Added Insulation	82
		5.3.3	Granular Fills	84
		5.3.4	Rigid Board Insulation and Insulation Inserts	84
		5.3.5	Insulation Location	86
	5.4	Vapor and Air Resistance		86

	5.5	Acoustical Properties	87
		5.5.1 Acoustical Characteristics of Masonry	87
		5.5.2 Sound Absorption	88
		5.5.3 Sound Transmission	89
		5.5.4 STC Ratings	89
	5.6	Green Buildings	90
		5.6.1 Resource Management, Recycled Content, and Embodied Energy	94
		5.6.2 Construction Site Operations	94
		5.6.3 Indoor Air Quality and Building Ecology	95
		5.6.4 Masonry and LEED	95
	5.7	Masonry Costs	95
		5.7.1 Factors Affecting Cost	96
		5.7.2 Value Engineering	96
6	**Expansion and Contraction**		**99**
	6.1	Movement Characteristics	99
		6.1.1 Brick Expansion and Contraction	99
		6.1.2 Concrete Masonry Expansion and Contraction	100
		6.1.3 Differential Movement	101
	6.2	Flexible Anchorage	103
	6.3	Movement Joints	103
		6.3.1 Joint Design	106
		6.3.2 Joint Locations	110
		6.3.3 Accommodating Movement Joints in Design	114
7	**Moisture and Air Management**		**121**
	7.1	Definitions	121
	7.2	Moisture Management	122
	7.3	Deflection: Limit Rain Penetration	123
		7.3.1 Wall System Concepts	123
		7.3.2 Masonry Wall Types	125
		7.3.3 Masonry Units and Mortar	127
		7.3.4 Parapets	130
		7.3.5 Coatings and Water Repellents	137
	7.4	Drainage: Prevent Moisture Accumulation	139
		7.4.1 Flashing	140
		7.4.2 Flashing Locations	146
		7.4.3 Weeps	150
	7.5	Drying: Evaporation and Venting	151
	7.6	Controlling Air and Vapor Movement	154
		7.6.1 Water-Resistive Barriers	155
		7.6.2 Vapor Retarders	155
		7.6.3 Air Barriers	156
8	**Single-Wythe Wall Details**		**159**

	8.1	Single-Wythe Concrete Masonry Unit Walls	159
	8.2	Insulation	161
	8.3	Interior Finishes	162
	8.4	Water Penetration Resistance	162
	8.5	Parapet Details	165
	8.6	Roof-to-Wall Details	168
	8.7	Window Head Details	168
	8.8	Window Sill Details	168
	8.9	Base Flashing Details	171
	8.10	Miscellaneous Details	172
9	**Multi-Wythe Wall Details**		**175**
	9.1	Multi-Wythe Walls	175
		9.1.1 Composite Walls	175
		9.1.2 Cavity Walls	176
	9.2	Insulation	178
	9.3	Interior Finishes	179
	9.4	Water Penetration Resistance	179
	9.5	Parapet Details	180
	9.6	Roof-to-Wall Details	183
	9.7	Shelf Angle Details	184
	9.8	Window Details	186
	9.9	Base Flashing Details	187
	9.10	Miscellaneous Details	189
10	**Anchored Veneer Details**		**191**
	10.1	Brick and Concrete Masonry Unit Veneer	191
	10.2	Insulation	196
	10.3	Water Penetration Resistance	197
	10.4	Parapet Details	197
	10.5	Roof-to-Wall Details	199
	10.6	Shelf Angle Details	199
	10.7	Window Details	200
	10.8	Base Flashing Details	201
11	**Adhered Veneer Details**		**205**
	11.1	Adhered Masonry Veneer	205
	11.2	Thin Brick Veneer	205
	11.3	Adhered Manufactured Stone Masonry Veneer (AMSMV)	206
	11.4	Installation Methods	206
	11.5	Water Penetration Resistance	208
	11.6	Parapet Details	209
	11.7	Roof-to-Wall Details	209
	11.8	Window Details	209
	11.9	Wall Base Details	212
	11.10	Miscellaneous Details	213

12 Special Wall Types ... 215
- 12.1 Interior Partitions ... 215
- 12.2 Screen Walls and Fences ... 216
- 12.3 Glass Block Panels ... 231

13 Lintels and Arches ... 237
- 13.1 Lintels ... 237
  - 13.1.1 Determining Loads ... 237
  - 13.1.2 Steel Lintels ... 239
  - 13.1.3 Concrete and Concrete Masonry Lintels ... 242
  - 13.1.4 Reinforced Brick Lintels ... 243
- 13.2 Arches ... 245
  - 13.2.1 Minor Arch Design ... 245
  - 13.2.2 Graphic Analysis ... 245

14 Structural Masonry ... 251
- 14.1 Masonry Structures ... 251
  - 14.1.1 Differential Movement ... 253
  - 14.1.2 Load Distribution ... 253
  - 14.1.3 Beams and Girders ... 255
  - 14.1.4 Connections ... 256
- 14.2 Empirical Design ... 257
- 14.3 Analytical Design ... 261
  - 14.3.1 Unreinforced Masonry ... 266
  - 14.3.2 Reinforced Masonry ... 269
  - 14.3.3 Wind and Seismic Loads ... 272

15 Installation and Workmanship ... 275
- 15.1 Moisture Resistance ... 275
- 15.2 Preparation of Materials ... 276
  - 15.2.1 Material Storage and Protection ... 276
  - 15.2.2 Mortar and Grout ... 277
  - 15.2.3 Masonry Units ... 279
  - 15.2.4 Accessories ... 280
  - 15.2.5 Layout and Coursing ... 282
- 15.3 Installation ... 284
  - 15.3.1 Mortar and Unit Placement ... 284
  - 15.3.2 Flashing and Weep Holes ... 288
  - 15.3.3 Control and Expansion Joints ... 296
  - 15.3.4 Accessories and Reinforcement ... 296
  - 15.3.5 Grouting ... 299
- 15.4 Construction Tolerances ... 303
  - 15.4.1 Masonry Size Tolerances ... 304
  - 15.4.2 Mortar Joints ... 304
  - 15.4.3 Wall Cavity Width Variations ... 305
  - 15.4.4 Grout and Reinforcement ... 306

	15.5	Cold Weather Construction	306
	15.6	Hot Weather Construction	307
	15.7	Moist Curing	309

16 Specifications — 311
- 16.1 Specification Guidelines — 311
  - 16.1.1 Mortar and Grout — 312
  - 16.1.2 Masonry Accessories — 312
  - 16.1.3 Masonry Units — 312
  - 16.1.4 Construction — 313
  - 16.1.5 Quality Control Tests — 313
  - 16.1.6 Sample Panels and Mock-Ups — 314
- 16.2 Specifying with the MSJC Code — 315
  - 16.2.1 General — 316
  - 16.2.2 Products — 317
  - 16.2.3 Execution — 317

17 Quality Assurance and Quality Control — 319
- 17.1 Standard of Quality — 319
- 17.2 Quality Assurance/Quality Control in Masonry — 319
  - 17.2.1 Industry Standards for Masonry — 319
  - 17.2.2 Standards for Clay Masonry Units — 320
  - 17.2.3 Standards for Concrete Masonry Units — 323
  - 17.2.4 Standards for Masonry Mortar and Grout — 324
  - 17.2.5 Standards for Masonry Accessories — 325
  - 17.2.6 Standards for Laboratory and Field Testing — 326
- 17.3 Masonry Submittals — 327
  - 17.3.1 Specifying Submittals — 327
  - 17.3.2 Submittal Procedures — 328
  - 17.3.3 Shop Drawings — 329
  - 17.3.4 Product Data — 330
  - 17.3.5 Samples — 330
  - 17.3.6 Quality Assurance/Quality Control Submittals — 330
  - 17.3.7 Closeout Submittals — 331
- 17.4 Sample Panels and Mock-Ups — 331
  - 17.4.1 Sample Panel — 332
  - 17.4.2 Mock-Up — 332
  - 17.4.3 Grout Demonstration Panel — 334
- 17.5 Field Observation and Inspection — 334
  - 17.5.1 Materials — 334
  - 17.5.2 Construction — 335
  - 17.5.3 Workmanship — 335
  - 17.5.4 Protection and Cleaning — 337
  - 17.5.5 Moisture Drainage — 337

18 Forensic Investigations — 339
- 18.1 Water Leakage — 339
  - 18.1.1 Diagnostic Water Testing — 340
  - 18.1.2 Types of Water Tests — 340

	18.2	Structural Performance	344
	18.3	Cracking	345
A	Glossary		349
B	ASTM Reference Standards		367
		Clay Masonry Units	367
		Cementitious Masonry Units	368
		Adhered Manufactured Stone Masonry Veneer	368
		Natural Stone	368
		Mortar and Grout	368
		Reinforcement and Accessories	369
		Sampling and Testing	369
		Assemblages	371
	Bibliography		373
	Index		377

Preface

This edition of *Masonry Design and Detailing* has been condensed to focus primarily on brick and concrete block masonry. Some of the less commonly used masonry materials such as adobe, terra cotta, and structural clay tile have been deleted, and a new section on adhered manufactured stone masonry veneer has been added.

I have also deleted some of the lesser used information on materials and manufacturing, fireplaces, retaining walls, and masonry paving. This effort has been toward making the book more concise in its coverage of the most popular and widely used masonry systems. A good deal of additional material has been added about water penetration resistance, and the chapters have been reorganized in what I hope will be a more accessible format.

The book is also available in electronic format, which provides the advantage of key word searches to locate information quickly.

CHRISTINE BEALL, NCARB, CCS
Austin, Texas

Masonry Design
and Detailing

CHAPTER 1
Introduction to Masonry

The unwritten record of history is preserved in buildings. The stone and brick of skeletal architectural remains date back long before the beginning of recorded history. Every country in the world has ancient or historic structures that survived because they were built of masonry. Masonry buildings are extremely durable to weathering and not easily damaged by physical forces. Contemporary reinforced masonry also resists the extremes of hurricane and seismic forces.

Stone is the oldest, most abundant, and perhaps the most important raw building material, and brick is the oldest man-made building product. Masonry construction today includes not only quarried stone and clay brick but also a host of other materials. Concrete block, cast stone, structural clay tile, terra cotta, glass block, adobe, and manufactured stone are also part of masonry's color and texture palette. Masonry can be designed as rustic, formal, residential, commercial, industrial, institutional, monumental, and even palatial. Masonry remains popular with both consumers and designers not only because of its durability but also because of its beauty and variety.

Moisture resistance is one of the primary concerns in building design and construction. Preventing water intrusion into buildings is a primary focus, but so is preventing damage to the materials of the building envelope itself.

Contemporary masonry walls are more water permeable than traditional masonry walls because of their relative thinness. Contemporary masonry is also more brittle because of the portland cement that is now used in masonry mortar. As is the case with any material or system used to form the building envelope, the movement of moisture into and through the envelope has a significant effect on the performance of masonry walls. Contemporary masonry systems are designed not with the intent of providing a barrier to water penetration but rather as drainage walls in which penetrated moisture is collected on flashing membranes and drained through a series of weep holes.

Materials, design, and workmanship are all important to the performance of masonry in resisting water penetration and moisture damage. The successful weather resistance of masonry walls depends on several basic requirements:

- Mortar joints must be full.
- Mortar must be compatible with and well bonded to the masonry units or stone.
- An appropriate flashing material must be selected for the expected service life of the building.

- Flashing details must be properly designed and installed at all necessary locations.
- Weep holes must be properly sized and spaced.
- Weep holes must be unobstructed and must provide effective drainage.

With adequate provision for moisture drainage, masonry wall systems provide long-term performance with little required maintenance. The chapters that follow describe materials, design, and workmanship with an eye toward achieving durability and weather resistance as well as adequate structural performance in masonry systems.

CHAPTER 2

Brick, Concrete Masonry Units, and Stone

There are many different types of masonry materials, but the most widely used in the United States are brick, concrete block, and stone. Each has its own aesthetic, economic, and performance characteristics that make it appropriate or desirable for any given project. Despite their many attributes, masonry materials are generally chosen because designers and building owners like the way they look.

2.1 Brick

Although concrete brick, sand-lime brick, and adobe brick can be of the same approximate size and shape as clay brick, it is usually clay brick that comes to mind for most people when we use the term *brick*. The variety of types, textures, and colors available give brick a wide range of style and appearance options from which architects, developers, and building owners may choose. Brick dates back to before recorded history and carries with it a sense of quality, durability, and permanence that appeals on many levels.

2.1.1 Types of Brick

ASTM standards prescribe minimum physical properties for building brick, facing brick, hollow brick, glazed brick, paving brick, and others. The most widely used is facing brick, or *face brick*, which it is commonly called. The requirements of ASTM C216, *Standard Specification for Facing Brick*, are related primarily to strength, durability, and resistance to weathering. The minimum compressive strength values listed in *Fig. 2-1* are substantially exceeded by most manufacturers. For standard-run extruded brick, average strengths for U.S.-made brick range from 4500 to 8000 psi.

Where the ASTM C216 Weathering index is greater than 50, Grade SW (severe weathering) is recommended for all applications. Grade MW (moderate weathering) may be used only in areas where the Weathering Index is less than 50. Grade NW (no weathering) is permitted only for interior work where there will be no weather exposure. The Weathering Index for almost all of the United States is above 50, so for all intents and purposes, Grade SW is used almost universally.

Face brick is typically selected for specific aesthetic criteria such as color, dimensional tolerances, uniformity, surface texture, and limits on the amount of cracks and defects that are permitted. Within each of the three weathering grades (SW, MW, and NW), face brick can be produced in three specific appearance types. Most of the brick used in commercial buildings in the United States is Type FBS (standard). Type FBX

Unit	Weathering Grade	Minimum Compressive Strength, Gross Area (psi)		Maximum Water Absorption by 5-Hour Boiling (%)	
		Average of 5 Tests	Individual Unit	Average of 5 Tests	Individual Unit
Face brick (ASTM C216)	SW	3000	2500	17	20
	MW	2500	2200	22	25
Building brick (ASTM C62)	SW	3000	2500	17	20
	MW	2500	2200	22	25
	NW	1500	1250	no limit	no limit
Hollow brick (ASTM C652)	SW	3000	2500	17	20
	MW	2500	2200	22	25
Glazed brick (ASTM C1405)	Exterior	6000	5000	—	7 (cold water)
	Interior	3000	2500	—	—
Glazed brick (ASTM C126)	—	3000	2500	—	—
	—	2000	1500	—	—

FIGURE 2-1 ASTM physical requirements for brick. (Copyright ASTM International, 100 Barr Harbor Drive, West Conshohocken, PA 19428. Reprinted with permission.)

(select) is for applications such as stack bond patterns where a crisp, linear effect and minimum size variation are desired. Type FBA (architectural) is manufactured with characteristic effects, such as distinctive irregularities in the individual units to simulate historic or hand-made brick (*see Figs. 2-2 and 2-3*). Extruded brick can be produced in any of the three types by progressively increasing the amount of texturing and roughening applied to the units. Types FBS, FBX, and FBA all meet the same physical requirements for strength and durability. Brick color and texture are not covered in

Type FBS Brick   Type FBA Brick   Type FBX Brick

Brick Types for Appearance

Type FBS and HBS	Brick for general use in masonry (traditional or contemporary styles of architecture)
Type FBX and HBX	Brick for general use in masonry where a higher degree of precision and lower permissible variation in size than permitted for Type FBS is required (crisp, linear, contemporary styles of architecture or stack bond patterns)
Type FBA and HBA	Brick for general use in masonry, selected to produce characteristic architectural effects resulting from non-uniformity in size and texture of the individual units (rustic styles of architecture)

FIGURE 2-2 Brick appearance Types.

	Tolerances on Dimensions				
	Maximum Permissible Variation (inches) Plus or Minus from				
	Specified Dimension	Specified Dimension		Average Brick Size in Job Lot Sample§	
Specified Dimension (in.)	ASTM 62	ASTM C216 Type FBX	ASTM C216 Type FBS	ASTM C216 Type FBX Smooth†	ASTM C216 Type FBX Rough‡
3 and under	3/32	1/16	3/32	1/16	3/32
Over 3 to 4	1/8	3/32	1/8	3/32	1/8
Over 4 to 6	3/16	1/8	3/16	3/32	3/16
Over 6 to 8	1/4	5/32	1/4	1/8	1/4
Over 8 to 12	5/16	7/32	5/16	3/16	5/16
Over 12 to 16	3/8	9/32	3/8	1/4	3/8

§ Lot size as determined by agreement between purchaser and seller. If not specified, lot size is understood to mean all brick of one size and color in the job order.

† Type FBS Smooth units have relatively fine texture and smooth edges, including wire cut surfaces.

‡ Type FBS Rough units have textured, rounded, or tumbled edges or faces.

FIGURE 2-3 Brick size tolerances. (Copyright ASTM International, 100 Barr Harbor Drive, West Conshohocken, PA 19428. Reprinted with permission.)

ASTM C216. Specific colors, color blends, and textures must be selected or specified for individual projects based on local or regional availability.

Glazed brick is fired with ceramic coatings that fuse to the clay body in the kiln and produce an impervious surface in clear or color, matte or gloss finish. Most glazed brick is fired at high temperatures in a single operation while firing the brick body itself. These are covered by ASTM C1405, *Standard Specification for Glazed Brick (Single Fired, Solid Brick Units)*. Some color glazes such as bright reds, yellows, and oranges must be fired at lower temperatures, which require a two-stage process. First, the brick is fired at normal kiln temperatures, then the glaze is applied and the units are fired again at a lower temperature. This double-fired process significantly increases the cost of the brick and usually limits such colors to accents and specialty applications. Some low-fired glazes are prone to crazing because they are not as hard as high-fired glazes. Standards for double-fired glazed brick are covered in ASTM C126, *Standard Specification for Ceramic Glazed Structural Clay Facing Tile, Facing Brick, and Solid Masonry Units*. Durability and weather resistance are not covered, so for exterior use, the body of the brick should be specified to conform to the requirements for ASTM C216 face brick, Grade SW, Type FBX, with the glaze in accordance with ASTM C126 standards.

Requirements for both ASTM C1405 and C126 glazed brick include unit strength and durability as well as properties of the glaze itself. Units are defined as Grade S (select) and Grade SS (select-sized, or ground edge), where a high degree of mechanical perfection, narrow color range, and minimum variation in size are required. Units may be either Type I, single-faced, or Type II, double-faced (opposite faces glazed). Type II units are generally special-order items and are not widely used. For weathering, units are designated as *Exterior Class* or *Interior Class*. Glazed brick may suffer severe freeze-thaw damage in cold climates if not adequately protected from moisture permeation. Glazed brick is not recommended for copings or other horizontal surfaces in any climate. Glazed brick is commonly available in several sizes and in stretchers, jambs, corners, sills, and other supplementary shapes (*see Fig. 2-4*).

Thin brick is similar to face brick except that it is only ½- to 1-in. thick. Thin bricks are made from fired clay or shale and are used as adhered veneer for both interior and exterior applications. There are also thin "brick" veneer units that are cast from a cementitious mix and made to look like clay brick (see Section 2.5), but they are very different products.

Standards for thin brick material properties and characteristics are covered in ASTM C1088, *Thin Veneer Brick Units Made From Clay or Shale*. The standard covers two

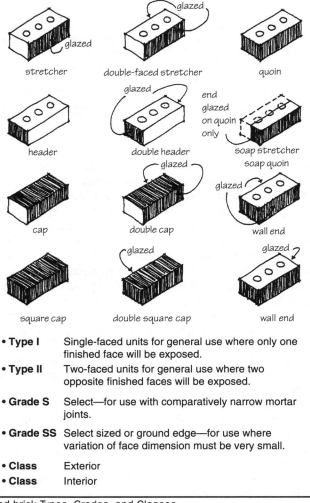

- **Type I**   Single-faced units for general use where only one finished face will be exposed.
- **Type II**   Two-faced units for general use where two opposite finished faces will be exposed.
- **Grade S**   Select—for use with comparatively narrow mortar joints.
- **Grade SS**   Select sized or ground edge—for use where variation of face dimension must be very small.
- **Class**   Exterior
- **Class**   Interior

FIGURE 2-4   Glazed brick Types, Grades, and Classes.

weathering or exposure grades designated as *Exterior* and *Interior*. Three appearance types—TBS, TBX, and TBA—are the equivalent of the face brick types discussed earlier. Thin brick is ideal for interior applications and for cladding or recladding the exteriors of buildings that do not have a supporting brick ledge. Thin brick can also be laid in concrete forms and cast as part of a precast element or tilt-up concrete wall or made into wall panels. Building codes set the maximum allowable weight for exterior, adhered masonry veneer at 15 lb/ft². The thickness of adhered masonry veneers must be between ¼ and 2-5/8 in., with a maximum dimension of 36 in. for either direction and a maximum face area of 5 sq ft.

2.1.2 Brick Sizes and Shapes

Face brick is available in widths or bed depths ranging from 3 to 12 in., heights from 2 to 8 in., and lengths of up to 16 in. Production includes both nonmodular and modular sizes conforming to the 4-in. grid system of structural and material coordination. Some typical units are illustrated in *Figs. 2-5 and 2-6*, which list several of the modular sizes, their recommended joint thicknesses, and coursing heights. Nominal dimensions

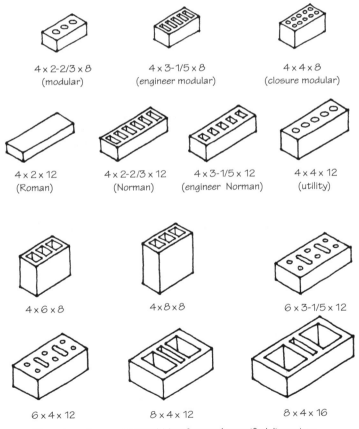

Dimensions shown are nominal. Manufacturer's specified dimensions are usually 3/8" less than the nominal dimension.

FIGURE 2-5 Examples of modular brick sizes.

Modular Brick Sizes and Coursing

Unit Designation	Nominal Dimensions (in.) W	H	L	Joint Thickness (in.)	Specified Dimensions (in.) W	H	L	Vertical Coursing
Modular	4	2-2/3	8	3/8 1/2	3-5/8 3-1/2	2-1/4 2-1/4	7-5/8 7-1/2	3 courses = 8 in.
Engineer modular	4	3-1/5	8	3/8 1/2	3-5/8 3-1/2	2-3/4 2-13/16	7-5/8 7-1/2	5 courses = 16 in.
Closure modular	4	4	8	3/8 1/2	3-5/8 3-1/2	3-5/8 3-1/2	11-5/8 11-1/2	1 course = 4 in.
Roman	4	2	12	3/8 1/2	3-5/8 3-1/2	1-5/8 1-1/2	11-5/8 11-1/2	2 courses = 4 in.
Norman	4	2-2/3	12	3/8 1/2	3-5/8 3-1/2	2-1/4 2-1/4	11-5/8 11-1/2	3 courses = 8 in.
Engineer Norman	4	3-1/4	12	3/8 1/2	3-5/8 3-1/2	2-3/4 2-13/16	11-5/8 11-1/2	5 courses = 16 in.
Utility	4	4	12	3/8 1/2	3-5/8 3-1/2	3-5/8 3-1/2	11-5/8 11-1/2	1 course = 4 in.
	4	6	8	3/8 1/2	3-5/8 3-1/2	5-5/8 5-1/2	7-5/8 7-1/2	2 courses = 12 in.
	4	8	8	3/8 1/2	3-5/8 3-1/2	7-5/8 7-1/2	7-5/8 7-1/2	1 course = 8 in.
	6	3-1/5	12	3/8 1/2	5-5/8 5-1/2	2-3/4 2-13/16	11-5/8 11-1/2	5 courses = 16 in.
	6	4	12	3/8 1/2	5-5/8 5-1/2	3-5/8 3-1/2	11-5/8 11-1/2	1 course = 4 in.
	8	4	12	3/8 1/2	7-5/8 7-1/2	3-5/8 3-1/2	11-5/8 11-1/2	1 course = 4 in.
	8	4	16	3/8 1/2	7-5/8 7-1/2	3-5/8 3-1/2	15-5/8 15-1/2	1 course = 4 in.

FIGURE 2-6 Modular brick size and coursing table.

include the actual size of the unit plus the thickness of mortar joint with which the unit was designed to be used. Face brick is laid with a 3/8- or ½-in. joint, and glazed brick is laid with a ¼-in. joint.

Brick core designs vary with the manufacturer. Coring is designed, among other things, to ease forming and handling and to improve grip and mortar bond. Cored brick is considered "solid brick" as defined by ASTM only if the voids do not exceed 25% of the area in the bearing plane (*see Fig. 2-7*). Most contemporary masonry units are designed to a 4-in. module to connect at 8- or 16-in. course heights. For example, two courses of concrete block with mortar joints is 16 in. vertically, which is the same height as three, five, or six courses of various size brick. This permits horizontal mechanical connection between the facing and backup wythes of a multi-wythe wall at the same vertical intervals.

Brick, Concrete Masonry Units, and Stone 9

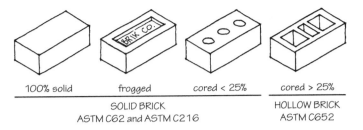

FIGURE 2-7 Solid brick and hollow brick.

In addition to the rectangular cut, brick is available in many special shapes for specific job requirements. Some of the more common shapes include square and hexagonal pavers, bullnose and stair tread units, caps, sills, special corner brick, and wedges for arch construction (*see Fig. 2-8*). The color of special-shape bricks may not exactly match the standard units of the same color because the special-shape bricks are typically fired

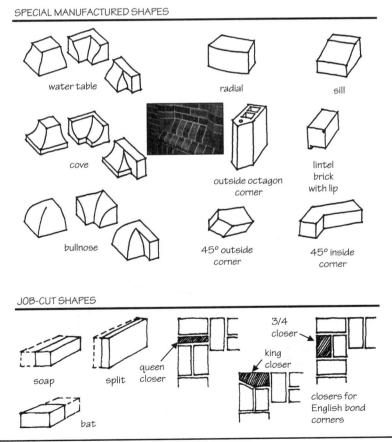

FIGURE 2-8 Job-cut and special manufactured brick shapes.

in a different run. The variations are usually minor but can be particularly noticeable at building corners and other vertical elements. Color variations in horizontal courses of special units blend into a wall better or, at worst, create a banding effect.

The lipped lintel brick shown in *Fig. 2-8* is made for use at window and door lintels to hide the front edge of the steel angle. When used at window or door lintels, lipped brick will not align with the mortar bed joints to either side. Lintel brick is sometimes also used at shelf angles, but detailing, coursing, and installation can be challenging, especially at corners. Only manufactured lip brick should be used. Field-cut brick is more prone to breakage because of overcutting.

The most unusual examples of customized masonry are sculptured pieces hand-crafted from the green clayware before firing. The unburned units are firm enough to allow the artist to work freely without damage to the brick body, but sufficiently soft for carving, scraping, and cutting. After execution of the design, the units are returned to the plant for firing, and the relief is permanently set in the brick face (*see Fig. 2-9*).

ASTM C652 covers hollow brick with core areas between 25 and 40% (Class H40V) and between 40 and 60% (Class H60V) of the gross cross-sectional area in the bearing plane (*see Fig. 2-10*). The two grades listed correspond to the same measure of durability as that used for building brick and face brick: Grade SW and Grade MW. The requirements for appearance Types HBX (select), HBS (standard), and HBA (architectural) are identical to those for face brick Types FBX, FBS, and FBA. Another fourth hollow brick type, HBB, is for general use in walls and partitions where color and texture are not a consideration and greater variation in size is permissible. Typical hollow brick sizes range from $4 \times 2\frac{1}{4} \times 12$ in. to $8 \times 4 \times 16$ in.

Figure 2-9 Sculptured brick is carved by the artist before the brick is fired. (Photo courtesy BIA.)

Brick, Concrete Masonry Units, and Stone

- Class H40V—hollow brick with voids 25 to 40%
- Class H60V—hollow brick with voids 40 to 60%

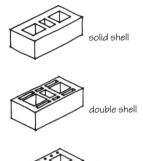

Nominal Width of Units	Class H60V Hollow Brick Minimum Thickness of Face Shells and Webs (in.)		
	Face Shell Thickness		End Shells or End Webs
	Solid	Cored or Double Shell	
3 and 4	3/4	—	3/4
6	1	1-1/2	1
8	1-1/4	1-1/2	1
10	1-3/8	1-5/8	1-1/8
12	1-1/2	2	1-1/8

FIGURE 2-10 ASTM C652 hollow brick. (Copyright ASTM International, 100 Barr Harbor Drive, West Conshohocken, PA 19428. Reprinted with permission.)

2.1.3 Brick Colors and Textures

Brick is available in a wide variety of colors depending on the clay or shale that is available in a given area. Natural colors can be altered or augmented by the introduction of various minerals in the mix and further enhanced by application of a clear, lustrous glaze. Ceramic glazed finishes range from bright primary colors to more subtle earth tones in solid, mottled, or blended shades. Glossy, matte, and satin finishes, as well as applied textures, add other aesthetic options.

2.1.4 Brick Absorption

The absorption of brick is defined as the weight of water taken up by the unit under given laboratory test conditions and is expressed as a percentage of the dry weight of the unit. Because highly absorptive brick exposed to weathering can retain a large quantity of moisture in a wall, ASTM standards limit face brick absorption to 17% for Grade SW and 22% for Grade MW units. Most bricks produced in the United States have absorption rates of only 4 to 10%. The durability of brick usually refers to its ability to withstand freezing in the presence of moisture, as this is the most severe test to which it is subjected. Compressive strength and absorption are indicators of freeze-thaw resistance, as a value cannot be assigned specifically for durability.

The suction, or *initial rate of absorption* (IRA), of brick affects the strength and intimacy of bond between units and mortar as well as the watertightness of the joints. Strong, water-resistant brick walls can be constructed with brick of any IRA if the brick is paired with the appropriate mortar or wetted prior to being laid. Optimum bond and minimum water penetration are achieved with ordinary handling and construction practices if the bricks have an IRA between 5 and 25 g·min^{-1}·30 sq in.$^{-1}$ (the approximate area of the bed surface of a modular brick) at the time they are laid (*see Fig. 2-11*). If high-suction brick absorbs excessive water from the mortar too quickly, it can retard cement hydration, weaken the bond, and result in water-permeable joints. Low-suction brick

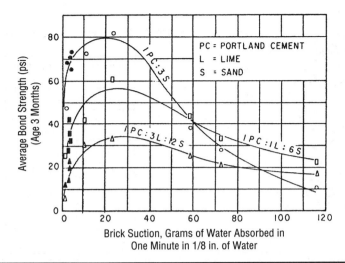

Figure 2-11 Relationship between bond strength and initial rate of absorption (IRA). (From T. Ritchie and J. Davison, Cement-Lime Mortars, National Research Council, Ottawa, Ontario, Canada, 1964.)

may not absorb enough cement paste from the mortar to form a good bond. ASTM standards do not specify IRA limits, but the IRA of a given brick is helpful in deciding which mortar to use or if the brick should be wetted before use. IRA information can be obtained from the brick manufacturer.

High-suction brick with an IRA greater than 30 $g \cdot min^{-1} \cdot 30$ sq in.$^{-1}$ should be thoroughly wetted but surface dry before the units are laid. Wetting can be done with an ordinary water hose and should take place a minimum of 3 hours but preferably 24 hours before use. If wetting the brick is impractical, the units should be laid with a mortar having high water retention (i.e., higher proportion of lime such as ASTM C270, *Mortar for Unit Masonry*, Type N or even Type O). Bricks with a low IRA should be covered at the job site to prevent wetting and used with a mortar with lower water retention (i.e., lower proportion of lime such as ASTM C270 Type S). Mortar for use with low-suction brick should be mixed with the minimum amount of water that will provide good workability.

Both high- and low-IRA brick can be accommodated by choosing a compatible mortar mix and by either wetting or protecting the brick as appropriate to its suction properties. The optimum range of IRA should be included in project specifications with allowance for wetting the brick or adjusting the mortar proportions within the ranges permitted by ASTM C270.

2.2 Concrete Masonry Units

Concrete masonry units (CMU) cure and harden by cement hydration. CMU include concrete brick and block as well as sand-lime brick. Cast stone and manufactured stone are also cementitious products, but they are not typically referred to as CMU. Cast stone and manufactured stone are discussed separately later in this chapter.

2.2.1 Types of CMU

Of the cementitious masonry products marketed in the United States, concrete block is the most familiar and by far the most widely used. Aggregates determine the weight of the block and affect different properties of the units (*see Fig. 2-12*). Lightweight aggregates reduce the weight by as much as 20 to 40% with little sacrifice in strength. Specifications for aggregates are covered in ASTM C33, *Standard Specification for Concrete Aggregates,* and ASTM C331, *Standard Specification for Lightweight Aggregates for Concrete Masonry Units.* There are three CMU weight classifications based on density of the concrete:

- Normal weight, 125 lb/cu ft or more
- Medium weight, between 105 and 125 lb/cu ft
- Lightweight, less than 105 lb/cu ft

All three weight classifications can be used in any type of construction, but lightweight units have higher fire, thermal, and sound resistance but they have lower water penetration resistance. The choice of unit weight will depend on project design

Average Weight of Concrete Masonry Units (lbs)							
Nominal Size (in.)	Concrete Density (pcf)						
	80	90	100	110	120	130	140
4 x 8 x 16	14.5	16.5	18.0	20.0	22.0	23.5	25.5
6 x 8 x 16	17.0	19.0	21.5	23.5	25.5	27.5	30.0
8 x 8 x 16	22.5	25.0	28.0	30.5	33.5	36.0	39.0
10 x 8 x 16	27.5	31.0	34.5	37.5	41.0	44.5	48.0
12 x 8 x 16	31.0	35.0	39.0	43.0	47.0	50.5	54.5

Effect of Aggregate on Weight and Physical Properties						
Classification	Aggregate	Unit Weight of Concrete (pcf)	Average Weight of 8 x 8 x 16 Unit (lb)	Net Area Compressive Strength (psi)	Water Absorption (lb/cu.ft of concrete)	Thermal Expansion Coefficient (per °F x 10^{-4})
Normal weight	Sand and gravel	135	38	2200-3400	7-10	5.0
	Crushed stone	135	38	2000-3400	8-12	5.0
Medium weight	Air-cooled slag	120	34	2000-2800	10-15	4.6
Light weight	Coal cinders	95	27	1300-1800	12-18	2.5
	Expanded slag	95	27	1300-2200	12-16	4.0
	Scoria	95	27	1300-2200	12-16	4.0
	Expanded clay, shale, and slate	85	24	1800-2800	12-15	4.5
	Pumice	75	22	1300-1700	13-18	4.0

Figure 2-12 CMU aggregate type, unit weight, and unit properties.

requirements. There are two kinds of concrete block: ASTM C90, *Standard Specification for Loadbearing Concrete Masonry Units*, and ASTM C129, *Standard Specification for Non-Loadbearing Concrete Masonry Units*.

Concrete brick is produced from a controlled mixture of portland cement and aggregates in sizes, colors, and proportions similar to those of clay brick. Concrete brick is governed by the requirements of ASTM C55, *Standard Specification for Concrete Building Brick*, and can be loadbearing or non-loadbearing. Aggregates include gravel, crushed stone, cinders, burned clay, and blast-furnace slag, producing both normal-weight and lightweight units. Coring, or "frogging," may be used to reduce weight and improve mechanical bond. Grading is based on strength and resistance to weathering. Grade N provides high strength and maximum resistance to moisture penetration and frost action. Grade S has only moderate strength and resistance to frost action and moisture penetration. Concrete mixes may be altered to produce a slight roll or slump when forms are removed, creating a unit similar in appearance to adobe brick. Color is achieved by adding natural or synthetic iron oxides, chromium oxides, or other pigments to the mix, just as in colored mortar. ASTM standards do not include color, texture, weight classification, or other special features. These properties must be covered in the project specifications.

Calcium silicate brick (also called *sand-lime brick*) is made with sand or other siliceous material and 5 to 10% hydrated lime, then steam-cured in high-pressure autoclaves. Calcium silicate brick is produced and used in the United States and widely used in industrial countries where suitable siliceous sands are more readily available than clay. The units have a natural near-white color with a slight yellow, gray, or pink tint, depending on the color of the sand used. With the addition of natural or synthetic pigments, other colors can be produced. Unit surfaces are smooth and uniform—the finer the sand particles, the smoother the surface. Texture is produced by sandblasting, mechanical brushing, or adding flint aggregates to the mix. Splitting hardened units produces a natural rock-face finish.

Most U.S. building codes permit the use of calcium silicate brick in the same manner as clay brick for both loadbearing and non-loadbearing applications. ASTM C73, *Standard Specification for Calcium Silicate Face Brick (Sand-Lime Brick)*, includes grading standards identical to those for clay face brick for severe weathering (Grade SW) and moderate weathering (Grade MW). Compressive strength minimums are 4500 and 2500 psi, respectively, and absorption rates are 10 and 13%, respectively. Strength and hardness are increased as carbon dioxide in the air slowly converts the calcium silicate to calcium carbonate.

Blocks made from *autoclaved aerated concrete* (AAC) have excellent insulating, sound-damping, and fire-resistive properties. A 4-in.-thick wall has a fire rating of 4 hours, and a 6-in. wall has a 6-hour rating. *R*-values are higher than for any other type of masonry. Compressive strength is relatively low, however, and moisture absorption is high. The exterior surface must be protected from wetting by a cladding (such as stucco) or a breathable acrylic coating (minimum 5 perms). The units weigh only one-fourth to one-third as much as normal concrete block, but not because of lightweight aggregates. The mix contains portland cement, lime, sand or fly ash, and aluminum powder, with water added to form a slurry. A chemical reaction causes the concrete to expand and set in cellular form. Smaller blocks are wire-cut or saw-cut from the large forms, and curing is completed under steam pressure in autoclave kilns.

AAC block has little size variation, so it can be laid with standard 3/8-in. mortar joints or with joints that are only 1/8-in. thick. It can be used for bearing walls in low-rise construction, for interior partition walls, as lightweight fireproofing for steel structural frames, and as acoustical partitions. AAC block can be cut or sawed with ordinary woodworking

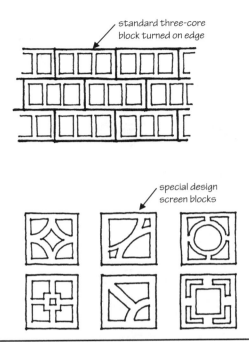

Figure 2-13 Concrete masonry screen block.

tools and is also nailable. Two ASTM standards for AAC have been published, and others are in development. ASTM C1386, *Specification for Precast Autoclaved Aerated Concrete (PAAC) Wall Construction Units,* and ASTM C1452, *Specification for Reinforced Autoclaved Aerated Concrete Elements,* cover both physical properties and testing methods.

Perforated *screen blocks* are available in several patterns and can be used as sun screens, ornamental partitions, and exterior sound baffles for damping low-frequency airborne noise (*see Fig. 2-13*). Ordinary concrete blocks are typically laid with the hollow cores oriented vertically. Screen blocks, however, are laid with the hollow cores oriented horizontally, which yields a lower compressive strength for axial loads. Screen blocks are non-loadbearing, but they must be strong enough to carry their own weight and the weight of the units above them.

2.2.2 CMU Colors and Textures

Standard utility block, or "gray block," is not typically used as an exterior finish. Gray block is most often used as the backing wythe in masonry cavity walls and as interior partitions and foundation walls. If it is exposed to the weather, it should be protected by an acrylic or cement-based paint or stucco finish.

Many decorative effects can be achieved through various CMU surface treatments. Architectural concrete blocks are made in colors, patterns, and textures more suitable for exterior finishes, but they absorb water as readily as gray block. Integral water-repellent admixtures in combination with field-applied water repellents are used to reduce absorption, but these treatments cannot protect against moisture intrusion through cracks or bond line separations at the mortar joints. Any wall with an exterior masonry facing must be designed to effectively drain moisture that penetrates the wall. Architectural block provides a unique appearance, with texture and scale very different from those of brick. The most common architectural blocks are the split-face, ribbed, and burnished (*see Fig. 2-14*).

FIGURE 2-14 Ribbed, split-faced, and burnished concrete block.

Requirements for Facing	
Property or Characteristic	Requirement
Resistance to crazing, cracking, or spalling	None when tested in accordance with specified method
Chemical resistance	No change after testing with specified chemicals
Adhesion of facing	No failure of adhesion of facing material at unit compression test failure
Abrasion resistance	Wear index shall exceed 130 when tested in accordance with specified method
Surface burning characteristics	Flame spread less than 25, smoke density less than 50
Color, tint, and texture	As specified by purchaser, change less than 5 Delta units when tested in accordance with specified method
Soiling and cleansability	Visible stain not to exceed trace when tested in accordance with specified method Spotting media completely removed when tested in accordance with specified method

Manufacturing Tolerances	
Type of Tolerance	Requirement
Size	Total variation from specified dimensions of finished face height and length shall not exceed ± 1/16 in.
Distortion	Distortion of plane and edges of facing from plane and edges of unit, maximum 1/16 in.
Cracking, chippage	Units shall be sound and free of cracks or other defects which interfere with placement or impair strength. Facing shall be free of chips, crazes, cracks, blisters, crawling, holes, and other imperfections which detract from appearance when viewed from a distance of 5 ft. perpendicular to facing surface using daylight without direct sunlight.

FIGURE 2-15 ASTM C744 pre-faced CMU requirements. (Copyright ASTM International, 100 Bar Harbor Drive, West Conshohocken, PA 19428. Reprinted with permission.)

Glazed surfaces may be applied to concrete brick or block and to sand-lime brick. All glazed surfaces must meet the requirements of ASTM C744, *Standard Specification for Prefaced Concrete and Calcium Silicate Masonry Units,* in tests of imperviousness, abrasion, stain resistance, chemical resistance, fire resistance, crazing, and adhesion of facing material to unit (*see Fig. 2-15*). Minimum requirements for strength and abrasion are lower for glazed CMU than for glazed clay masonry units. Like glazed clay units, prefaced CMU combine the functionality of masonry with a hygienic, cleanable surface and a wide palette of color choices. Manufacturing tolerances are only ± 1/16 in. so that narrow joints can be used to minimize mortar exposure.

CMU colors can be varied by using different aggregates, cements, or the integral mixing of natural or synthetic pigments. Penetrating stains may also be applied to the finished wall to achieve a uniform color. These stains are typically mixed with field-applied penetrating water repellents.

2.2.3 CMU Coring

CMU having less than 25% core area in relation to the gross cross section in the bearing plane are classified as "solid." This would include concrete brick and calcium silicate brick. Concrete block typically has 40 to 50% coring and is classified as "hollow." Coring design and percent of solid volume vary, depending on the unit size, the equipment, and the methods of the individual manufacturers. For structural reasons, ASTM standards for loadbearing units specify minimum face shell and web thickness, but these stipulations do not apply for non-loadbearing units.

Block is produced in two-core and three-core designs and with smooth or flanged ends (*see Fig. 2-16*). Two-core designs offer several advantages, including a weight reduction of approximately 10% and larger cores for the placement of vertical reinforcing steel and

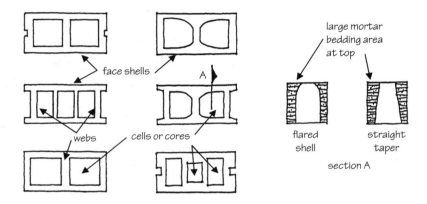

Nominal Width of Unit (in.)	Minimum Face Shell Thickness† (in.)	Web Thickness	
		Minimum Web Thickness† (in.)	Minimum Equivalent Web Thickness‡ (in./linear ft.)
3 and 4	3/4	3/4	1-5/8
6	1	1	2-1/4
8	1-1/4	1	2-1/4
10	1-3/8 1-1/4 §	1-1/8	2-1/2
12	1-1/2 1-1/4 §	1-1/8	2-1/2
Any width unit solidly grouted	5/8	—	—

Minimum Required Face Shell and Web Thickness for Loadbearing (ASTM C90 Only) Concrete Masonry Units

† Average of measurements on three units, taken at thinnest point.
§ Allowable design load must be reduced in proportion to reduction in face shell thickness.
‡ Sum of measured thickness of all webs, multiplied by 12, and divided by length of unit.

(Copyright ASTM International, 100 Barr Harbor Drive, West Conshohocken, PA 19428. Reprinted with permission.)

FIGURE 2-16 Unit coring and minimum face shell thickness.

conduit. Accurate vertical alignment of both two-core and three-core designs is important in grouted and reinforced construction. End designs of block may be smooth or flanged, and some have a mortar key or groove for control joints. Units with a flat face and ends must be used for corner construction, piers, pilasters, and so on. *Minimum face shell thickness* required by ASTM standards for loadbearing units refers to the narrowest cross section. Because compressive strengths of hollow units are established on the basis of gross cross-sectional area and fire-resistance ratings on equivalent solid thickness, these details of unit design become important in determining actual ratings for a particular unit.

2.2.4 CMU Grading and Moisture Content

Concrete block no longer has grade classifications. Two types of concrete block were formerly recognized, based on moisture content of the units. The limits on moisture content for some units were based on efforts to minimize shrinkage cracking. Because moisture content was difficult to control at the construction site, the *National Concrete Masonry Association* (NCMA) has developed new guidelines for crack control joints, and ASTM has eliminated the moisture-type designation from its standards. Refer to Chapter 6 for a discussion on controlling movement and cracking in masonry construction.

2.2.5 CMU Sizes and Shapes

CMU are manufactured to the same 4-in. module as clay masonry products. The basic concrete block size is derived from its relationship to modular brick. A nominal $8 \times 8 \times 16$-in. block is the equivalent of one modular brick in width, two modular bricks in length and three modular brick courses in height. Horizontal ties may be placed at 8- or 16-in. vertical intervals with a CMU backing and brick facing. These are nominal dimensions that include allowance for a standard 3/8-in. mortar joint. Concrete block widths include 4, 6, 8, 10, and 12 in., with a standard face size of 8×16 in. (*see Fig. 2-17*). Half-length and half-high units are available for special conditions at openings, corners, base courses, and so on. A number of special shapes have been developed for specific structural functions, such as lintel blocks, sash blocks, pilaster units, and control joint blocks (*see Fig. 2-18*). Terminology is not fully standardized, and availability will vary, but

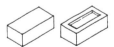

CONCRETE AND CALCIUM SILICATE (SAND LIME) BRICK

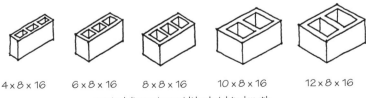

STANDARD CONCRETE BLOCK STRETCHER UNITS

FIGURE 2-17 Basic CMU sizes and shapes.

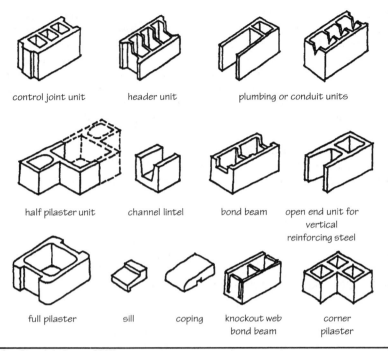

FIGURE 2-18 Special-shape CMU.

most manufacturers produce and stock at least some of the more commonly used special items. In the absence of such shapes, however, standard units can be field cut to accommodate many functions.

Proprietary specialty units include flashing blocks, angled units for making 45° corners and intersections, and special block for laying curved walls. Another proprietary design incorporates channels in the block webs to accommodate reinforcing bars and hold them in place without the need for spacers. Still others offer cornice, sill, and water table units, inspection blocks for grouting, angled keystone blocks for arches, and others (*see Fig. 2-19*). Specialty blocks are usually patented designs and may not be available in all areas.

2.2.6 CMU Strength

CMU physical properties and characteristics fall into a number of structural, aesthetic, and functional categories. Strength and absorption have the greatest influence on overall performance. Compressive strength varies with the type and gradation of the aggregate, the water-cement ratio, and the degree of compaction achieved in molding. In general, the lighter-weight aggregates produce slightly lower strength values and have increased rates of absorption (*see Fig. 2-20*).

Aggregate size and gradation and water-cement ratio affect compaction and consolidation and are important determinants of strength. Higher compressive strengths are generally associated with wetter mixes, but manufacturers must individually determine optimum water-cement ratio to obtain a balance among moldability, handling, breakage, and strength. For special applications, higher-strength units may be obtained

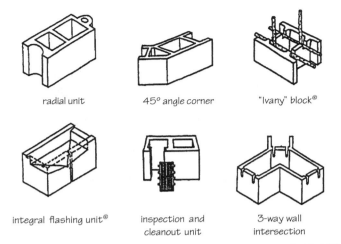

FIGURE 2-19 Special-purpose CMU.

Unit	Minimum Compressive Strength (psi)		Maximum Water Absorption (pcf), Average of 3 Units, Weight Classification—Oven Dry Weight of Concrete		
	Average of 3 Units	Individual Unit	Light-weight (less than 105 pcf)	Medium Weight (105 to 125 pcf)	Normal Weight (125 pcf or more)
Loadbearing CMU (ASTM C90)	1900 (net area)	1700 (net area)	18	15	13
Non-Loadbearing CMU (ASTM C129)	600 (net area)	500 (net area)	—	—	—
Concrete Brick (ASTM C55) Grade N Grade S	3500 (gross area) 2500 (gross area)	3000 (gross area) 2000 (gross area)	15 18	13 15	10 13
Calcium Silicate Brick (Sand Lime Brick) (ASTM C73) Grade SW Grade MW	4500 (gross area) 2500 (gross area)	3500 (gross area) 2000 (gross area)	— —	— —	10 13

FIGURE 2-20 Strength and absorption requirements for CMU. (Copyright ASTM International, 100 Barr Harbor Drive, West Conshohocken, PA 19428. Reprinted with permission.)

from the same aggregates by careful design of the concrete mix and slower curing, increasing net strength ratings to as much as 4000 psi.

2.2.7 CMU Absorption

Water absorption properties are an indication of durability in resistance to freeze-thaw cycles. Highly absorptive units, if frozen when saturated with water, can be fractured by the expanding ice crystals. A drier unit can accommodate some expansion into empty pore areas without damage. Minimum ASTM requirements differentiate between unit weights because of the effect of unit density and aggregate characteristics on this property. Absorption values are measured in pounds of water per cubic foot of concrete. They may range from as little as 4 or 5 lb/cu ft for heavy sand and gravel materials to 20 lb/cu ft for the most porous, lightweight aggregates.

Porosity influences other properties, such as thermal insulation and sound absorption. Pore structure varies for different aggregates and material types and has varying influence on these values and their relationships to one another. Relatively large interconnected pores readily absorb air and sound as well as water but are less likely to suffer damage from freeze-thaw cycles because the freezing water has room to expand. Unconnected or closed pores such as those in structural-grade expanded aggregate offer good insulating qualities and reduced absorption of water and sound. A high initial rate of water absorption, or suction, adversely affects the bond between mortar and unit just as it does in clay masonry. Unlike brick, however, concrete products should never be prewetted at the job site to control suction because of the moisture shrinkage inherent to concrete. Prewetting of CMU can increase shrinkage cracking in the wall. Suction can be controlled to a certain extent through the use of highly water retentive mortars (i.e., maximum proportion of lime), which can enhance the integrity of the bond.

Integral water-repellent admixtures are used in architectural CMU where the exterior weathering face of a building is constructed of masonry without additional finishes or coatings. They can also improve the performance of CMU backing walls that will not be exposed to the weather. If water penetrates past the exterior cladding, the water repellent limits migration of the water through the backing wall to the interior of the building.

Integral water repellents reduce the wicking characteristics of CMU. Increasing the water repellency of the block reduces its ability to extract mortar or grout paste in forming a bond. Research, however, shows that bond is primarily influenced by the mechanical interlock of the mortar or grout to the small voids in the block face. The rougher the unit face, the better the mechanical bond.

Because the admixture manufacturers have no control over the surface texture of the block, they develop admixtures that will maximize performance on any type, weight, or finish of block. Although mortar-to-unit bond is primarily mechanical, it is customary to add a compatible water-repellent admixture to mortar that will be used with treated block. The ingredients and formulas of water-repellent admixtures vary from one manufacturer to the next, so both the block and mortar admixtures should be from the same manufacturer to ensure chemical compatibility.

2.3 Natural Stone

Despite their abundant variety, relatively few types of natural stone are suitable for building. In addition to accessibility and ease of quarrying, the stone must also satisfy the requirements of strength, hardness, workability, porosity, durability, and appearance.

2.3.1 Granite

Some of the natural stones that satisfy the requirements of building construction are granite, limestone, sandstone, slate, and marble. Many others, such as quartzite and serpentine, are used locally or regionally but to a much lesser extent.

Physical Properties and Minimum ASTM Requirements					
Stone Type	Maximum Absorption by Weight, ASTM C97 (%)	Minimum Density, ASTM C97 (%)	Minimum Compressive Strength, ASTM C170 (psi)	Minimum Abrasion Resistance, ASTM C241 (hardness)	Thermal Expansion Coefficient (10^{-6}/°F)
Marble					
ASTM C503					3.69-12.30
I. Calcite	0.20	162	7500	10	
II. Dolomite	0.20	175	7500	10	
III. Serpentine	0.20	168	7500	10	
IV. Travertine	0.20	144	7500	10	
Limestone					
ASTM C568					2.4-3.0
I. Low Density	12.0	110	1800	10	
II. Medium Density	7.5	135	4000	10	
III. High Density	3.0	160	8000	10	
Granite					
ASTM C615	0.40	160	19000	—	6.3-9.0
Quartz-based stone					
ASTM C616					5.0-12.0
I. Sandstone	20.0	135	2000	8	
II. Quartzite Sand-stone	3.0	150	10000	8	
III. Quartzite (Bluestone)	1.0	160	20000	8	
Slate					
ASTM C629					9.4-12.0
I. Exterior	0.25	—	—	8	
II. Interior	0.45	—	—	8	

FIGURE 2-21 Properties of building stone.

Granite is very hard, strong, and durable and is noted for its hard-wearing qualities. Compressive strength ranges from 7700 to 60,000 psi, depending on the source. ASTM C615, *Standard Specification for Granite Dimension Stone,* requires a minimum of 19,000 psi for acceptable performance in building construction (*see Fig. 2-21*). Granite is classified as fine, medium, or coarse grained. Although the hardness of the stone lends itself to a highly polished surface, it also makes sawing and cutting very difficult. Carving or lettering on granite, which was formerly done by hand or pneumatic tools, is now done by sandblasting and can achieve a high degree of precision.

2.3.2 Limestone

Limestone is a sedimentary rock that is durable and easily worked. The compressive strength of limestone varies from 1800 to 28,000 psi, depending on the silica content, and the stone has approximately the same strength in all directions. ASTM C568, *Standard Specification for Limestone Dimension Stone,* classifies limestone in three categories: I (low density), II (medium density), and III (high density), with minimum required compressive strengths of 1800, 4000, and 8000 psi, respectively. Limestone is much softer, more porous, and has a higher water absorption capacity than granite, but is a very attractive and widely used building stone. Although soft when first taken from the ground, limestone weathers hard upon exposure. Its durability is greatest in drier climates.

Limestone textures are graded as A, statuary; B, select; C, standard; D, rustic; E, variegated; and F, old Gothic. Grades A, B, C, and D come in buff or gray and vary in grain from fine to coarse. Grade E is a mixture of buff and gray and is of unselected grain size. Grade F is a mixture of D and E and includes stone with seams and markings.

When quarried, limestone contains groundwater (commonly called *quarry sap*) that includes varying amounts of organic and chemical matter. Gray stone generally contains more natural moisture than does buff-colored stone. As the quarry sap dries and stabilizes, the stone lightens in color and is said to "season." Buff stone does not usually require seasoning beyond the 60 to 90 days it takes to quarry, saw, and fabricate the material. Gray stone, however, may require seasoning for as long as 6 months. If unseasoned stone is placed in the wall, it may be very uneven in color for several months or even as long as a year. No specific action or cleaning procedure will noticeably improve the appearance during this period, nor can it reduce the seasoning time. Left alone to weather, the stone eventually attains its characteristic light neutral color. No water repellents or other surface treatments should be applied until after the stone is seasoned.

Limestone is used as cut stone for veneer, copings, lintels, and sills with either rough or finished faces. Naturally weathered or fractured fieldstone is often used as a rustic veneer on residential and low-rise commercial buildings. Veneer panels may be sliced in thicknesses ranging from 2 to 6 in. and face sizes from 3×5 ft to 5×14 ft.

Travertine is a porous limestone characterized by small pockets or voids. This natural and unusual texturing presents an attractive decorative surface highly suited to facing materials and veneer slabs.

The denser varieties of limestone, including travertine, can be polished and for that reason are sometimes classed as marble in the trade. The dividing line between limestone and marble is often difficult to determine.

2.3.3 Marble

Marble texture is naturally fine, permitting a highly polished surface. The crystalline structure of marble adds depth and luster to the colors as light penetrates a short distance and is reflected back to the surface by the deeper-lying crystals. Pure marbles are white, and Brecciated marbles are made up of angular and rounded fragments embedded in a colored paste or cementing medium. Marble is characterized by grains, streaks, or blotches throughout the stone.

Marble often has compressive strengths as high as 20,000 psi, and when used in dry climates or in areas protected from precipitation, the stone is quite durable. Some varieties, however, are decomposed by weathering or exposure to industrial fumes and are suitable only for interior work. ASTM C503, *Standard Specification for Marble Dimension Stone (Exterior)*, covers four marble classifications, each with a minimum required compressive strength of 7500 psi: I, calcite; II, dolomite; III, serpentine; and IV, travertine. Marbles are classified as A, B, C, or D on the basis of working qualities, uniformity, flaws, and imperfections. For exterior applications, only Group A, highest-quality materials, should be used. The other groups are less durable and will require maintenance and protection. Group B marbles have less favorable working properties than those of Group A and will have occasional natural faults requiring limited repair. Group C marbles have uncertain variations in working qualities; contain flaws, voids, veins, and lines of separation; and will always require some repair (known as sticking, waxing, filling, and reinforcing). Group D marbles have an even higher proportion of natural structural variations requiring repair and have great variation in working qualities.

2.3.4 Sandstone

Sandstone is softer than granite and marble and is categorized by grain size and cementing media.

- Siliceous sandstone is strong and is durable if it is cemented with silica. Many siliceous sandstones contain iron, which is oxidized by acidic pollutants (or acidic cleaners), and turns the stone brown. If the cementing medium is largely iron oxide, the stone is naturally red or brown.
- Ferruginous sandstone is cemented together with iron oxide, so it is naturally red to deep brown in color and is softer and more easily cut than siliceous sandstone.
- Calcareous sandstone is cemented together with calcium carbonate, which is sensitive to acids and can deteriorate rapidly in a polluted urban environment.
- Dolomitic sandstone is cemented together with dolomite, which is more resistant to acid.
- Argillaceous sandstone contains large amounts of clay, which can quickly deteriorate simply from exposure to rain.

ASTM C616, *Standard Specification for Quartz-Based Dimension Stone*, recognizes three classifications of sandstone:

- Type I, sandstone, is characterized by a minimum of 60% free silica content.
- Type II, quartzite sandstone, is characterized by 90% free silica.
- Type III, quartzite, is characterized by 95% free silica content.

As a reflection of these varying compositions, minimum compressive strengths are 2000, 10,000, and 20,000 psi, respectively. Absorption characteristics also differ significantly, ranging from 20% for Type I to 3% for Type II and 1% for Type III. When first taken from the ground, sandstone contains a lot of water, which makes it easy to cut. When the moisture evaporates, the stone becomes considerably harder.

There are both fine and coarse textures of sandstone, some of which are highly porous and therefore low in durability. The structure of sandstone lends itself to textured finishes and to cutting and tooling for ashlar and dimension stone in veneers, moldings, sills, and copings. Sandstone is also used in rubble masonry as fieldstone. Bluestone is a form of sandstone split into thin slabs for flagging.

2.3.5 Selecting Stone

Design and aesthetics will determine the suitability of a particular stone's color, texture, aging characteristics, and general qualities appropriate for the type of building under consideration. Stone for building construction is judged on the basis of

- Appearance
- Durability

- Strength
- Economy
- Ease of maintenance

Stone is used in building construction in two forms: rubble stone and dimension stone. Rubble stone is irregular in size and shape (*see Fig. 2-22*). *Fieldstone rubble* is harvested from fields in its natural form. It is weathered smooth but is irregular and uneven. *Quarried rubble* comes from the fragments of stone left over from the cutting and removal of large slabs at the stone quarry. It has freshly broken faces, which may be sharp and angular. Rubble may be either broken into suitable sizes or roughly cut to size with a hammer. *Dimension stone,* such as ashlar, decorative elements, and thin veneer slabs, is delivered from stone fabricators cut and dressed to a specific size and thickness and squared to dimension each way. Ashlar is a flat-faced dimension stone, generally in small squares or rectangles, with sawed or dressed beds and joints.

Rubble and ashlar dimension stone can be laid in a variety of pattern bonds (*see Figs. 2-23 and 2-24*). Dimension stone is also available in thin slabs. Depending on the type of stone, surface treatments include smooth, rough, or natural split face and textured, polished, honed, or flamed finishes. Carved elements and lettered panels require stones of fine grain to produce and preserve the detail.

For practicality and long-term cost, durability is the most important consideration in selecting building stones today. Stones of the same general type such as limestone, sandstone, and marble differ greatly in durability, depending on their softness and porosity. Suitability will depend not only on the characteristics of the stone but also on local environmental and climatic conditions. Soft, porous stones are more liable to absorb water and to flake or disintegrate from freeze-thaw cycles and may not be suitable in cold, rainy climates. In warm, dry climates, any building stone may be used.

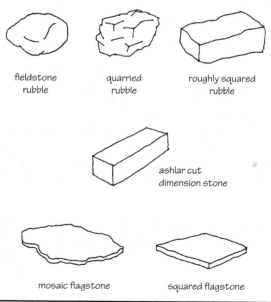

FIGURE 2-22 Building stone is used in several forms.

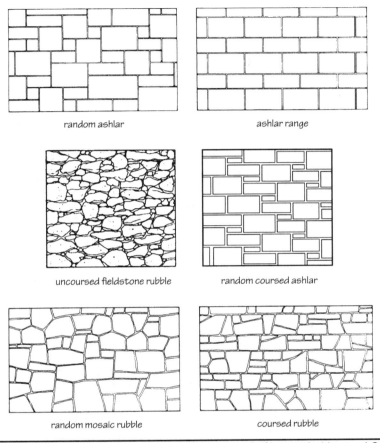

FIGURE 2-23 Stone bonding patterns. (From C. Ramsey and R. Sleeper, Architectural Graphic Standards, 6th edition, ed. Joseph Boaz. John Wiley & Sons, Inc., 1970. Reprinted by permission of John Wiley & Sons, Inc.)

Building stone is relatively volume-stable, returning to its original dimensions after undergoing only slight thermal expansion and contraction. Some fine-grained, uniformly textured, relatively pure marbles, however, retain small incremental volume increases after each heating cycle. Marble becomes less dense when it expands during warm weather but does not return to its prior density during cool weather. This irreversible expansion is called *hysteresis*. In relatively thick veneers, the greater expansion on the exposed exterior surface is restrained or accommodated by the unaffected interior mass. In thin veneers, however, dilation of the exterior surface can easily overcome the restraint of the inner layers, causing a dishing effect.

2.4 Cast Stone

What is the difference between cast stone and precast concrete? Both precast concrete and cast stone are made with aggregates, portland cement, and water and may also contain mineral pigments and chemical admixtures. Precast concrete is typically

FIGURE 2-24 Rubble and ashlar stone pattern bonds.

thought of as large structural elements or wall panels. Functional pieces such as beams or girders are usually left unfinished and have the appearance of ordinary gray concrete. *Architectural precast* is an aesthetic element of colored concrete often cast with textured finishes through the use of special form liners.

Cast stone is a type of precast concrete or CMU manufactured specifically to simulate the appearance of natural cut stone. It can be fabricated in smaller elements to be set in mortar such as sills and copings, but more elaborate decorative pieces can also be made (*see Fig. 2-25*). A high percentage of fine aggregate gives cast stone mixes a consistency that lends itself to simulating the texture of different natural stones and the profiles of hand-carved elements. Cast stone can be precisely colored using white portland cement and selecting aggregates and mineral pigments to simulate the color of different types of stone.

FIGURE 2-25   Cast stone decorative elements.

Cast stone is used in new construction, to replicate original stone elements in restoration projects, and to match existing stone for additions or renovations. Typical cast stone elements include sills, jambs, heads, keystones, door surrounds, balustrades, stair treads and risers, quoins, and columns (*see Fig. 2-26*). Cast stone should meet the minimum requirements of ASTM C1364, *Standard Specification for Architectural Cast Stone*, for permissible color variations and dimensional tolerances.

In addition to a savings in cost over natural cut stone, cast stone has the advantage that it can be fabricated with steel reinforcing if needed for additional strength in specific applications or to facilitate shipping and handling without damage to the units. Reinforcement should not be routinely incorporated into cast stone unless there is a specific need.

ASTM C1364 requires that reinforcement in weather-exposed pieces have a minimum concrete cover of two times the bar diameter. If coverage is less than 1½ in., the reinforcement must be corrosion resistant. Water penetration can occur if there is inadequate concrete cover, and if the steel is not corrosion resistant, the expansive forces of rust formation can accelerate deterioration and failure of cast stone in the same way that it can with precast or cast-in-place concrete (*see Fig. 2-27*). Although stone masonry anchors are typically stainless steel, reinforcing bars are more commonly hot-dip

Chapter Two

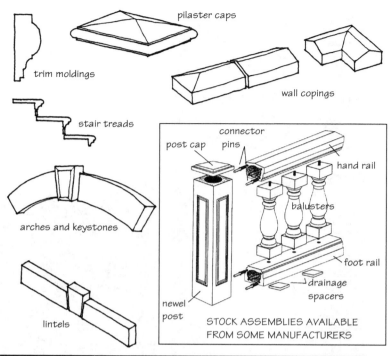

FIGURE 2-26 Typical cast stone elements used with stone and unit masonry.

FIGURE 2-27 The expansive force of rusting steel can accelerate deterioration.

Brick, Concrete Masonry Units, and Stone

unacceptable surface cracking, corrosion of reinforcing steel

FIGURE 2-28 Light surface crazing does not affect integrity or durability, but a porous surface can allow water absorption that corrodes the steel and breaks the stone.

galvanized or epoxy coated when corrosion resistance is required. Corrosion-protected steel should be touched up with zinc or epoxy paint wherever bars have been cut to length for fabrication.

Surface crazing does not affect cast stone durability, and ASTM C1364 stipulates that surface crazing does not by itself constitute a cause for product rejection. Cracking that is more than superficial can allow excessive water penetration and contribute to the deterioration of the cast stone itself and to the wall system and/or building contents (*see Fig. 2-28*).

2.5 Adhered Manufactured Stone Masonry Veneer

Thin, manufactured stone for adhered veneer applications has been called many things including simulated stone, cultured stone, precast stone, faux stone, cast stone, and engineered stone. It is essentially a type of CMU manufactured as a lightweight, non-loadbearing architectural veneer product. The veneer units are cast into random sizes and shapes in a variety of colors simulating the look of undressed quarried stone. Even though they can be made similar in appearance to brick, units of this type are a cementitious product distinctly different from the thin clay brick described earlier.

In an effort to standardize the name and the minimum properties, an ASTM subcommittee has been formed to develop standard specifications and installation guidelines. Although no standard has yet been published, the product has been provisionally titled *"adhered manufactured stone masonry veneer"* (AMSMV) and provisionally defined as being manufactured from a mixture of cements, normal and lightweight aggregates, iron oxide pigments, additives, water, and other components.

The economy of thin stone veneer is in the reduced weight of the product and the corresponding reduction in support and anchorage requirements. Manufactured stone veneer can simulate the appearance of virtually any type of natural stone in color, texture, and pattern. Anything from river rock to rubble stone, to ashlar and dry-stack ledgestone can be produced (*see Fig. 2-29*). Trim and accessory pieces are typically available in matching colors and textures from any given manufacturer. Available items may include keystones and quoins, window sills and surrounds, coping stones, and trim or belt courses. The unit types, colors, textures, and shapes will vary from one manufacturer to the next.

FIGURE 2-29   Examples of adhered manufactured stone masonry veneer.

2.6   Glass Block

Glass block is considered a masonry material because it is typically laid up in mortar. Glass block can be used as security glazing or to produce special daylighting effects. There are a variety of glass block sizes in both solid and hollow form. Decorative blocks are produced in clear, reflective, or color glass with smooth, molded, fluted, etched, or rippled textures. Glass block can diffuse or direct light for different illuminating requirements and provides security and energy efficiency for glazed areas.

The majority of glass block is made of clear glass that admits the full spectrum of natural light (*see Fig. 2-30*). Hollow block with patterns pressed into the interior face distorts images creating visual privacy and can be manufactured to either diffuse or reflect light. Glass block can reduce solar heat gain, and because of their large air cavity, hollow glass blocks have more thermal resistance than that of ordinary flat glass. Solar reflective block is coated on the inside with a metal oxide that can reduce heat gain by as much as 70%.

Glass block comes in nominal face sizes of 6 × 6-, 8 × 8-, and 12 × 12-in. square units and 4 × 8- and 6 × 8-in. rectangular units. Actual dimensions vary by manufacturer and style. Units made in the United States are ¼-in. to 3/8-in. less than nominal dimensions. Most hollow blocks are 3-7/8 in. thick (nominal 4 in.), but some manufacturers also make thin blocks that measure only 3-1/8 in. and weigh 20% less than standard units

Unit Type	Light Transmitted (%)	U-Value	Shading Coefficient	Heat Gain (Btu/hr/sq.ft)
Solid	80	—	—	—
Hollow	50-75	—	—	—
Diffusion	28-40	—	—	—
Reflective	5-20	—	—	—
8 x 8 Reflective	20	0.51	0.25	42
8 x 8 Clear	62	0.51	0.65	140
1/4" Clear sheet glass	90	1.04	1.00	215

FIGURE 2-30 Light transmission and thermal performance of glass block.

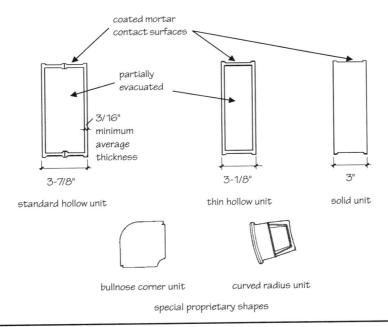

FIGURE 2-31 Glass block sizes and shapes.

(see Fig. 2-31). Solid glass blocks are used for high-security glazing. They come in 3 × 8-in. rectangular units and 8 × 8-in. square units.

The weight of glass block requires limiting the number of courses laid at one time so that fresh mortar is not extruded from the joints before it has begun to set. There are several proprietary types of spacers that help speed construction. Unit weight is transferred directly from block to block by the spacers, allowing work to progress rapidly without waiting for substantial mortar cure to support the weight of the units. Mortar adhesion to glass block is limited.

CHAPTER 3
Mortar and Grout

Although mortar may account for as little as 7% of the volume of a masonry wall, it influences performance far more than the percentage indicates. Mortar adds color and texture to masonry, binds the individual masonry units together, resists air and moisture penetration, and bonds with anchors, ties, and reinforcing to join the building components together. In loadbearing applications, mortar strength and performance are as critical as unit strength and workmanship.

3.1 Mortar Ingredients

There are three basic components in masonry mortar: cementitious materials, aggregates, and water. There are, however, numerous types of cementitious materials that meet the requirements of ASTM C270, *Standard Specification for Mortar for Unit Masonry*. Choices of these materials affect properties of the plastic mortar and the hardened mortar.

3.1.1 Cementitious Materials

Portland Cement

There are five types of portland cement described in ASTM C150, *Standard Specification for Portland Cement*. Each of the five types has different physical and chemical characteristics.

Because the properties required for mortar are significantly different from the qualities called for in concrete, not all of the five types of cement are suitable for masonry construction. For most ordinary mortars, Type I all-purpose portland cement is used. In some instances, such as masonry catch basins or underground drainage structures where mortar may come in contact with sulfates in the soil, Type II portland cement can be used to resist chemical attack. A more common substitute for Type I is Type III, high-early-strength cement. This mixture attains ultimate compressive strength in a very short period of time and generates greater heat during the hydration process. For use in cold weather construction, these properties help keep the wet mortar from freezing and permit a reduction in the period of time required for protection against low temperatures.

Air-entraining portland cement, Types IA, IIA, and so on, is made by adding selected chemicals to produce minute, well-distributed air bubbles in hardened concrete or mortar. Increased air content improves workability, increases resistance to frost action and the scaling caused by chemical removal of snow and ice, and enhances moisture, sulfate, and abrasion resistance. Air-entrained mortar mixes are not as strong as ordinary portland cement mortar mixes, and excessive air is detrimental in mortar because it can impair the bond to masonry units and reinforcing steel.

Air-entrained cements are used primarily in horizontal concrete applications where exposure to ponded water, ice, and snow is greatest. Entrained air produces voids in the concrete into which freezing water can expand without causing damage. Rigid masonry paving applications installed with mortared joints may also enjoy some of the benefits of air-entrained cements in resisting the expansion of freezing water. Although industry standards for masonry mortar limit the air content of mortar, the benefits of higher air content for masonry paving systems in resisting freeze-thaw damage may outweigh the decrease in bond strength. Because rigid masonry paving systems are generally supported on concrete slabs, the flexural bond strength of the masonry is less important than its resistance to weathering. In these applications, lower bond strength might be tolerated in return for increased durability against freezing and thawing.

Other Cements

ASTM C270 permits the use of hydraulic cement, blended hydraulic cement, and portland blast-furnace slag cement, but these are typically used only in laboratory mixed designs.

Lime

The term *lime* when used in reference to building materials means a burned form of lime derived from the calcination of sedimentary limestone. Powdered, hydrated lime is the most common and convenient form of lime used today. Of the two types of hydrated lime permitted by ASTM C207, *Standard Specification for Hydrated Lime for Masonry Purposes*, only Type S is suitable for masonry mortar.

Lime improves workability by helping mortar to permeate tiny surface indentations, pores, and irregularities in the masonry units and develop a strong physical bond. Lime also improves water retention. The mortar holds its moisture longer, resisting the suction of dry, porous units so that sufficient water is maintained for proper curing and development of a good bond. Its slow setting quality facilitates retempering to replace evaporated moisture before the mortar begins to stiffen. Lime undergoes less volume change or shrinkage than other cementitious ingredients so it helps prevent shrinkage cracking in mortar joints. It contributes to mortar integrity and bond through autogenous healing, which is the ability to combine with moisture and carbon dioxide to fill small cracks. Some manufacturers preblend portland cement and lime and sell bagged mixes that require only the addition of sand and water at the job site.

Masonry Cement

Proprietary mixes of cement and workability agents, or masonry cements, are popular with masons because of their convenience and good workability. However, ASTM C91, *Standard Specification for Masonry Cement*, places no limitations on chemical composition, and the ingredients as well as the properties and performance vary widely among the many brands available. Although the exact formula is seldom disclosed by the manufacturer, masonry cements generally contain combinations of portland cement, plasticizers, and air-entraining additives. Finely ground limestone, clay, and lime hydrate are often used as plasticizers because of their ability to adsorb water and thus improve workability. Air-entraining additives protect against freeze-thaw damage and provide some additional workability.

Like all proprietary products, different brands of masonry cements will be of different qualities. Because of the latitude permitted for ingredients and proportioning, the properties of a particular masonry cement cannot be accurately predicted solely on the basis of compliance with ASTM C91. They must be established through performance records and laboratory tests.

Mortar Cement

Mortar cements are also proprietary products, but they must meet higher performance standards than masonry cements. ASTM C1329, *Standard Specification for Mortar Cement*, permits a maximum air content of 16% for mortars made with mortar cement and also prescribes minimum flexural bond strength. Mortar cement mortars are considered to be equivalent to cement-lime mortars in bond and weather resistance.

3.1.2 Sand

Sand accounts for at least 75% of the volume of masonry mortar. Manufactured sands have sharp, angular grains, whereas natural sands obtained from banks, pits, and riverbeds have particles that are smoother and more round. Natural sands produce mortars that are more workable than those made with manufactured sands.

Sand must be clean, sound, and well graded according to requirements set by ASTM C144, *Standard Specification for Aggregate for Masonry Mortar*. The sand in masonry mortar acts as a filler. The cementitious paste must completely coat each particle to lubricate the mix. Sands that have a high percentage of large grains produce voids between the particles and will make harsh mortars with poor workability and low resistance to moisture penetration. When the sand is well proportioned of both fine and coarse grains, the smaller grains fill these voids and produce mortars that are more workable and plastic. If the percentage of fine particles is too high, more cement is required to coat the particles thoroughly, more mixing water is required to produce good workability, and the mortar will be weaker, more porous, and subject to greater volume shrinkage.

3.1.3 Mortar Admixtures

Although admixtures are often used in concrete construction, they can have adverse effects on the properties and performance of masonry mortar. A variety of proprietary admixtures are available that are reported by their manufacturers to increase workability or water retentivity, lower the freezing point, and accelerate or retard the set. Although they may produce some effects, they can also reduce compressive strength, impair bond, contribute to efflorescence, increase shrinkage, or corrode metal accessories and reinforcing steel. If admixtures are permitted to produce or enhance some special property in the mortar, the specifications should require that they meet the requirements of ASTM C1384, *Standard Specification for Modifiers for Masonry Mortar*.

Set accelerators are sometimes used in winter construction to speed cement hydration, shorten setting time, increase 24-hour strength, and reduce the time required for cold weather protective measures. Set accelerators are sometimes mistakenly referred to as "antifreeze" compounds. Set accelerators sometimes include calcium chloride, which can contribute to efflorescence and cause corrosion of embedded steel anchors and reinforcement. Nonchloride accelerators are a little more expensive but less damaging to the masonry. Chlorides should be prohibited in mortars that contain embedded metals such as anchors, ties, or joint reinforcement. *Water-reducing accelerators* increase early strength and ultimate strength by reducing the water-cement ratio needed to produce a workable mix.

Set retarders extend the board life of fresh mortar for as long as 4 to 5 hours by helping to retain water for longer periods of time. Set retarders are sometimes used during hot weather to counteract the effects of rapid set and high evaporation rates. With soft, dry brick or block, set retarders are also sometimes used to counteract rapid suction and help achieve a better bond. Mortar with set retarders cannot be retempered.

Extended-life retarders slow the hydration of the cement and water to give the mortar a 12- to 72-hour board life, depending on the dosage rate. The extended workability allows the mortar to be premixed at a central batching plant and then shipped to the job site. The admixture has little or no effect on setting time because the retarder is absorbed by the masonry units on contact, allowing normal cement hydration to begin. Most extended-life retarders increase air content in the mortar slightly, so use with other air-entrained mortar ingredients should be very carefully controlled or avoided entirely.

Bond enhancers are intended to improve adhesion to smooth, dense-surfaced units such as glass block. Bond enhancers cannot be used with air-entraining agents or air-entrained cement or lime.

Integral water repellents reduce the water absorption or wicking of hardened mortar by as much as 60%. They are used in conjunction with concrete masonry units that have also been treated with an integral water-repellent admixture. Integral water repellents reduce the capillarity of the mortar while still permitting moisture vapor transmission. Using water-repellent-treated mortar with untreated masonry units, or vice versa, can reduce mortar-to-unit bond and the flexural strength of the wall. Reduced bond can negate the effects of the water repellent by allowing moisture to penetrate the wall freely at the joint interfaces. Mortars and block treated with integral water repellents achieve better bond and better moisture resistance only if the admixtures are chemically compatible, so both the mortar and the block admixtures should be from the same manufacturer.

3.1.4 Coloring Pigments

Colored mortar can dramatically change the appearance of a masonry wall. Natural and synthetic pigments used to color masonry mortar should meet the requirements of ASTM C979, *Standard Specification for Pigments for Integrally Colored Concrete*. Most mortar colorants are made from natural or synthetic iron oxide pigments. Iron oxides are nontoxic, colorfast, chemically stable in mortar, and resistant to ultraviolet radiation. Carbon black and lampblack (used to make blacks and browns) are less weather resistant than the iron oxides used to make the same colors. When pigments are used in mortar, bond strength is reduced by 3 to 5%.

Colored mortar can be made at the job site from powdered or liquid pigments. Powdered pigments are used most frequently, and the majority are packaged so that one bag contains enough pigment to color 1 cu ft of cementitious material (i.e., for each 1 cu ft bag of masonry cement, portland cement, or lime, one bag of color is added). Pigment manufacturers supply charts that identify the exact number of bags or bottles of pigment required for various mortar proportions.

The color of a finished mortar joint is affected by the sand and cement, the workmanship, curing conditions, cleaning procedures, joint type, and joint tooling techniques. White cement and white sand produce cleaner, brighter colors. Consistency is critical in the use of colored mortar. Material suppliers, preparation, mixing, handling, workmanship, and cleaning must be consistent from batch to batch to produce consistent mortar colors. When colored mortar is used, it is best to evaluate and select materials on the basis of samples that closely approximate job-site materials and design and to incorporate the colored mortar into a job-site sample panel before acceptance.

3.2 Mortar Properties

Although concrete, masonry mortar, and masonry grout share some common ingredients (*see Fig. 3-1*), they are quite different from one another. The methods and materials used to produce strong, durable concrete do not apply to masonry mortar and grout.

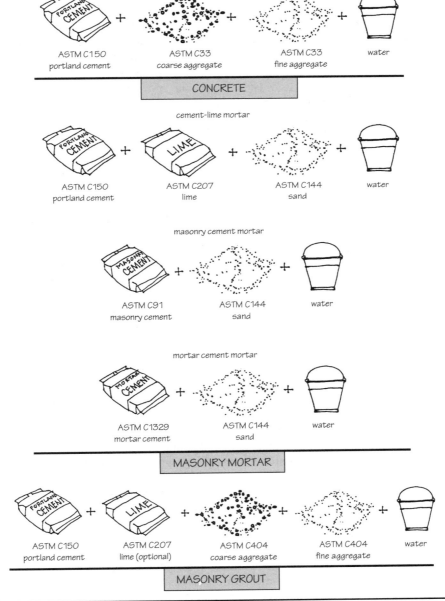

FIGURE 3-1 Comparison of ingredients used to make concrete, mortar, and grout. (*From C. Beall and R. Jaffe,* Concrete and Masonry Databook, *McGraw-Hill, 2003.*)

The most important physical property of concrete is compressive strength, but compressive strength is only one of several properties important to mortar and grout, such as bond strength and durability. These qualities are influenced by two distinct sets of properties, which interact to affect overall performance: (1) *properties of the plastic mortar*, including workability and water retention; (2) *properties of the hardened mortar*, including bond strength, and durability, as well as compressive strength. No single mortar property defines mortar quality.

3.2.1 Workability

Workability is the most important property of wet mortar. Research has shown that workability is more important than compressive strength in producing good masonry for most applications. Workability is a complex property including adhesion, cohesion, density, plasticity, and viscosity. A "workable" mortar has a smooth consistency, is easily spread with a trowel, readily adheres to vertical surfaces, and extrudes readily from the joints without smearing or dropping when the units are placed. Well-graded, smooth aggregates enhance workability, as do lime, air entrainment, and proper amounts of mixing water.

Variations in unit absorption and in environmental conditions affect optimum mortar workability. Warmer summer temperatures and units with a high IRA (initial rate of absorption), for example, require a softer, wetter mix to compensate for evaporation and unit suction. Colder winter temperatures and low IRA units require a drier mix. But at the same time, dry mixes can lose too much water to dry masonry units, and the mortar will not cure properly. Excessively wet mixes cause units to float and will decrease mortar bond strength. The "proper" amount of mixing water is universally agreed upon as the maximum compatible with "workability." A mortar that is workable makes better contact with the unit surface and maximizes the extent of bond. The mason is the best judge of workability for any given mortar mix and job-site conditions. Project specifications should not dictate water/cement ratios or the amount of mixing water for masonry mortar. This should be left up to the mason in the field.

3.2.2 Water Retention

Water retention allows mortar to resist the loss of mixing water by evaporation and the suction of dry masonry units to retain enough moisture for proper cement hydration.

Water retention generally increases as the proportion of lime in the mix increases (*see Fig. 3-2*). At one extreme, a mortar made with only portland cement and sand, without any lime, would have a high compressive strength but low water retention. At the other extreme, a mortar made with only lime and sand, without portland cement, would have low compressive strength but high water retention. High-suction units, especially if laid in hot or dry weather, should be used with a mortar that has high water retention (i.e., a higher proportion of lime). Low-suction units, especially if laid in cold or wet weather, should be used with a mortar that has low water retention (i.e., a lower proportion of lime).

Mortar is subject to water loss by evaporation, particularly on hot, dry days. *Retempering* (the addition of mixing water to compensate for evaporation) is not an acceptable practice in concrete construction, but it is essential in masonry. Because highest bond strengths are obtained with moist mixes, a partially dried and stiffened mortar is less effective if the evaporated water is not replaced. Mortar that has begun to harden as a result of cement hydration, however, should be discarded. Evaporative drying is related

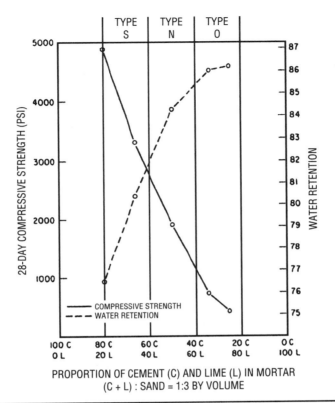

Figure 3-2 As more lime is substituted for portland cement in mortar, water retention and workability increase and compressive strength decreases. (*From T. Ritchie and J. Davison,* Cement-Lime Mortars, *National Research Council, Ottawa, Ontario, Canada, 1964.*)

to both time and temperature. When ambient temperatures are above 80°F, mortar may be safely retempered as needed during the first 1½ to 2 hours after mixing. When temperatures are below 80°F, mortar may be retempered for 2 to 2½ hours after mixing before it should be discarded. ASTM C270 requires that all mortar be used within 2½ hours without reference to weather conditions and permits retempering as frequently as needed within this time period. It is beneficial to the performance of the masonry to maximize workability and bond by replacing evaporated moisture.

3.2.3 Bond Strength

For the majority of masonry construction, the single most important property of hardened mortar is bond strength. For durability, weather resistance, and resistance to loads, it is critical that this bond be strong and complete. The term *mortar bond* includes:

- Extent of bond or area of contact between unit and mortar
- Bond strength or adhesion of mortar to units

The strength and extent of the bond are affected by many variables of material and workmanship. Complete and intimate contact between the mortar and the unit is essential, and the mortar must have sufficient workability to spread easily and wet the contact surfaces. The moisture content, absorption, pore structure, and surface characteristics of the units, the water retention of the mortar, and curing conditions such as temperature, relative humidity, and wind combine to influence the completeness and integrity of the mortar-to-unit bond. Voids at the mortar-to-unit interface offer little resistance to water infiltration and may cause disintegration and failure if repeated freezing and thawing occur while the wall is saturated.

All other factors being equal, mortar bond strength increases slightly as compressive strength increases, although there is no direct proportion. Although higher cement ratios in the mix increase both compressive and bond strength, high-cement–low-lime mortars are less workable. This can leave voids and gaps at the mortar-unit interface that decrease the extent of bond and therefore also decrease bond strength. Because of the trade-off between compressive strength and workability, there is actually little gain in bond strength, if any, when compressive strength is increased. To obtain optimum bond:

- Use mortars with properties compatible with those of the masonry units that will be used.
- Use mortars with the appropriate water retention characteristics for the absorption characteristics of the units.
- Mix mortar with the *maximum* water content compatible with workability.
- Allow retempering of the mortar within recommended time limits.
- Minimize the time between spreading mortar and placing masonry units.
- Apply pressure in forming the mortar joint.
- Do not subsequently disturb units that have been placed.
- Moist-cure the masonry (*see* Chapter 15).

A simple field test to check extent of bond can be made by lifting a unit from its fresh mortar bed to see if the mortar has fully adhered to all bedding surfaces. The extent of bond is good if the mortar sticks to the masonry unit and shows no air pockets or dry areas.

3.2.4 Compressive Strength

The compressive strength of mortar is determined by the amount of cement in the mix. Compressive strength increases with the proportion of cement in the mix and decreases as the lime content is increased. Mortar compressive strengths may be as high as 5000 psi but need not exceed either the design requirements or the strength of the units themselves. Mortar with a high compressive strength does not produce masonry with a proportionately higher compressive strength.

Although compressive strength is less important than bond strength in masonry, simple and reliable testing procedures make compressive strength a widely accepted basis for comparing one mortar with another.

For veneer construction and for low-rise loadbearing construction, mortar compressive strength is rarely a critical design factor because both the mortar and the masonry are usually much stronger in compression than necessary. Compressive strength is important in high-rise loadbearing construction, but structural failure due to compressive loading is rare. More critical properties such as flexural bond strength are usually given higher priority.

3.3 Three Kinds of Mortar

There are three kinds of masonry mortar based on the cementitious ingredients used in the mix:

- Cement-lime mortar
- Masonry cement mortar
- Mortar cement mortar

3.3.1 Cement-Lime Mortar

Mortars made of portland cement and sand have high compressive strengths but are stiff and unworkable and have low water retention and poor bond. Mortars made of lime and sand have low compressive strength but excellent workability and bond. The combination of lime and portland cement in masonry mortar results in the best combination of compressive strength, workability, water retention, and bond strength. The relative proportions of portland cement and lime in a *cement-lime mortar* can be adjusted to enhance some properties and minimize others:

- The more cement in the mortar, the higher the compressive strength, the more rapid the set, and the shorter the board life.
- The more lime in the mortar, the better the workability, water retention, and bond and the longer the board life.

The minimum requirements for portland cement are prescribed in ASTM C150 and those for lime in ASTM C207.

3.3.2 Masonry Cement Mortar

Proprietary masonry cements offer convenience, consistency, and economy because they come in a single bag and are mixed only with sand and water at the job site. *Masonry cement mortar* does not require the proportioning and mixing of lime into the mortar at the job site. Masonry cements generally contain cement, plasticizers, and entrained air.

Air-entraining agents contribute to mortar workability by introducing tiny air bubbles that act as lubricants in the mix. Although the voids created by these bubbles usually reduce bond strength and increase water permeability, they also increase freeze-thaw durability by providing interstitial spaces that accommodate the expansion of ice crystals without damage to the structure of the mortar. This can be particularly important for mortars used in horizontal paving applications. The minimum requirements for masonry cements are listed in ASTM C91.

3.3.3 Mortar Cement Mortar

The third kind of masonry mortar is called *mortar cement mortar*. The minimum requirements for mortar cement are covered in ASTM C1329.

ASTM C1329 essentially sorts out masonry cements with high flexural bond strength capabilities from those that can only provide lower bond strengths. The mortar cements that meet ASTM C1329 are capable of producing mortars with flexural bond strengths equivalent to those of portland cement-lime mortars under identical laboratory test conditions. When high flexural bond strengths are required on a project and it is also desirable to use a single bag mix for its advantageous convenience and economy, a mortar cement conforming to ASTM C1329 should be specified.

3.3.4 Choosing Cement-Lime or Masonry Cement Mortar

Historically, portland cement-lime mortars have exhibited higher flexural strengths than those of masonry cement mortars. Higher flexural strengths increase resistance to lateral loads and to moisture penetration as well. According to one industry survey, both contractors and architects preferred portland cement-lime mortars for water penetration resistance, bond strength, and durability.

Masonry cements are more widely used than either portland cement and lime or mortar cement for masonry mortars. The vast majority of projects that use masonry cements perform quite satisfactorily. On projects that have experienced flexural bond failures or excessive moisture penetration, the cause is seldom attributable solely to the use of masonry cement versus portland cement and lime in the mortar. Usually, there are other defects that contribute more to the problems, such as poor workmanship or inadequate flashing and drainage. Both portland cement-lime mortars and masonry cement mortars can allow water penetration through masonry walls. The amount of water entering the wall is generally higher with masonry cement mortars, but when workmanship is poor, joints are unfilled, and flashing and weeps are not functional, both cement-lime and masonry cement mortars can produce a leaky wall. A wall system with well-designed and properly installed flashing and weeps will allow tolerance of a much greater volume of water penetration without damage to the wall, the building, or its contents than that of a wall system without such safeguards. Ultimately, workmanship and the flashing and drainage system rather than the cementitious materials in a mortar will determine the success or failure of most masonry installations.

3.4 Mortar Types

ASTM C270 outlines requirements for five different mortar types, designated as M, S, N, O, and K. No single mortar type is universally suited to all applications. Variations in proportioning the mix will always enhance one or more properties at the expense of others. Each of the five basic mortar types has certain applications to which it is particularly suited and for which it may be recommended.

3.4.1 Type M Mortar

Type M mortar has high compressive strength and is recommended for both reinforced and unreinforced masonry that may be subject to high compressive loads. Type M is an alternative mix recommendation for various purposes but is not a primary recommendation for any single purpose.

3.4.2 Type S Mortar

Type S mortar is recommended for structures at or below grade and in contact with the soil, such as foundations, retaining walls, pavements, sewers, and manholes. Type S should also be used for adhered veneer such as manufactured stone.

3.4.3 Type N Mortar

Type N is a good general-purpose mortar for use in above-grade masonry. It is recommended for exterior masonry veneers and for interior and exterior loadbearing walls. This "medium-strength" mortar represents the best balance of compressive and flexural strength, workability, and economy and is, in fact, recommended for most masonry applications.

3.4.4 Type O Mortar

Type O is a "high-lime," low-compressive-strength mortar. It is recommended for exterior above-grade and interior non-loadbearing walls and veneers that will not be subject to freezing in the presence of moisture. Type O mortar is often used in one- and two-story residential work and is a favorite of masons because of its excellent workability and economical cost.

3.4.5 Type K Mortar

Type K mortar has a very low compressive strength and a correspondingly low tensile bond strength. It is seldom used in new construction and is recommended in ASTM C270 only for tuckpointing historic buildings constructed originally with lime-sand mortar.

3.4.6 Choosing the Right Mortar Type

There are several rules of thumb in selecting a mortar type. Use mortar with the lowest compressive strength that meets structural requirements because the lower the compressive strength, the more flexible the mortar in accommodating movements in the wall. In areas exposed to significant freeze-thaw cycling, and in particular for horizontal applications in those areas, specify mortars with a higher cement content and/or entrained air. For low-suction clay masonry units, use mortars with a lower lime content, and for high-suction clay masonry units, use mortars with a higher lime content.

The appendix to ASTM C270 contains guidelines on the selection and use of the various masonry mortar types (see *Fig. 3-3*). Type N and Type O are recommended for most applications, including glass block. Type S is recommended only for exterior masonry at or below grade. Although there are alternative recommendations, there is seldom any good reason to specify the alternative rather than the recommended mortar type.

For the vast majority of projects, a Type N mortar is not only adequate in compressive and bond strength but also the best and most appropriate choice. The unnecessary specification of a Type S mortar when a Type N is adequate in strength sacrifices workability in the wet mortar and a degree of elasticity in the finished wall. John H. Matthys, director of the Construction Research Center at the University of Texas at Arlington, has estimated that:

- Type N mortar is the most appropriate choice in 95% of masonry work.
- Type O mortar is the most appropriate choice in 80% of masonry work.
- Type S mortar is the most appropriate choice in only 5 to 10% of masonry work.
- Type M mortar is the most appropriate choice in only 1% of masonry work.

Type S or Type M mortar should not be specified unless there is a compelling reason to do so. A Type N mortar is the best choice for most loadbearing and non-loadbearing masonry because it provides the best balance of strength and workability.

ASTM C270 Recommended Mortar Type Applications				
		Mortar Type	
Location	Building Segment	Recommended	Alternative
Exterior, above grade	Loadbearing walls Non-loadbearing walls Parapet walls	N O§ N	S or M N or S S
Exterior, at or below grade	Foundation walls, retaining walls, manholes, sewers, pavements†, walks† and patios†	S	M or N
Interior	Loadbearing walls Non-loadbearing partitions	N O§	S or M N

§ Type O mortar is recommended for use where the masonry is unlikely to be frozen when saturated and unlikely to be subjected to high winds or other significant lateral loads. Type N or S should be used in other cases.

† Masonry exposed to weather in a nominally horizontal surface is extremely vulnerable to weathering. Mortar for such masonry should be selected with due caution.

(*Copyright ASTM International, 100 Barr Harbor Drive, West Conshohocken PA, 19428. Reprinted with permission.*)

RULE OF THUMB

Always select the mortar type with the lowest compressive strength appropriate to its location and use.

- Type N is most appropriate 95% of the time
- Type O mortar is most appropriate 80% of the time
- Type S is most appropriate 5% of the time
- Type M is most appropriate only 1% of the time

(*Statistics from Dr. John H. Matthys, Construction Research Center, University of Texas at Arlington.*)

FIGURE 3-3 Mortar types and recommended applications.

3.4.7 Proportion Versus Property Method of Specifying Mortar

Conformance with ASTM C270 may be based either on volume proportions or on minimum property requirements (*see Fig. 3-4*). The *proportion specification* prescribes by volume a range of proportions of cementitious materials and aggregate for each mortar type. The *property specifications* are based on minimum compressive strength, minimum water retention, maximum air content, and aggregate ratio. The strength requirements and strength testing of mortar is probably the most misunderstood aspect of masonry construction.

The proportion requirements of ASTM C270 are conservative and will generally yield compressive strengths greater than the minimums required in the property specification. Conversely, the minimum compressive strengths required by the property specification generally can be achieved with a smaller proportion of cement and larger proportion of lime than those prescribed under the proportion specification. There are no minimum property requirements or compressive strengths for proportion-specified mortars. Empirical experience has shown that mortar mixed with the recommended proportions will have more than adequate compressive strength and bond.

Mortar and Grout 47

Mortar Proportions (by Volume)									
Mortar	Type	Portland Cement or Blended Cement	Masonry Cement Type			Mortar Cement Type			Hydrated Lime or Lime Putty	Aggregate (Sand) Measured in a Damp, Loose Condition
			M	S	N	M	S	N		
Cement-Lime	M	1							¼	
	S	1							over ¼ to ½	
	N	1							over ½ to 1½	
	O	1							over 1¼ to 2½	
Mortar Cement	M	1						1		not less than 2¼ and not more than 3 times the sum of the separate volumes of cement and lime
	M					1				
	S	½						1		
	S						1			
	N							1		
	O							1		
Masonry Cement	M	1			1					
	M		1							
	S	½			1					
	S			1						
	N				1					
	O				1					

Mortar Properties§ (ASTM C270 Test Methods)				
Mortar	Type	Minimum Average Compressive Strength at 28 Days (psi)	Minimum Water Retention (%)	Maximum Air Content† (%)
Cement-Lime	M	2500	75	12
	S	1800	75	12
	N	750	75	14
	O	350	75	14
Mortar Cement	M	2500	75	12
	S	1800	75	12
	N	750	75	14
	O	350	75	14
Masonry Cement	M	2500	75	18
	S	1800	75	18
	N	750	75	20
	O	350	75	20

§ The aggregate ratio, measured in a damp, loose condition, shall not be less than 2¼ and not more than 3 times the sum of the separate volumes of cement and lime.
† When structural reinforcement is incorporated in cement-lime or mortar cement mortars, maximum air content shall not exceed 12%. When structural reinforcement is incorporated in masonry cement mortars, maximum air content shall not exceed 18%.

FIGURE 3-4 Use either the proportion specification or the property specification for mortar. (*Copyright ASTM, 100 Barr Harbor Drive, West Conshohocken, PA 19428. Reprinted with permission.*)

The property specifications encourage preconstruction testing of a mix design to gain the economic advantage of meeting minimum property requirements at lower cost. On larger projects, the savings in mortar costs will more than offset the cost of the laboratory mix designs. Because it is generally recommended to use the mortar type with the minimum necessary compressive strength, specifying mortar by the property requirement method ensures that the mortar is not any stronger than it needs to be. There are no "standard" proportions for property-specified mortars. Using the test methods in ASTM C270, the laboratory compares one mix design with another and ultimately provides to the masonry contractor the proportions of ingredients that should be used in the field to achieve the minimum property requirements.

The big difference between the laboratory mix design and the actual field mortar is the amount of mixing water. The field mortar is mixed with a much larger quantity of water. The higher water-cement ratio of the field mortar reduces its compressive strength below that which was obtained with the same proportions of materials tested with a much smaller quantity of water during the ASTM C270 laboratory mix design process. The larger quantity of water in the field mortar is necessary for workability and to satisfy the initial absorption of the masonry units. Absorption characteristics vary from brick to concrete masonry unit (CMU), from one brick to another and one block to another, and even from one day or time of day to the next. The ASTM C270 laboratory mix design does not consider and is independent of these unit absorption characteristics and environmental conditions. The mason adjusts the amount of mixing water in the field but not the proportions of cementitious materials and sand that were determined by the laboratory design mix. Once the initial absorption of the units has been satisfied, the water-cement ratio is rapidly reduced and the strength of the field mortar more closely *approximates* that of the ASTM C270 minimum property requirements.

The minimum strength requirements for property specification mortars are usually lower than those actually produced by proportion specification mortars. If we follow the rule of thumb that the mortar should not be any stronger than required by the project design, then a laboratory mix design by the property specification will provide the best and most economical mortar. On smaller projects where the volume of mortar is much less, using the proportion specification saves the cost of laboratory mix designs. However, it will usually yield mortars with higher strengths than needed (at the sacrifice of other properties). If ASTM C270 is referenced in project specifications without indication as to whether the property or proportion method should be used, the proportion method always governs.

The property specification method in ASTM C270 includes minimum requirements for compressive strength, water retention, air content, and aggregate ratio. The property specifications (including minimum compressive strengths) in ASTM C270 are for laboratory-prepared samples only, and the values _will not_ correlate with those obtained from field samples. ASTM C270 laboratory samples are made with a very low water-cement ratio to simulate the moisture content of field mixed mortar *after* unit suction and evaporation have occurred. Mortars mixed at the job site are made with much higher water-cement ratios because the units are absorptive and will immediately extract much of the mixing water from the mortar paste (*see Fig. 3-5*). Field-sampled mortars yield a much lower compressive strength than that of laboratory-prepared mortar because of the difference in water content.

ASTM C270 *does not* require field testing for mortar, but field testing for consistency from batch to batch may be desirable for quality assurance. Field testing for consistency may include simple visual observation that the specified materials and mix proportions are being used.

Testing of field-sampled mortar, when desired, is appropriate only for mortars specified by the property method, but *never* for mortars specified by the proportion method. If the project will require field sampling of mortar during construction for laboratory

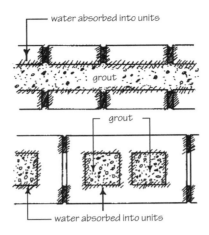
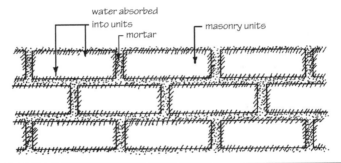

FIGURE 3-5 Excess water in mortar and grout is absorbed by the units. (*From Masonry Institute of America, "Grout . . . The Third Ingredient."*)

testing, ASTM C780, *Standard Test Method for Preconstruction and Construction Evaluation of Mortars for Plain and Reinforced Unit Masonry*, must be used both to set the preconstruction benchmark and to perform the construction phase testing. The goal of the testing is only to verify that the field mix matches the properties of the preconstruction benchmark mix.

The ASTM C780 benchmark test of the specified project mortar will not (and is not supposed to) match, meet, or exceed the minimum compressive strength requirements in ASTM C270. ASTM C270 test methods are not appropriate for job-site-mixed mortar. Results from ASTM C780 tests cannot be compared with results from ASTM C270 tests or with the minimum property requirements listed in ASTM C270. It will not be an apples-to-apples comparison. Because of the different water-cement ratios in the ASTM C270 and ASTM C780 test methods, the compressive strength values obtained from ASTM C780 tests of job-site-mixed mortars are neither required nor expected to meet the property requirements of ASTM C270. ASTM C780 tests of job-site-mixed mortar can be compared *only* with ASTM C780 preconstruction benchmark tests based on the same materials, design mix, and *quantity of water*.

Even ASTM C780 tested compressive strength is not the same as the compressive strength of the mortar in the wall. There is no ASTM test method for determining conformance of a *hardened* mortar to the ASTM C270 property requirements or the ASTM C780 preconstruction benchmark test. Proportions of materials in hardened mortar can be verified by ASTM C1324, *Examination and Analysis of Hardened Masonry Mortar*, which uses simple petrographic methods. Aggregate ratio for plastic mortar can be tested and verified by ASTM C780. (*See* Chapter 17 for quality assurance/quality control recommendations.)

3.5 Masonry Grout

Grout is a fluid, flowable mixture of cementitious material and aggregate with enough water added to allow the mix to be poured or pumped into masonry cores and cavities (*see Fig. 3-6*).

In unreinforced loadbearing construction, unit cores or cavities are sometimes grouted to give added strength. In both loadbearing and non-loadbearing construction, grout may be added to unit cores and cavities to increase fire or sound resistance, security, and thermal storage capacity. In reinforced masonry, grout must bond the masonry units and reinforcing bars together for resisting superimposed loads. Grout requirements are prescribed in ASTM C476, *Standard Specification for Grout for Masonry*.

3.5.1 Grout Materials

The same cementitious materials that are used in mortar are also used in masonry grout. Cement and lime are combined with sand and gravel aggregate to produce either "fine" or "coarse" grout. Fine grout is made with sand aggregate, and coarse grout is

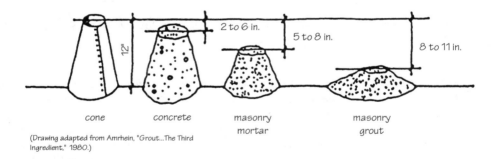

(Drawing adapted from Amrhein, "Grout...The Third Ingredient," 1980.)

Concrete	Masonry Mortar	Masonry Grout
Concrete is generally mixed with the minimum amount of water required to produce workability appropriate to the method of placement. The amount of water is determined by laboratory mix design.	Masonry mortar is generally mixed with the maximum amount of water required to produce good workability with a given unit. The amount of water is determined by the mason based on masonry unit absorption and field conditions.	Masonry grout is usually mixed with the maximum amount of water required to produce good flow properties. The amount of water is determined by the mason based on masonry unit absorption and field conditions.

FIGURE 3-6 Relative consistency of concrete, mortar, and grout as measured by slump test.

made with a combination of sand and gravel aggregates. ASTM C404, *Standard Specification for Aggregates for Masonry Grout*, defines both fine and coarse aggregates based on sieve analysis.

3.5.2 Grout Admixtures

Pumping aids (commonly called grouting aids or shrinkage-compensating admixtures) are the most common grout additives. Grout typically shrinks 5 to 10% after placement as the surrounding masonry units absorb water. To minimize volume loss, maintain good bond, and give workers more time to vibrate the grout before it stiffens, these specially blended admixtures expand the grout, retard its set, and lower the water requirements. Pumping aids can also be used to accelerate set in cold weather or retard set in hot weather. Superplasticizers may be used in hot weather to increase slump without adding water or reducing strength. Superplasticizers can also be advantageous when units have low initial absorption rates.

Air-entraining admixtures are permitted for increased freeze-thaw durability but are typically not used in grout because grout is not exposed to moisture saturation and freezing. Antifreeze liquids, salts, or other substances cannot be used to lower the freezing point of masonry grout. Set retarders slow grout set in hot weather, which extends fluidity and allows more time for placement and consolidation. Set accelerating admixtures will speed up the set time during cold weather and reduce the length of time that the wall must be protected from freezing. Set accelerators are sometimes mistakenly referred to as "antifreeze" compounds. Set accelerators sometimes include calcium chloride, which can contribute to efflorescence and cause corrosion of embedded steel anchors and reinforcement. Nonchloride accelerators are a little more expensive but less damaging to the masonry. Chlorides should be prohibited in mortars and grouts that contain embedded metals such as anchors, ties, or joint reinforcement.

Water-repellent admixtures are not typically used in grout. CMU made with integral water-repellent admixtures are less absorptive than untreated block, but the bond between treated units and grout is not inhibited. Integral water-repellent admixtures only make the block *water repellent*, not *waterproof*. Block treated with water-repellent admixtures are designed to withstand a wind-driven rain that is equivalent to about 2 in. of hydrostatic pressure. Grout exerts significantly higher hydrostatic pressure, which overpowers the repellent properties of the CMU and pushes moisture and grout paste into the block surface to form a tight bond.

During cold weather construction, grouted walls made with water-repellent CMU admixtures will maintain higher moisture contents for longer periods of time than those of walls with untreated block. This will require that weather protection measures be maintained for longer periods of time as well. Using a Type III High Early Strength Portland Cement instead of a Type I accelerates setting, as do nonchloride accelerators, which can shorten the time required for protection. Water-reducing admixtures or superplasticizers reduce water content while maintaining adequate flow, which can also help reduce protection time. These additional measures can be used *with* code-required cold weather protection requirements, *not instead* of them.

Admixtures should be specified in the project specifications or approved by the design professional. All grout mixes that contain admixtures should be tested in advance of construction to ensure proper performance. Grout mix designs that meet project requirements and ASTM guidelines can be determined in the laboratory by preconstruction testing of trial batches.

3.5.3 Grout Properties

The fluid consistency of grout is important in determining compressive strength, in ensuring that the mix will pour or pump easily without segregation, and in ensuring that the mix will flow into corners and recesses and encapsulate the reinforcing bars without leaving voids. Performance records indicate that a minimum slump of 8 in. is necessary for units with low absorption and as much as 11 in. for units with high absorption.

The slump and amount of water required for proper fluidity should be adjusted according to:

- Initial rate of absorption of the masonry units
- Size of the grout space
- Lift and pour height
- Weather conditions

Masonry units with a high initial rate of absorption will absorb more water than that of units with a low initial rate of absorption, so water content and slump of the grout should be at the higher end of the permissible range. Concrete block and mortar made with integral water repellents have lower absorption, so water content and slump should be at the lower end of the permissible range. Small grout spaces have larger surface-to-volume ratios and will absorb more water than wide grout spaces, so water content and slump should be on the higher end of the permissible range. High lifts and pours require grout that is more fluid than low lifts and pours, so water content and slump should be on the highest end of the permissible range. Masonry constructed in hot, dry climates will absorb water at a faster rate than masonry constructed in a colder and more moist climate. In drier climates with high suction units, grout with a 10- to 11-in. slump is recommended. In wet climates with low absorption units, grout with an 8- to 9-in. slump is recommended.

3.5.4 Grout Types and Proportions

ASTM C476 includes both fine and coarse grout types based on aggregate size and grading. Selection of a fine or coarse grout is based on the size of the core or cavity and the height of the pour or lift to be grouted. If the maximum aggregate size is 3/8 in. or larger, the grout is classified as *coarse*. If the maximum aggregate size is less than 3/8 in., it is classified as *fine*. The smaller the grout space, the smaller the maximum aggregate size allowed. Although ASTM C404 limits the maximum aggregate size to 3/8 in., some engineers allow up to ¾-in. aggregate for grouting large voids such as columns and pilasters. The larger aggregate takes up more volume, reduces grout shrinkage, and requires less cement for equivalent strength. The table in *Fig. 3-7* shows the recommended grout type for various grout spaces.

ASTM C476 permits specifying grout either by mix proportions or by compressive strength. Grout proportion requirements are prescribed in ASTM C476 (*see Fig. 3-8*). When the compressive strength method is used, the minimum 28-day compressive strength must be 2000 psi when tested in accordance with ASTM C1019, *Standard Test Method for Sampling and Testing Grout*. When the design requires grout with a higher compressive strength, the compressive strength method should be used to develop an appropriate design mix. Optimum water content, consistency, and slump will depend on the absorption rate of the units as well as job-site temperature and humidity

Mortar and Grout

Minimum Grout Space Requirements for ASTM C476 Grout (with tolerance of +3/8" or −1/4")			
Grout Type	Maximum Grout Pour Height (ft.)	Minimum Width of Grout Space Between Wythes of Masonry§ (in.)	Minimum Grout Space Dimensions for Grouting Cells or Cores of Hollow Units§† (in. x in.)
Fine	1	¾	1½ x 2
Fine	5	2	2 x 3
Fine	12	2½	2½ x 3
Fine	24	3	3 x 3
Coarse	1	1½	1½ x 3
Coarse	5	2	2½ x 3
Coarse	12	2½	3 x 3
Coarse	24	3	3 x 4

§ Grout space dimension is the clear dimension between any masonry protrusion and shall be increased by the diameters of the horizontal bars within the cross section of the grout space.
† Area of vertical reinforcement not to exceed 6% of the area of the grout space.

FIGURE 3-7 Minimum grout space requirements for fine or coarse grout. (*From Masonry Standards Joint Committee,* Building Code Requirements for Masonry Structures, *ACI 530/ASCE 5/TMS 402.*)

- Specify masonry grout either by mix proportions or by minimum compressive strength.
- Select grout type based on the size of the space to be grouted.
- ASTM C476 and most building codes require a minimum grout compressive strength of 2000 psi when tested in accordance with ASTM C1019.

Grout Type	Parts by Volume of Portland Cement or Blended Cement	Parts by Volume of Hydrated Lime or Lime Putty	Aggregate Measured in Damp, Loose Condition	
			Fine	Coarse
Fine	1	0 to 1/10	2-1/4 to 3 times the sum of the volumes of the cement and lime	—
Coarse	1	0 to 1/10	2-1/4 to 3 times the sum of the volumes of the cement and lime	1 to 2 times the sum of the volumes of the cement and lime

FIGURE 3-8 Requirements for ASTM C476 masonry grout. (Copyright ASTM International, 100 Bar Harbor Drive, West Conshohocken, PA 19428. Reprinted with permission.)

conditions. Like mortar, grout mixed by the prescribed proportions typically produces higher compressive strengths than those mixed by the compressive strength method. The compressive strength method is only used on large projects where a laboratory mix design will be more economical.

Water absorbed by the units is retained for a period of time, providing a moist condition for curing. Unit absorption is affected not only by the characteristics of the brick or block but also by the size of the cavity and its shape. The greater the surface area, the more water will be absorbed, so water content and slump limits should be adjusted accordingly by the mason.

3.5.5 Self-Consolidating Grout

Self-consolidating grout is a highly fluid mix that is easy to place and eliminates the need to vibrate the grout for consolidation. Self-consolidating grout is a carefully controlled mix of cement, aggregate, and water with high-range water-reducing admixtures. Self-consolidating grout may be mixed on the job site, but proportioning of the grout on site is not permitted. Mix proportions are critical, so self-consolidating grout is often delivered to the job site as a premixed product.

Self-consolidating grout is divided into the same two types as conventional grout: fine and coarse. It should meet the requirements of ASTM C404 for aggregate and ASTM C476 for proportion or property requirements.

CHAPTER 4

Masonry Accessories

Most masonry construction includes accessory items that perform a variety of functions from attachment and reinforcement to moisture collection and drainage. Both metallic and nonmetallic accessories must be durable for their purpose and remain serviceable for the life of the masonry. The premature deterioration of embedded accessories can cause performance failure of the masonry and cost millions of dollars in repairs. Saving a few dollars of initial cost by "value engineering" the accessories can be an extremely expensive mistake.

4.1 Metals and Corrosion

Metal corrosion is caused by weathering, exposure to certain chemicals, and galvanic action. When metal accessories are only partially embedded in masonry, corrosion may be caused by prolonged exposure to the combination of air and moisture. Exposure to air occurs in the drainage cavities behind veneers and between wythes of masonry and in the ungrouted cores of hollow masonry units. The source of moisture may be condensation, water that penetrates through the masonry, and even atmospheric humidity higher than 75%. Chemical attack can be caused by chlorides, so set-accelerating admixtures that contain calcium chloride should not be used in masonry mortar or grout. Deep carbonation of mortar caused by carbon dioxide intrusion through cracks or voids at the mortar-to-unit interface may also accelerate the corrosion of steel anchors, ties, and joint reinforcement embedded in the mortar.

Some metal corrosion in masonry is caused by galvanic action. Galvanic corrosion occurs between dissimilar metals in contact with each other *and* in the presence of moisture. To protect against galvanic corrosion when dissimilar metals are used, isolation can be provided by an electrical insulator such as asphalt felt or membrane flashing.

The density of the galvanic corrosion current is important; that is, the size of the current relative to the anode surface. If a fastener has a surface that is small compared with the metal to which it will be fastened, its current density will be high and therefore subject to rapid corrosion. To prevent this (as a general rule), a fastener in a given environment should be equal to or higher on the galvanic scale (more noble) than the material to which it will be fastened. This allows, for instance, the use of stainless steel screws (more noble) to attach hot-dip galvanized anchors to a cold galvanized metal stud (less noble) without galvanic corrosion. The same is true for metal drip edges with metal flashing or on top of steel shelf angles or lintels. The drip edge, which is the smaller of the contiguous elements, must be of the same metal or of a metal that is more noble than the flashing or angle itself or it must be isolated to prevent galvanic corrosion. Galvanized drip edges on galvanized or ungalvanized angles or stainless steel drip edges on galvanized or ungalvanized angles do not corrode. Copper flashing used with stainless or galvanized steel drip edges will cause corrosion if they are not isolated. The drip edge must be of copper or rubber to prevent galvanic corrosion.

Like the reinforcing bars used in concrete construction, those used in masonry are made of billet steel (ASTM A615, *Deformed and Plain Billet-Steel Bars for Concrete Reinforcement*), rail steel (ASTM A616, *Rail-Steel Deformed and Plain Bars for Concrete Reinforcement*), or axle steel (ASTM A617, *Axle-Steel Deformed and Plain Bars for Concrete Reinforcement*). Corrosion of reinforcing bars is minimal because it is limited to that which occurs during the grout curing process. Once the moisture in the grout is gone, the steel is protected from further moisture exposure by its complete encapsulation within the grouted masonry. Because of the low risk of deterioration, steel reinforcing bars are not typically galvanized. Epoxy-coated steel rebar is not typically used in masonry because, unlike the concrete highway paving and bridge construction to which it is more common, moisture exposure is expected to be much less.

The metals used for masonry anchors, ties, and joint reinforcement are stainless or galvanized steel. Nonferrous metals such as copper and copper alloys are used for masonry flashing. Copper and copper alloys are essentially immune to the corrosive action of uncured mortar. Because of this immunity, copper can be safely embedded in fresh mortar even under saturated conditions. Aluminum is deteriorated by the alkalinity of fresh mortar. Because it is deteriorated by the mortar itself, aluminum is never used in masonry construction.

With the exception of wire fabric, the steel accessories used in masonry should be galvanized or stainless steel. Although zinc is also susceptible to corrosive attack, it is used in the galvanizing process both to provide a barrier coating to isolate the steel from corrosive elements and as a sacrificial anodic coating that is consumed to protect the steel at uncoated areas such as scratches and cut ends. If the masonry is absorbing excessive moisture because of design or construction defects, corrosion of the steel may continue after the sacrificial zinc is consumed and the expansive pressures increase substantially over time. As this "rust jacking" continues, the masonry cracks, allowing even more moisture to enter the wall (*see Fig. 4-1*).

Figure 4-1 The expansion of corroding steel can cause severe displacement and cracking of the surrounding masonry (rust jacking).

Accessory Item	ASTM Standard†	Class	Weight or Thickness of Coating
Galvanized Coatings			
Joint reinforcement, interior walls	A641	1	0.10 oz/sq.ft.
Wire ties or anchors in exterior walls completely embedded in mortar or grout	A641	3	0.80 oz/sq.ft.
Wire ties or anchors in exterior walls *not* completely embedded in mortar or grout	A153	B2	1.50 oz/sq.ft.
Joint reinforcement in exterior walls or interior walls exposed to a mean relative humidity exceeding 75% (e.g., food processing or swimming pool)	A153	B2	1.50 oz/sq.ft.
Sheet metal ties or anchors in exterior walls or interior walls exposed to a mean relative humidity exceeding 75% (e.g., food processing or swimming pool)	A153	B2	1.50 oz/sq.ft.
Sheet metal ties or anchors in interior walls	A653	G60	0.60 oz/sq.ft.
Steel plates and bars (as applicable to size and form indicated)	A123 A153	— B	— —
Epoxy Coatings			
Joint reinforcement	A884	B2	18 mils
Wire ties and anchors	A899	C	20 mils
Sheet metal ties and anchors	—	—	20 mils‡

Masonry Standards Joint Committee Specification for Masonry Structures (ACI 530.1/ASCE 6/TMS 602) Requirements for Corrosion Protection§

§ Corrosion protection may also be provided by using AISI Type 304 stainless steel as follows:
  joint reinforcement, ASTM A580 *Stainless and Heat-Resisting Steel Wire*
    plate and bent bar anchors, ASTM A666 *Austenitic Stainless Steel Sheet, Strip, Plate and Flat Bar for Structural Applications*
  sheet metal ties and anchors, ASTM A167 *Stainless and Heat-Resisting Chromium-Nickel Steel Plate, Sheet and Strip*
  wire ties and anchors, ASTM A580 *for Stainless and Heat-Resisting Steel Wire*

† ASTM A641 *Zinc Coated (Galvanized) Carbon Steel Wire*
  ASTM A153 *Zinc Coating (Hot-Dipped) on Iron and Steel Hardware*
  ASTM A653 *Steel Sheet, Zinc-Coated (Galvanized) or Zinc-Iron Alloy-Coated (Galvanealed) by the Hot-Dip Process*
  ASTM A884 *Epoxy-Coated Steel Wire and Welded Wire Fabric for Reinforcement*
  ASTM A899 *Steel Wire Epoxy Coated*

‡ Per surface or manufacturer's specification.

FIGURE 4-2 Required corrosion protection for masonry accessories. (*From MSJC*, Specifications for Masonry Structures, *ACI 530.1/ASCE 6/TMS 602.*)

The table in *Fig. 4-2* lists the corrosion protection requirements for masonry accessories found in the *Masonry Standards Joint Committee* (MSJC) *Specifications for Masonry Structures* (ACI 530.1/ASCE6/TMS 602). Masonry accessories in exterior walls and interior walls exposed to relative humidities of 50% or higher should be hot-dip galvanized after fabrication in accordance with ASTM A153, *Standard Specification for Zinc Coating (Hot-Dip) on Iron and Steel Hardware,* Class B. Mill galvanizing can be used only for interior walls exposed to lower humidity. The life expectancy of zinc galvanizing is directly proportional to its thickness. Stainless steel and epoxy-coated steel accessories are less susceptible to corrosion and provide greater long-term durability for masonry construction. Stainless steel will provide the highest corrosion protection in severe exposures and should conform to Series 300, ASTM A167, *Standard Specification for Stainless Steel and Heat-Resisting Chromium-Nickel Steel Plate, Sheet, and Strip*.

4.2 Joint Reinforcement

Joint reinforcement is used to control shrinkage cracking in *concrete masonry unit* (CMU) walls. It does not *stop* cracks from occurring; it just reduces the widths of the cracks that do occur, producing many small hairline cracks instead. The total width of all the hairline cracks approximate the width of a crack that would occur without the reinforcement. Joint reinforcement can also be used to tie the wythes of multi-wythe walls together and to bond intersecting walls. Joint reinforcement is laid in the horizontal bed joints of masonry walls at specified intervals. The basic types of joint reinforcement available are shown in *Fig. 4-3*. Some designs are better for certain applications than others.

- In single-wythe walls, two-wire ladder or truss-type reinforcement is most appropriate. Under most circumstances, the ladder type provides adequate restraint against shrinkage cracking. The truss type is stronger and provides about 35% more area of steel, but the ladder type interferes less with grout flow and vertical bar placement in structurally reinforced walls. Specify ladder type unless there is an engineering need for truss type.
- For multi-wythe walls in which the backing and facing wythes are of the same type of masonry, three-wire joint reinforcement of either the truss- or ladder-type design is suitable. If the wythes are laid up at different times, however, the three-wire design makes installation awkward. Three-wire truss-type reinforcing should never be used when insulation is installed in the cavity between wythes because it is too stiff to allow for differential thermal movement between the backing and facing wythes.
- For walls in which the backing and facing wythes are laid at different times or walls that combine clay and concrete masonry in the facing and backing wythes, joint reinforcement with adjustable ties allows differential movement between wythes and facilitates the installation of the outer wythe after the backing wythe is already in place. The adjustable ties may be either a tab or eye-and-pintle design.
- For uninsulated cavity walls of block and brick where the backing and facing wythes are laid at the same time, truss- or ladder-type reinforcement with fixed welded-wire tab ties can be used. It is less expensive than reinforcement with adjustable ties but also allows less differential movement. If the cavity is insulated, the tabs restrain differential thermal movement between the backing and facing wythes.
- For construction in seismically active areas, joint reinforcement with seismic anchors is available from several manufacturers.

The table in *Fig. 4-4* summarizes the general recommendations for using various types of joint reinforcement in various applications.

Horizontal joint reinforcement is available in stainless steel, hot-dip galvanized steel, mill galvanized steel, and epoxy-coated steel. Stainless steel and epoxy-coated

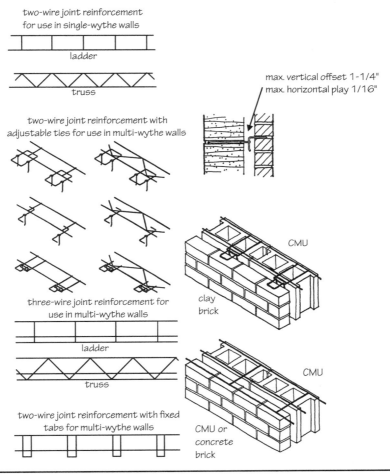

FIGURE 4-3 Prefabricated joint reinforcement.

steel are for extreme conditions. Stainless steel or epoxy-coated steel joint reinforcement is seldom used in typical masonry construction because it is not necessary and because it adds unnecessary expense. Joint reinforcement is usually made of galvanized steel wire. Spacing of the welded lateral ties should not exceed 16 in. for deformed wire or 6 in. for smooth wire. If used as structural reinforcing, the longitudinal chords *must* be of deformed wire. Joint reinforcement should conform to the requirements of ASTM A951, *Standard Specification for Joint Reinforcement for Masonry*. For exterior walls and for interior walls exposed to a relative humidity of 75% or higher, joint reinforcement should be hot-dip galvanized after fabrication in accordance with ASTM A153, Class B2. For interior walls exposed to lower humidity, joint reinforcement can be zinc coated in accordance with ASTM A641, *Zinc Coated (Galvanized) Carbon Steel Wire*. Stainless steel or epoxy-coated steel joint reinforcement will provide the highest corrosion protection in severe exposures and should conform to ASTM A167, Type 304 stainless steel, or ASTM A884 or ASTM A889 epoxy-coated steel.

Wall configuration	2-Wire ladder	2-Wire truss	2-Wire ladder or truss with adjustable ties	2-Wire ladder or truss with fixed tabs	2-Wire ladder or truss with seismic ties
Single-wythe • With vertical reinforcing steel	•				
Single-wythe • Without vertical reinforcing steel		•			
Multi-wythe • Wythes laid at different times • Backing wythe CMU, facing wythe clay masonry or stone			•		
Multi-wythe • Uninsulated cavity • Both wythes laid at same time • Backing wythe CMU, facing wythe clay masonry or stone				•	
Multi-wythe • Wythes laid at different times • Backing wythe CMU, facing wythe clay masonry or stone • Seismic Performance Category C					•

Note: Three-wire joint reinforcement is not recommended unless both wythes are laid up at the same time and there is no insulation between the wythes.

Figure 4-4 Joint reinforcement selection guide. (*Adapted from Mario Catani, "Selecting the Right Joint Reinforcement for the Job,"* The Magazine of Masonry Construction, *January 1995.*)

Joint reinforcement is available in several wire diameters and in standard lengths of 10 to 12 ft. Longitudinal wires are available in standard 9 gauge (W1.7) and extra-heavy 3/16 in. (W2.8). Standard 9-gauge wire provides better fit and more practical constructability in 3/8-in. mortar joints. Heavy-gauge joint reinforcement should be used only when there is compelling engineering rationale. Cross-wires are typically either 9 or 12 gauge. Fabricated joint reinforcement widths are approximately 1-5/8 in. less than the actual wall thickness to ensure adequate mortar coverage. Mortar cover at the exterior wall face should be at least 5/8 in. to prevent corrosion from exterior moisture (*see Fig. 4-5*). Prefabricated "L" and "T" sections are used at corners and intersecting walls to prevent cracking and separation (*see Fig. 4-6*).

4.3 Connectors

There are two different types of masonry connectors. *Anchors* attach masonry to a structural support such as an intersecting wall, a floor, a beam, or a column. This type of connector includes anchor bolts and veneer anchors used to attach masonry veneers to backing walls of nonmasonry construction. *Ties* connect multiple wythes of masonry together in cavity wall or composite wall construction.

exposed joint reinforcement

rusted anchor

FIGURE 4-5 Inadequate mortar cover at the outside face of the wall can cause even galvanized joint reinforcement and accessories to corrode.

FIGURE 4-6 Prefabricated "L" and "T" sections of joint reinforcement are used at corners and intersecting walls to prevent this type of cracking.

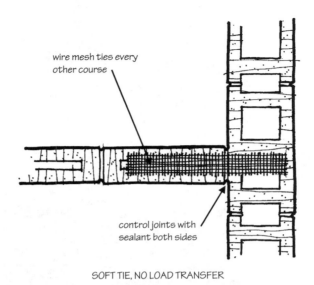

FIGURE 4-7 Wire mesh tie at wall intersection does not transfer loads.

4.3.1 Ties

Joint reinforcement can provide longitudinal strength in addition to lateral connection between wythes, but individual *corrugated* or *wire ties* function only in the lateral direction, providing intermittent rather than continuous connection. There are several shapes and configurations, different wire gauges, and various sizes to suit the wall thickness. *Woven wire mesh* is sometimes used to connect intersecting masonry walls when no load transfer is desired. This is a soft connection and requires the installation of control joints at the wall intersection (*see Fig. 4-7*). Wire ties should be used in open-cavity walls and grouted multi-wythe walls. Wire ties may be rigid for laying in bed joints at the same height or adjustable for laying in bed joints at different levels (*see Figs. 4-8 and 4-9*). Adjustable ties also permit differential expansion and contraction between backing and facing wythes of cavity walls. This is particularly important when connecting between clay and concrete masonry because the thermal and moisture movement characteristics of the materials are so different. Crimped ties that form a water drip in the cavity are not recommended because the deformation reduces their strength in transferring lateral loads. Crimped ties, in fact, are prohibited under some building codes because their compressive strength is only about half that of uncrimped ties. Drips are incorporated by some manufacturers by installing a plastic ring at the midsection of the wire. Adjustable tab and eye-and-pintle ties have a natural drip. Building codes prescribe maximum tie spacing.

Rigid wire ties may be rectangular or Z-shaped, in lengths of 4, 6, or 8 in. (*see Fig. 4-8*). Z-ties should have at least a 2-in., 90° leg at each end. Rectangular ties should have a minimum width of 2 in. and welded ends if the width is less than 3 in. Either type may be used for solid masonry (core area less than 25%), but Z-ties are less expensive. Only rectangular ties should be used in ungrouted walls of hollow masonry. Corrugated steel ties should have 0.3- to 0.5-in. wavelength, 0.06- to 0.10-in. amplitude, 7/8-in. width,

Masonry Accessories

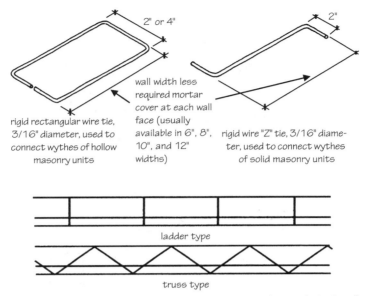

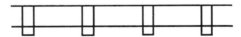

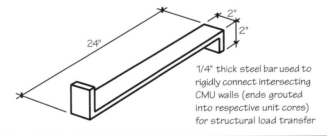

FIGURE 4-8 Rigid masonry ties.

and minimum 22-gauge thickness. Wire mesh ties should be formed of unwelded, woven wire, 16 gauge or heavier. A minimum width of 4 in. is required and a ½ × ½ in. or finer mesh.

For best performance, metal ties should be fabricated of the following materials:

- Stainless steel, ASTM A167, Series 300
- Steel plate, headed and bent bar ties, ASTM A36
- Sheet metal, ASTM A366

64   Chapter Four

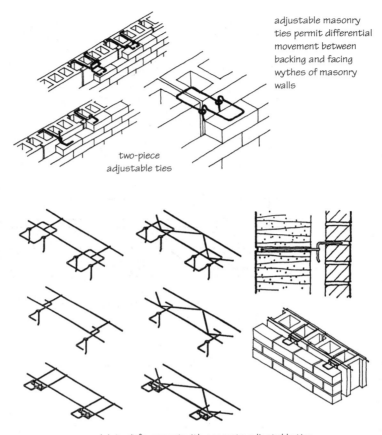

Figure 4-9 Adjustable masonry ties.

- Wire mesh, ASTM A185
- Wire ties, ASTM A82

For exterior walls, ties that are not fabricated of stainless steel are required by code to be hot-dip galvanized after fabrication in accordance with ASTM A153, Class B2.

4.3.2 Anchors

Masonry veneer anchors provide connections that can resist compressive, tensile, and shear stresses. Anchors may be of either wire or sheet metal for attaching masonry veneer to steel, concrete, or stud backing (*see Figs. 4-10 through 4-12*). Anchors must allow differential movement between the masonry and the backing wall.

Corrugated sheet metal anchors should meet the same physical requirements as corrugated ties (0.3- to 0.5-in. wavelength, 0.06- to 0.10-in. amplitude, 7/8-in. width, and

Masonry Accessories

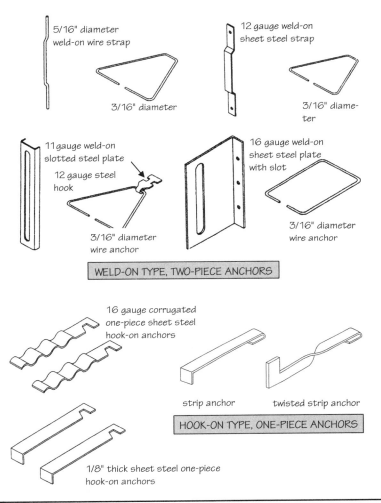

FIGURE 4-10 Masonry-to-steel veneer anchors.

minimum 22-gauge thickness). Corrugated sheet metal anchors may be used with solid or hollow units where the distance between the veneer and supporting frame is 1 in. or less and are typically limited to residential or one-story light commercial construction. One end of the anchor is nailed or screwed to a stud, and the other end is embedded in a mortar joint (*see Fig. 4-11*). Performance is greatly reduced if the attaching nail or screw is not located within ½ in. of the bend. Corrugated dovetail anchors are fabricated to fit a dovetailed slot in a concrete structural frame and are usually at least 16 gauge (*see Fig. 4-12*).

Building codes require special anchorage of masonry veneers in seismic areas. Seismic anchors typically consist of a single or double continuous reinforcing wire attached to a plate for connection to different types of backing walls (*see Fig. 4-13*).

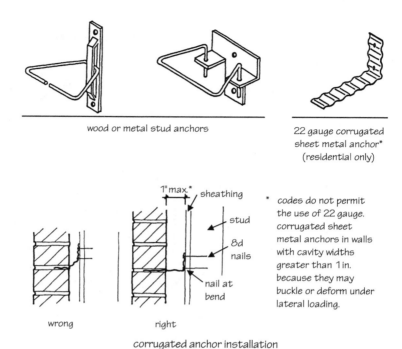

FIGURE 4-11 Masonry-to-stud veneer anchors.

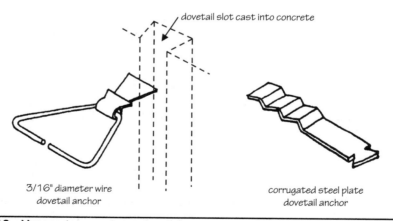

FIGURE 4-12 Masonry-to-concrete veneer anchors.

4.4 Reinforcing Bars and Bar Positioners

Proper structural function of reinforced masonry and proper interaction between grout and reinforcement require that the reinforcing bars be located in the position required by the design. Accurate positioning requires the use of special accessories or special units (*see Fig. 4-14*) that are capable of holding the reinforcement in place

Masonry Accessories

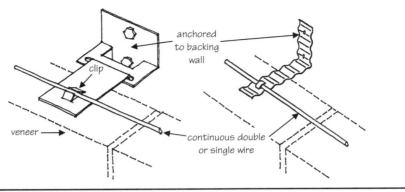

FIGURE 4-13 Seismic veneer anchors.

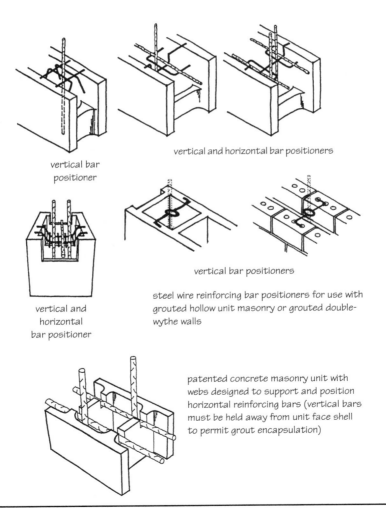

FIGURE 4-14 Bar positioners for masonry reinforcement.

during grouting operations. Although reinforcing bars in masonry construction are not typically galvanized, bar positioners typically are and should be hot-dip galvanized after fabrication in accordance with ASTM A153, Class B2, for exterior walls and mill galvanized for interior walls.

4.5 Flashing Materials

Flashing in masonry may be used as a barrier against the intrusion of water and as a moisture-collection device. All flashing materials must be impervious to moisture and resistant to corrosion, abrasion, and puncture. *Stainless steel* flashings are highly resistant to corrosion and provide the highest long-term durability. *Copper sheet* resists ordinary corrosive action, provides an excellent moisture barrier, and is easily shaped. Both stainless steel and copper flashing can be "saw-toothed" or "dovetailed" in section to provide a mechanical bond with the mortar (*see Fig. 4-15*).

Galvanized steel flashings for exterior exposures require a 26-gauge thickness, and concealed installations require 28 gauge. *Aluminum* flashing should not be used in masonry.

Copper is commonly used in 3-, 5-, or 7-oz sheets laminated on both sides with asphalt, kraft paper, asphalt-saturated cotton fabrics, or glass fiber fabrics. Laminated copper flashing is also available with a self-adhering rubberized asphalt facing on one side to eliminate the need for a separate mastic adhesive. Some copper flashings are made with glass fiber fabrics laminated to the copper with non-asphaltic adhesives for better joint sealant compatibility. The non-asphaltic materials also reduce the possibility of "bleed" at the outside face of the wall. Laminated copper flashings provide good protection at lower cost than that of plain copper by allowing thinner metal sections.

Plastic sheet flashing of *PVC* membrane may also be used in concealed locations but deteriorates and becomes brittle over time. In general, PVC flashing is not recommended because of its short service life. Various types of thin, inexpensive sheet flashings are used in residential construction, but some higher-end custom homes now use commercial-grade flashing.

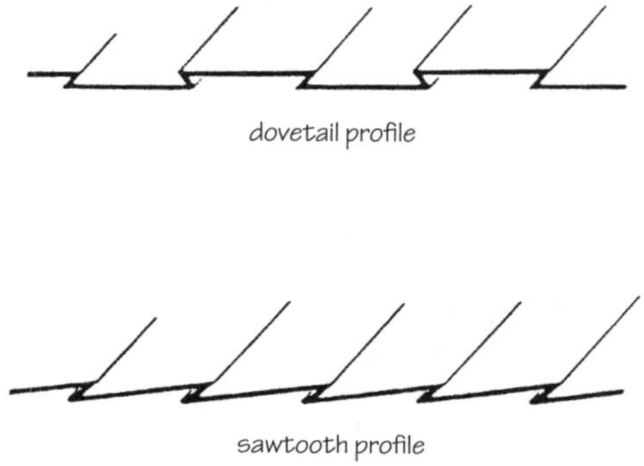

dovetail profile

sawtooth profile

FIGURE 4-15 Deformed metal flashing forms a mechanical bond with the mortar.

Rubberized asphalt flashing is self-adhering and self-healing of small punctures. It installs quickly and easily and is relatively forgiving of uneven substrates. However, good adhesion depends on a clean, dry substrate and temperatures that are relatively warm. Primers help ensure good adhesion of rubberized asphalt flashing to concrete, sheathing, and other substrates and can make cold-weather adhesion easier to achieve than with a heat gun. Primers perform a critical function and are recommended for all rubberized asphalt flashing installation. Rubberized asphalt flashing cannot tolerate ultraviolet exposure. Rubberized asphalt flashing has become very popular for commercial construction, in part because it is easily installed and prevents the flow of water underneath the membrane. *Figure 4-16* lists the advantages and disadvantages of the most commonly used flashing materials.

Flashing must be extended to or beyond the face of the wall, so materials such as rubberized asphalt and laminated copper that cannot tolerate ultraviolet exposure typically use separate drip edges. Drip edges prevent water from flowing around the front edge of the flashing membrane and underneath it onto supporting angles or into the cores of the masonry units below. Preformed stainless steel or galvanized steel drips are available from most accessory manufacturers, but drips may also be custom fabricated. In the absence of drips, flashing that is brought to the face of the wall and trimmed off flush should be fully adhered to prevent water from flowing under it and back into the wall.

\multicolumn{4}{c}{Masonry Flashing Materials}			
Material	Minimum Thickness or Gauge	Advantages	Disadvantages
Stainless steel	26 gauge/0.018 in.	Very durable, non-staining	Difficult to solder and form
Cold-rolled copper	16 oz.	Flexible, durable, easily formed and jointed	Damaged by excessive flexing, can stain surfaces below where water runs off, bitumen and fire-retardant treated wood containing salts are corrosive to copper
Galvanized steel	28 gauge/0.015 in.	Durable and easy to paint	Difficult to solder, corrodes early in acidic and salty air
Copper laminates	5 oz. (copper)	Easy to form and join	Fabric degrades in UV light, more easily torn than full copper
Rubberized asphalt	40 mil	Fully adhered, separate lap adhesive not needed, self-healing, flexible, easy to form and join	Full support required, degrades in UV light, metal drip edge required, difficult adhesion in cold weather, surfaces must be clean and some require priming
PVC	30 mil	Easy to form and join, non-staining, low cost	Easily damaged, full support required, metal drip edge required, questionable durability, embrittled and cracked by age and thermal cycling. Not recommended

FIGURE 4-16 Flashing types and properties.

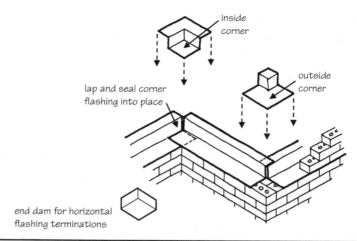

Figure 4-17 Prefabricated flashing corners and end dams.

Prefabricated corners and end dams make flashing installation easier but must be compatible with the flashing material (*see Fig. 4-17*). All joint sealant installed in contact with masonry flashing must be compatible so that adequate adhesion is developed. If specific types of joint sealants are required, the flashing material should be selected with compatibility in mind, and vice versa. If necessary, field adhesion tests can be performed to verify adhesive capability, but manufacturers should be consulted regarding long-term performance.

The cost of flashing is minimal compared with the overall construction budget, and it is counterproductive to economize on flashing materials at the expense of durability. Flashing material selection should take into account the function, environment, and expected service life of the building. For institutional buildings and others that will be in service for long periods of time, only the most durable materials should be used.

4.6 Weeps

Masonry cavity walls and veneers are designed to drain moisture. Without effective weep holes in the course above flashings, walls collect moisture and hold it like a reservoir. The most common type of weep hole in brick masonry is the open-head joint, which provides the largest open area and thus the most effective evaporation and drainage. Mortar is left out of head joints every 24 in. for brick masonry, leaving open channels that are 3/8-in. wide × course height × veneer depth. For concrete block, weeps should be spaced every 32 in. The height of an open-head joint in CMU walls is not typically the full 8-in. height of a standard block. An opening about 3-in. high is sufficient. The primary drawback to open-joint weeps is appearance and the possible intrusion of insects and vermin. A dark shadow is created at each opening, particularly with light-colored units and mortar. The openings are so large, in fact, that building maintenance crews all too often caulk the weep holes shut, mistakenly thinking they are the source of leaks. Some products camouflage the open joints but still allow them to work properly. One is a vinyl or aluminum cover with louver-type

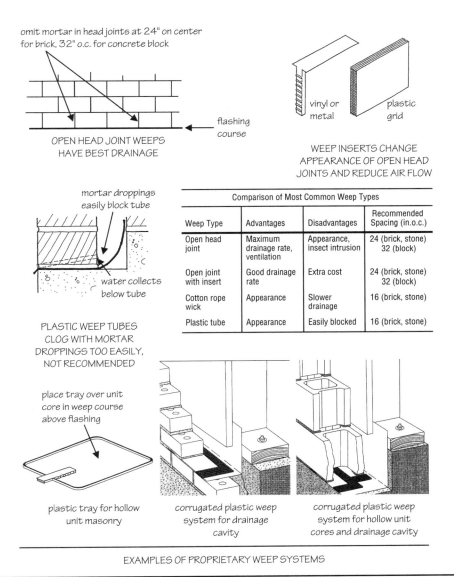

FIGURE 4-18 Weep accessories.

slots, and another is a plastic grid 3/8 in. wide × brick course height × veneer depth (*see Fig. 4-18*). A third type is made of woven filament. All three disguise the openings and still permit effective drainage and evaporation, but they do reduce airflow in the drainage cavity.

Small-diameter hollow plastic weep tubes are typically spaced at 16" on center because of their reduced drainage capacity. Weep tubes are easily clogged by mortar, debris, or insects and thus rendered ineffective. Some manufacturers make larger, rectangular tubes that measure 3/8 × 1½ × 3½ in. Because the opening is much larger, blockage is less likely, but the weep holes are more noticeable.

Cotton wicks are used to form another type of weep system. A ¼- to 3/8-in.-diameter rope is installed in the joints at 16 in. on center. The rope should be at least 12 in. long and extend through the exterior wythe and up into the cavity well above the height of possible mortar droppings. The top end of the rope should be fastened to the backing wall to prevent it from falling to the bottom of the cavity where it might be buried by mortar droppings. Moisture in the cavity is absorbed by the cotton material and wicked to the outside face of the wall, where it evaporates. This is a slower process than that of open weeps, and nylon or hemp rope will not perform well. The cotton will be wet throughout its service life and eventually will rot, leaving an open drainage hole. Using cotton wicks, however, ensures that drainage is not inadvertently blocked by mortar. Wicks are also inconspicuous in the wall and are particularly popular in rubble stone veneers with unevenly spaced head joints.

Another alternative is oiled rods or ropes, which are mortared into the head joints at 16 in. on center and then removed when the mortar has set. The rods function much the same as plastic tubes and share some of the same disadvantages. The 3/8-in.-diameter rods used are generally 3½ to 4 in. long, oiled slightly to prevent mortar bond, and extended through the veneer thickness to the core or cavity. The opening left after removal is a full 3/8 in. as the thickness of the tube shell is eliminated, but the hole is still small and easily blocked by mortar droppings. The oiled rope technique is similar to that of the wick system in that an unobstructed drainage path is provided. After the wall is completed to story height, the rope can be removed. The rope should be a minimum of 12 in. long to allow adequate height in the cavity. By removing the rope instead of using it as a wick, the hole provides more rapid drainage at the outset of construction, and its size makes it less noticeable than open-head joint weeps. The remaining hole, although it is not subject to blockage from mortar droppings, can nevertheless be penetrated by insects and vermin.

4.7 Drainage Accessories

Weep holes are not effective if the flow of moisture in the wall cavity is obstructed by mortar droppings. Although we want drainage cavities as free of mortar droppings as possible, it is unrealistic to expect perfection, so there will always and inevitably be some mortar buildup. It was once common practice to use a layer of pea gravel in the bottom of the cavity to promote drainage and keep mortar droppings away from the weeps. However, research proved that, rather than ensuring drainage in the presence of mortar droppings, the gravel merely raised the elevation of the mortar droppings but did not prevent them from obstructing water flow to the weeps. There are a number of proprietary products on the market, including several woven filament drainage mats, which are more effective than pea gravel in breaking up mortar droppings in the bottom of the cavity (*see Fig. 4-19*). Some flashing is available with mesh drainage mat or fabric attached to the surface, and some insulation boards include a drainage mat or drainage channels similar to that used for below-grade drainage systems. These products are intended to maintain a moisture flow path to the weeps but should be used in conjunction with the recommended workmanship techniques described in Chapter 15 for minimizing the amount of mortar droppings that fall into the wall cavity.

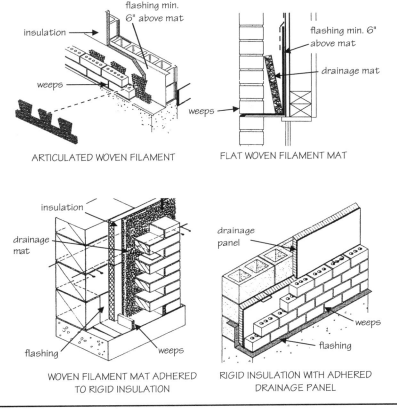

FIGURE 4-19 Drainage accessories.

4.8 Flashing and Drainage Systems

Flashing, weeps, and drainage accessories have traditionally been produced and marketed by separate manufacturers. There are now several manufacturers who produce integrated *systems* for moisture collection and drainage designed specifically for either solid or hollow masonry units, cavity walls and veneers, or single-wythe walls. These systems offer effective performance that includes all of the elements required for moisture control in masonry cavity, veneer, and single-wythe walls. In combination with units that are manufactured with an integral water-repellent admixture and supplemented with a field-applied surface water repellent, the appropriate drainage systems are particularly effective for single-wythe, hollow concrete block construction.

CHAPTER 5
Masonry Properties

Masonry is noncombustible and, in its various forms, can be used as both structural and protective elements in fire-resistive construction. Masonry is durable against wear and abrasion, and most types weather well without protective coatings. The mass and density of masonry also provide efficient thermal and acoustical resistance.

5.1 Durability and Freeze-Thaw Resistance

Masonry units and mortar are not susceptible to rust, rot, or decay from exposure to sun, wind, or water. Saturated units and mortar subjected to freezing temperatures can spall or flake at the surface, but good design and workmanship can minimize or eliminate this risk.

Freezing and thawing is not harmful to brick, concrete masonry unit (CMU), stone, or mortar, but freezing and thawing when the units or mortar are saturated with water can cause surface spalling and flaking. Water expands about 7% when it freezes. If the masonry units or mortar are saturated, there are no open pores or interstitial spaces into which the freezing water can expand, so the face of the unit or the mortar joint may eventually pop off. Freeze-thaw cycling does not affect a CMU because the pore structure is so much larger that there is plenty of room for the expanding ice crystals. Glazed brick are especially susceptible to freeze-thaw damage because the glazing traps moisture just under the surface and makes the spalling worse.

ASTM C216, *Standard Specification for Facing Brick,* Grade SW (severe weathering) bricks are required to have either a saturation coefficient less than 0.78 or a total average cold water absorption of less than 8%. If the units meet either of these criteria, they are classified as Grade SW and presumed to be sufficiently durable to withstand freeze-thaw cycling. For bricks that do not meet either of these two criteria, there is an alternate freeze-thaw test in ASTM C67, *Sampling and Testing Brick and Structural Clay Tile.* The freeze-thaw test is not a requirement for Grade SW brick and should be used only to qualify units that otherwise do not meet either the saturation coefficient or the maximum cold water absorption requirement. Bricks pass the ASTM C67 freeze-thaw test if, after 50 freeze-thaw cycles, they do not exhibit significant weight loss, breaking, or cracking.

There are other factors that affect a given brick's ability to withstand cyclical freezing and thawing. There is a complex interaction of the moisture content of the brick at the time of freezing, the rate of cooling, the temperature, and the number of freeze-thaw cycles. Material properties can also affect the performance of brick. The rate of water absorption, surface drying characteristics, pore size distribution and interconnectedness, and material strength all affect how a brick will perform during freezing and thawing. Consequently, classification as a Grade SW brick alone does not guarantee good freeze-thaw resistance.

The severity of freeze-thaw exposure and its effect on the durability of a brick wall can also be affected by workmanship, mortar type, mortar air content, and the degree of saturation in a wall. Remember that freezing and thawing in itself is not harmful to brick, but freeze-thaw cycling when the brick is saturated with water can cause surface spalling. Researchers have recently proposed new test protocols and methodologies to predict the critical degree of saturation at which freeze-thaw damage will begin. The theory is as follows:

- There is a critical degree of saturation.
- Below the critical degree of saturation, no freeze-thaw damage will occur regardless of the number of freeze-thaw cycles.
- Above the critical degree of saturation, freeze-thaw damage is measurable after only a few cycles.

The best defense against freeze-thaw damage in brick masonry is good design and workmanship. It is impossible to prevent a wall from becoming saturated in wet climates, but it is possible to design a wall that can dry quickly and effectively. Attention to proper flashing and weep design, water-shedding details, and good workmanship promote drying and decrease the possibility of freeze-thaw damage even in harsh climates.

Cast stone elements can suffer freeze-thaw damage at the surface of the units. The Cast Stone Institute recommends specifying that the cast stone be tested using a modified version of ASTM C666, *Test Method for Resistance of Concrete to Rapid Freezing and Thawing*. Under controlled laboratory procedures, the stone samples are limited to a 5% cumulative percent weight loss due to spalling after 300 cycles of freezing and thawing.

5.2 Fire-Resistance Characteristics

Building fires are a serious hazard to life and property, and fire safety in construction is therefore a primary consideration of every architect, building owner, and code authority.

Fire regulations are concerned primarily with the safety of occupants, the safety of fire fighters, the integrity of the structure, and the reduction of damage. Construction must (1) limit the spread of fire within a building, (2) prevent fire spread to adjacent buildings, (3) maintain the integrity of occupant evacuation routes, and (4) allow for attack by fire services. The overall risk is reduced when noncombustible construction is used to construct or protect structural elements and to divide a building into compartments for the containment of fire. Noncombustible masonry and concrete construction provide the highest level of protection through fire wall containment and structural integrity. Noncombustible construction is not required in low-rise multifamily buildings, and standard fire ratings are misleading about the relative fire safety of different types of construction. "Fire-resistive" construction with combustible materials is an oxymoron, and sprayed fireproofing on steel framing is subject to abrasion and delamination, which leave the structure essentially unprotected.

5.2.1 Fire-Resistance Ratings

Fire-resistance ratings for masonry are based on the rate of heat transmission through the wall and the rate of temperature rise on the opposite face. Fire-resistance ratings for masonry are not based on time of structural failure like gypsum drywall assemblies because masonry walls do not fail structurally in a fire. Standard fire testing according to ASTM E119, *Standard Test Method for Fire Tests of Building Construction and Materials,* or *Underwriters Laboratories* (UL) standard 263, *Standard for Fire Tests of Building Construction and Materials,* involves controlled fire exposure followed by a hose stream test. The hose stream induces thermal shock and tests the structural integrity of the wall.

Prescriptive fire ratings in the *International Building Code* (IBC) and other codes are based on tables of component requirements for various assemblies to attain specific rating periods. Fire-resistance ratings for masonry may also be calculated based on the equivalent solid thickness of the units exclusive of open cores or cells. The calculation method provides a way to determine ratings for composite masonry assemblies other than those specific combinations listed in the code tables.

Code tables list the minimum thickness of a particular material or combination of materials required for ratings of 1, 2, 3, and 4 hours. The tables in *Figs.* 5-1 and 5-2 list

Material Type	Minimum Wall Thickness (inches) for Fire Resistance Rating of [§±†]			
	1-hr.	2-hr.	3-hr.	4-hr.
Solid brick of clay or shale[‡]	2.7	3.8	4.9	6.0
Hollow brick or tile of clay or shale, unfilled	2.3	3.4	4.3	5.0
Hollow brick or tile of clay or shale, grouted or filled with sand, pea gravel, crushed stone, slag, pumice, scoria, expanded clay, shale or fly ash, cinders, perlite, or vermiculite	3.0	4.4	5.5	6.6

[§] Equivalent thickness determined by the formula
$$T_E = V_n/LH$$
where:
- T_E = equivalent thickness of the clay masonry unit (inches)
- V_n = net volume of the clay masonry unit (inch^3)
- L = specified length of clay masonry unit (inches)
- H = specified height of clay masonry unit (inches)

[±] Calculated fire resistance between hourly increments listed may be determined by linear interpolation.

[†] Where combustible members are framed into the wall, the thickness of solid material between the end of each member and the opposite face of the wall, or between members set in from opposite sides, shall be not less than 93% of the thickness shown.

[‡] For units in which the net cross-sectional area of cored brick in any plane parallel to the surface containing the cores is at least 75% of the gross cross-sectional area measured in the same plane.

(*From* International Building Code)

FIGURE 5-1 Fire resistance of loadbearing and non-loadbearing clay masonry walls.

Minimum Required Equivalent Thickness For CMU Made Of Various Aggregates					
Type of Aggregate	Fire Resistance Rating (hours)				
	1	1½	2	2½	3
Pumice or expanded slag	2.1	2.7	3.2	3.6	4.0
Expanded shale, clay, or slate	2.6	3.3	3.6	4.0	4.4
Limestone, cinders, or unexpanded slag	2.7	3.4	4.0	4.5	5.0
Calcareous or siliceous gravel	2.8	3.6	4.2	4.8	5.3

Notes:
- Where combustible members are framed into the wall, the thickness of solid material between the ends of each member and the opposite face of the wall, or between members set in from opposite sides, shall not be less than 93% of the thickness shown.
- Minimum required equivalent thickness corresponding to the hourly fire resistance rating for units with a combination of aggregates shall be determined by linear interpolation based on the percent by volume of each aggregate used in manufacture.

(*From* International Building Code)

FIGURE 5-2 Fire resistance of loadbearing and non-loadbearing concrete masonry walls.

fire-resistance ratings for clay and concrete masonry walls from the IBC. Ratings for brick walls are a function of wall mass or thickness and depend to some extent on the percent of cored area in the individual units. Units with less than 25% cored area are considered solid, and units with more than 25% cored area are classified as hollow. An 8-in. hollow brick wall contains less mass than an 8-in. solid brick wall, and it therefore offers less resistance to fire and heat transmission.

For walls made up of combinations of masonry units or masonry units and plaster for which there is no listed rating, fire resistance can also be calculated based on the equivalent solid thickness of the units and the known fire-resistance characteristics of the materials. Code requirements in the IBC are based on a minimum required equivalent thickness for each rating. Equivalent solid thickness is the average thickness of solid material in the wall or unit and is calculated from the actual thickness and the percentage of solid material in the unit. Equivalent thickness is found by taking the total volume of a masonry unit, subtracting the volume of core or cell spaces, and dividing by the area of the exposed face of the unit.

Volume characteristics and equivalent thickness for some typical CMU are shown in *Fig. 5-3*. The fire resistance of concrete masonry is a function of both aggregate type and equivalent thickness. Aggregates have a significant effect on fire resistance. Lightweight aggregates such as pumice, expanded slag, clay, or shale offer greater resistance to the transfer of heat in a fire because of their increased air content. Units made with these materials require less thickness to achieve the same fire rating as a heavyweight aggregate unit. The table in *Fig. 5-2* lists aggregate types and equivalent thicknesses that will satisfy specific fire-rating requirements.

The application of plaster to one or both sides of a clay or concrete masonry wall increases the fire rating of the assembly. For portland cement plaster, the plaster thickness may be added to the actual thickness of solid units or to the equivalent thickness of hollow units in determining the rating. The methods used for calculating fire resistance are fully described in *Standard Method for Determining Fire Resistance of Concrete and Masonry Assemblies* (ANSI/ACI 216.1/TMS 0216) and in Section 721 of the IBC.

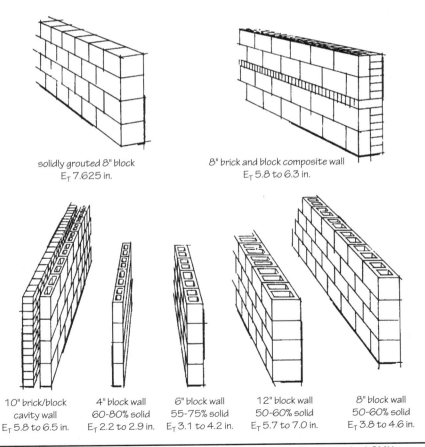

FIGURE 5-3 Volume characteristics and equivalent thickness (ET) of some typical CMU.

Glass block manufacturers typically offer units that have a 45-minute, 60-minute, or 90-minute fire-resistance rating. The 45-minute and 60-minute blocks can be used for "protected openings" in 1-hour–rated masonry and steel stud walls. These units can be installed with panel anchors or channel restraints (*see* Chapter 12). For 90-minute–rated construction, the 90-minute–rated block can be used only in masonry walls and only with channel restraints. Codes list size, area, and usage limitations that may apply.

5.2.2 UL Ratings

UL design numbers apply only to a specific proprietary product or assembly manufactured by a specific manufacturer or manufacturers. The fire-resistance ratings of clay and concrete masonry, in contrast, are generic. They apply to all products made from the same types of raw materials. Consequently, UL identifies masonry products by their classification rather than by design numbers. For example, Class B-4 CMU have a 4-hour rating, Class C-3 CMU have a 3-hour rating, and Class D-2 units have a 2-hour rating. The *UL Fire Resistance Directory* then lists CMU manufacturers who are eligible to issue a UL certificate for one or more of these classifications. The directory also gives UL numbers for several tested masonry wall assemblies. All of these assemblies were tested so that a specific manufacturer could show that a particular product (mortar mix or insulation insert, for example) could be added to or substituted in a "standard" masonry assembly and still achieve the same fire rating. Because they are proprietary materials, glass block manufacturers often have their units tested and listed in the UL directory.

Most of the masonry wall assemblies "listed" in the UL directory are too proprietary to apply to masonry construction in general. The UL numbers for these other assemblies are not appropriate if any of the component materials vary from the specific brand or type of products identified, including such items as veneer anchors or lime. For masonry, the more appropriate way to note construction documents is to reference the building code and table from which the rating requirement is taken and require that unit manufacturers provide test reports or certifications attesting to the fire-endurance ratings of their products.

5.2.3 Steel Fireproofing

Steel frame construction is vulnerable to fire damage and must be protected from heat and flame. Fire test results from the National Bureau of Standards form the basis of modern code requirements for protection of steel structural elements. Tables in the IBC show protective masonry coverings that are acceptable for various fire ratings.

5.2.4 Compartmentation

A key element in fire control and balanced fire protection is compartmentation of a building to contain fire and smoke. Codes require that a building be subdivided by fire walls into areas related in size to the danger and severity of fire hazard involved. Fire walls must be constructed of noncombustible materials, have a minimum fire rating of 4 hours, and have sufficient structural stability under fire conditions to allow collapse of construction on either side without collapse of the wall. Masonry fire walls may be designed as continuously reinforced cantilevered sections. They are self-supporting without depending on connections to adjacent structural framing. For additional lateral stability, free-standing cantilever walls may be stiffened by integral masonry pilasters

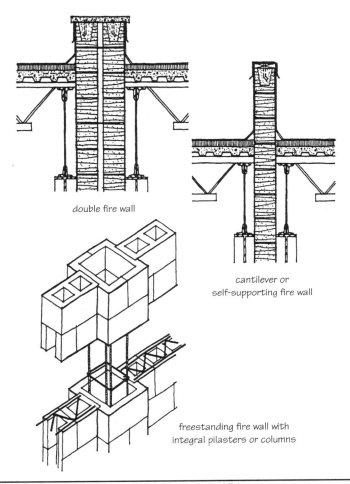

Figure 5-4 Masonry fire walls. (*From NCMA TEK Bulletin 95.*)

or columns with vertical reinforcing steel (*see Fig. 5-4*). Double fire walls can also be used, so that if the building frame on one side collapses, half the wall can be pulled over while the other half still protects adjacent areas. Masonry walls also provide a barrier against the spread of smoke and toxic gases.

5.3 Thermal Properties

The thermal efficiency of a building material is normally judged by its resistance to heat flow. A material's *R*-value is a measure of this resistance taken under laboratory conditions with a constant temperature differential from one side to the other.

Thermal resistance depends on the density of the material. By this measure, masonry is a poor insulator. Urethane insulation, in contrast, has a very high resistance because it incorporates closed cells or air pockets to inhibit heat transfer. The thermal resistance characteristics of concrete masonry vary with aggregate type and density. Exact values may be easily determined from basic information.

5.3.1 Heat Capacity

Heat flows from hot to cold. As the temperature rises on one side of a wall, heat begins to migrate toward the cooler side. Before heat transfer from one space to another can be achieved, the wall itself must undergo a temperature increase. The amount of thermal energy necessary to produce this increase is directly proportional to the weight of the wall. Masonry is heavy, so it can absorb and store heat and substantially retard its migration. This characteristic is called *heat capacity*. One measure of heat capacity is the elapsed time required to achieve equilibrium between inside- and outside-wall surface temperatures. The midday solar radiation load on the south face of a building will not completely penetrate a 12-in. solid masonry wall for approximately 8 hours.

The effects of wall mass on heat transmission are dependent on the magnitude and duration of temperature differentials during the daily cycle. Warm climates with cool nights benefit most. Seasonal and climatic conditions with only small daily temperature differentials tend to diminish the benefits.

Thermal lag and heat capacity are of considerable importance in calculating heat gain when outside temperature variations are great. During a daily cycle, walls with equal *U*-values but unequal mass will produce significantly different peak loads. The greater the heat capacity, the lower the total heat gain. Increased mass reduces actual peak loads in a building, thus requiring smaller cooling equipment. Building envelopes with more heat capacity will also delay the peak load until after the hottest part of the day, when solar radiation through glass areas is diminished and, in commercial buildings, after lighting, equipment, and occupant loads are reduced. This lag time decreases the total demand on cooling equipment by staggering the loads.

5.3.2 Added Insulation

The thermal performance of masonry walls and their resistance to heat flow can be improved by adding insulation. In severe winter climates where diurnal temperature cycles are of minimum amplitude, the thermal inertia of brick and block walls can be complemented by the use of resistance insulation such as loose fill or rigid board materials (*see Fig. 5-5*). Hollow units can easily be insulated with loose fill or granular materials, and multi-wythe cavity walls and veneer walls over wood or metal frame construction have open cavities for rigid insulating boards (*see Fig. 5-6*). The proper selection of insulating materials for masonry walls depends on more than just thermal performance:

- The insulation must not interfere with proper cavity drainage.
- Thermal insulating efficiency must not be impaired by retained moisture from any source (e.g., rain penetration or vapor condensation within the cavity).
- Granular fill materials must be able to support their own weight without settlement to ensure that no portion of the wall is without insulation.
- Insulating materials must be inorganic, or be resistant to rot, fire, and vermin.
- Granular insulating materials must be "pourable" in lifts of at least 4 ft for practical installation.

Material	Density (lb/ft^3)	R-Value per Inch of Thickness	Water Vapor Permeability (perm-In.)	Water Absorption (% by weight)	Dimensional Stability
Molded polystyrene	0.9-1.8	3.6-4.4	1.2-5.0	2-3§	no change
Extruded polystyrene	1.6-3.0	4.0-6.0	0.3-0.9	1-4	no change
Polyurethane, unfaced	1.7-4.0	5.8-6.2	2.0-3.0	negligible	0-12% change
Polyisocyanurate, unfaced	1.7-4.0	5.8-7.8	2.5-3.0	negligible	0-12% change
Perlite, loose fill	5.0-8.0	2.63	100	low	settles 0-10%
Vermiculite, loose fill	4.0-10.0	2.4-3.0	100	none	settles 0-10%

§ Water absorption given as percent by volume for molded polystyrene only.

FIGURE 5-5 Properties of insulation materials. (*From* Architectural Graphic Standards, 9th edition, *John Wiley & Sons.*)

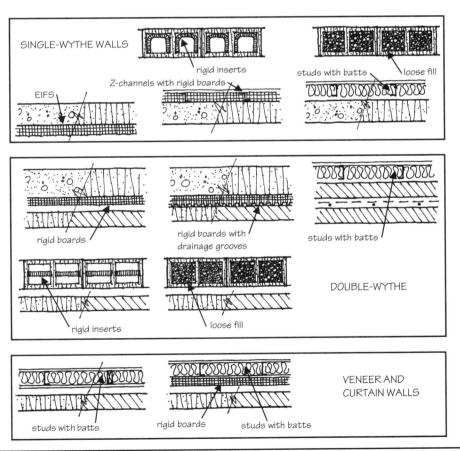

FIGURE 5-6 Methods of insulating concrete and masonry walls. (*From C. Beall and R. Jaffe,* Concrete and Masonry Databook, *McGraw-Hill, 2003.*)

5.3.3 Granular Fills

Two types of granular fill insulation have been tested by researchers at the Brick Industry Association and found to comply with these criteria: water-repellent–treated *vermiculite* and *perlite* fills.

Vermiculite is an inert, lightweight insulating material made from aluminum silicate expanded into cellular granules about 15 times their original size. Perlite is a white, inert, lightweight granular insulating material made from volcanic siliceous rock expanded up to 20 times its original volume. Specifications for water-repellent–treated vermiculite and perlite are published by the Vermiculite Association and the Perlite Institute, Inc. Each of these specifications contains limits on density, grading, thermal conductivity, and water repellency. Loose fill insulation should not settle more than 0.5% after placement; if it does, a thermal bridge will be created at the top of the wall.

Cavity wall construction permits natural drainage of moisture or condensation. If insulating materials absorb excessive moisture, the cavity can no longer drain effectively, and the insulation acts as a bridge to transfer moisture across the cavity to the interior wythe. Untreated vermiculite and perlite will accumulate moisture and suffer an accompanying decrease in thermal resistance.

Loose fill insulation is usually poured directly into the cavity from the bag or from a hopper placed on top of the wall. Pours can be made at any convenient interval, but the height of any pour should not exceed 20 ft. Rodding or tamping is not necessary and may in fact reduce the thermal resistance of the material. The insulation in the wall should be protected from weather during construction, and weep holes should be screened to prevent the granules from leaking out or from plugging the drainage path.

5.3.4 Rigid Board Insulation and Insulation Inserts

Rigid board insulations can be used in masonry cavity and veneer walls. Extruded polystyrene is the most moisture resistant and the most widely used. Some proprietary insulation products have drainage mats or drainage grooves designed to prevent mortar extrusions from obstructing the flow of moisture to weep holes (*see Fig. 5-7*). Air circulation behind the insulation will reduce effectiveness, so rigid insulation should be well adhered or fastened tightly to the backing wall. Gaps between boards will also reduce effectiveness.

Generally, rigid insulation is installed against the cavity face of the backing wall. A minimum of 2 in. should be left between the cavity face of the exterior wythe and the insulation board to facilitate construction and allow for drainage of the cavity. Insulation boards that are intended for use in masonry cavity walls are available in widths that will fit between the protruding joint reinforcement "eyes" in the backing wythe. Once the insulation is wedged between the eyes, the pintle ties are attached as the outer wythe of the wall is laid.

Some concrete block manufacturers produce units with rigid insulation inserts installed at the plant prior to shipment. These inserts may be of polystyrene or polyurethane and vary in shape and design for different proprietary products (*see Fig. 5-8*). Hollow unit cores are sometimes filled on site with foamed-in-place insulation, but the foam will prevent free drainage of moisture to the weeps in a single-wythe wall. Some foam insulations also absorb and hold huge amounts of water if there are any water penetration sources in the wall.

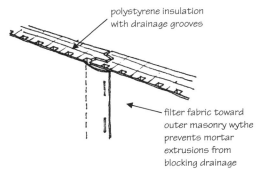

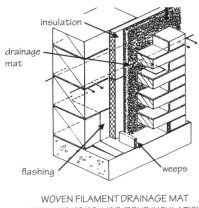

Figure 5-7 Proprietary rigid insulation designed to maintain moisture drainage from masonry wall cavity.

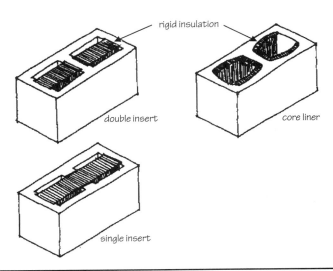

Figure 5-8 Rigid insulation inserts for hollow CMU.

R-Value of Insulated and Uninsulated Single-Wythe CMU Walls						
Nominal Wall Thickness (in.)	Unit Cores	Concrete Unit Weight				
		60 pcf	80 pcf	100 pcf	120 pcf	140 pcf
4	insulated	3.36	2.79	2.33	1.92	1.14
	uninsulated	2.07	1.68	1.40	1.17	0.77
6	insulated	5.59	4.59	3.72	2.95	1.59
	uninsulated	2.25	1.83	1.53	1.29	0.86
8	insulated	7.46	6.06	4.85	3.79	1.98
	uninsulated	2.30	2.12	1.75	1.46	0.98
10	insulated	9.35	7.45	5.92	4.59	2.35
	uninsulated	3.00	2.40	1.97	1.63	1.08
12	insulated	10.98	8.70	6.80	5.18	2.59
	uninsulated	3.29	2.62	2.14	1.81	1.16

Figure 5-9 Aggregate weight affects thermal resistance of concrete masonry. (*From NCMA* TEK Bulletin 38A.)

Foamed-in-place, loose fill, and most insulation inserts leave thermal bridges at the webs of hollow masonry units because the insulation is not continuous. Thermal bridging in single-wythe walls not only affects heat transfer and energy use but also may cause condensation. Lightweight CMU has higher thermal resistance than that of units made with heavy aggregate, so the effect of thermal bridging is somewhat modified and the benefits of added insulation somewhat greater (*see Fig. 5-9*).

5.3.5 Insulation Location

The most effective thermal use of massive construction materials is to store and re-radiate heat, so insulation location should be based on climatic exposure. In the thermal research conducted by the National Institute of *Standards and Technology* (NIST) and the *National Concrete Masonry Association* (NCMA), the effects of variable insulation location were studied. It was found that indoor winter temperature fluctuations were reduced by half when insulation was placed on the outside rather than the inside of the wall, and that the thermal storage capacity of the masonry was maximized. In cavity walls, performance in hot and cold climates is improved if the insulation is placed in the cavity rather than on the inside surface. Insulation location can affect the potential for condensation, so vapor flow as well as heat flow should be considered in optimizing wall performance.

5.4 Vapor and Air Resistance

Under certain conditions of design, it may be necessary to add a vapor retarder or air barrier to building walls to control both surface and interstitial condensation. Masonry construction is neither vapor nor air resistant. Masonry walls have a relatively high vapor permeance rating but typically allow less air penetration than light-framed metal or wood stud walls. The open pore structure of lightweight CMU allows some air

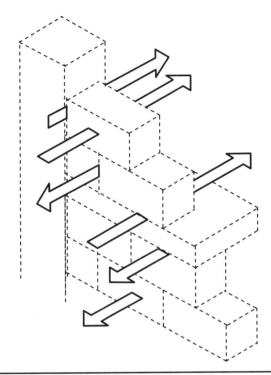

FIGURE 5-10 Air permeance of a CMU wall.

movement directly through single wythes of units (*Fig. 5-10*). Air penetration also occurs at the edges of masonry walls where they are built as in-fill between the beams and columns of steel or concrete building frames.

To achieve air or vapor diffusion resistance, masonry walls can be supplemented as needed by a variety of liquid, mastic, membrane, or sheet materials as vapor retarders and/or air barriers. Every project should be analyzed to determine the risk of condensation, including climate and building orientation. Vapor- and air-resistance characteristics can then be developed as needed for any type of masonry wall. (Chapter 7 describes condensation control in greater detail.)

5.5 Acoustical Properties

Steps can be taken in masonry design to absorb sound or prevent its transmission through walls, floors, and ceilings. Clay and concrete masonry walls and partitions have been tested and found to provide good sound insulation.

5.5.1 Acoustical Characteristics of Masonry

Noise is transmitted through building walls and partitions in three ways:

- As airborne sound through open windows or doors, through cracks around doors, windows, and wall penetrations

- As airborne sound through walls and partitions
- By vibration of the structure

Acoustical control includes absorbing the sound hitting a wall so that it will not reverberate and preventing sound transmission through walls into adjoining spaces.

Sound absorption involves reducing the sound emanating from a source within a room by diminishing the sound level and changing its characteristics. Sound is absorbed through dissipation of the sound-wave energy. The extent of control depends on the efficiency of the room surfaces in absorbing rather than reflecting these energy waves. *Sound transmission* deals with sound traveling through barriers from one space into another. To prevent sound transmission, walls must have enough density to stop the energy waves. With insufficient mass, the sound energy will penetrate the wall and be heard beyond it.

The acoustical characteristics of masonry may be subdivided into two categories: (1) sound absorption and reflectance, which depend primarily on surface texture, and (2) sound transmission, which is a function of density and mass (*see Fig. 5-11*).

The density of clay masonry determines its acoustical characteristics. Although sound absorption is almost negligible, the mass of a brick wall provides excellent resistance to the transmission of sound. This suggests best use as partitions or sound barriers between areas of different occupancy.

Normal-weight or heavyweight CMU also have high resistance to sound transmission. They will produce walls with higher STC ratings than those of lightweight units. Sound absorption is higher for coarse, open-textured surfaces with large pores. CMU can absorb from 18 to 68% of the sound striking the face of the wall, with lightweight units having the higher values.

5.5.2 Sound Absorption

Sound absorption relates to the amount of airborne sound energy absorbed on the wall adjacent to the sound. Sound is absorbed by porous, open-textured materials and by carpeting, furniture, draperies, or anything else in a room that resists the flow of sound and keeps it from bouncing around. If the room surfaces were capable of absorbing all

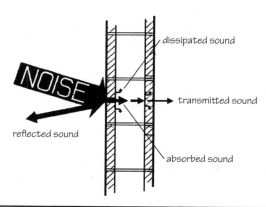

FIGURE 5-11 Sound absorption, reflectance, and transmission. (From National Concrete Masonry Association.)

sound generated within the room, they would have a sound absorption coefficient (SAC) of 1.0. If only 50% of it were absorbed, the coefficient would be 0.50.

Masonry, wood, steel, and concrete all have low sound absorption, ranging from 2 to 8%. Dense brick and heavyweight concrete block will have 1 to 3%, whereas lightweight block may be as high as 5%. Painting of the surface of a masonry wall effectively closes the pores of the material and reduces its absorptive capability even further. Conventional masonry products absorb little sound because of their density. Specially designed structural clay tile and concrete block units have higher sound absorption, with little or no sacrifice of strength or fire resistance. Acoustical clay tile units have a perforated face shell with the adjacent hollow cores filled at the factory with a fibrous glass pad. Perforations may be circular or slotted, uniform or variable in size, and regular or random in pattern. CMU sound blocks have NRC ratings from 0.45 to as high as 0.85.

5.5.3 Sound Transmission

Although it is an important element in control of unwanted noise, sound absorption cannot take the place of sound insulation or the prevention of noise transmission through building elements. Sound transmission loss is the total amount of airborne sound lost as it travels through a wall or floor. Each type may be identified at a particular frequency or by class.

Sound energy is transmitted to one side of a wall by air. The impact of the successive sound waves on the wall sets it in motion like a diaphragm. Through this motion, energy is transmitted to the air on the opposite side. The amount of energy transmitted depends on the amplitude of vibration of the wall, which in turn depends on four things:

- The frequency of the sound striking the surface
- The mass of the wall
- The stiffness of the wall
- The method by which the edges of the wall are anchored

The *sound transmission loss* (STL) of a wall is a measure of its resistance to the passage of noise or sound from one side to the other. If a sound level of 80 dB is generated on one side and 30 dB is measured on the other, the reduction in sound intensity is 50 dB. The wall therefore has a 50-dB STL rating. The higher the transmission loss of a wall, the better is its performance as a sound barrier.

5.5.4 STC Ratings

STC ratings represent the overall ability of an assembly to insulate against airborne noise. The higher the STC rating a wall has, the better the wall performs as a sound barrier. For homogeneous walls, resistance to sound transmission increases with unit weight. When surfaces are impervious, sound is transmitted only through diaphragm action. The greater the inertia or resistance to vibration, the greater is the ability to prevent sound transfer. The initial doubling of weight produces the greatest increase in transmission loss.

Porosity, as measured by air permeability, significantly reduces transmission loss through a wall. STC values vary inversely with porosity. Unpainted, open-textured

CMU, for instance, will have lower STC values than would be expected on the basis of unit weight alone. Porosity can be reduced, and STC values increased, by sealing the wall surface. The STC value is increased by about 10% with two coats of paint or plaster.

Cavity walls have greater resistance to sound transmission than that of solid walls of equal weight. Having two wythes separated by an air space interrupts the diaphragm action and improves sound loss. Up to about 24 in., the wider the air space, the more sound efficient the wall will be. Cavity walls are very effective where a high transmission loss, of the order 70 to 80 dB, is required. If the wythes are only an inch or so apart, the transmission loss is less because of the coupling effect of the tightly enclosed air. For maximum benefit, the wythes should be farther apart.

Some building codes incorporate standards for sound transmission characteristics in buildings. The standards generally specify minimum STC ratings for party wall and floor-ceiling separations. Party walls generally require an STC of 45 to 50. STC ratings for brick and block masonry walls are determined based on the standard methodology described in *The Masonry Society* (TMS) 0302, *Standard Method for Determining the Sound Transmission Class for Masonry Walls*. Tabulated values are shown in *Figs. 5-12 through 5-15*. STC ratings for glass block will vary slightly among manufacturers but generally range from about 35 to 50 depending on the type of block.

5.6 Green Buildings

The *American Society for Testing and Materials* (ASTM) Subcommittee E50.06 on Green Buildings defines the term *green building* as "building structures . . . that are designed, constructed, operated and demolished in an environmentally enhanced manner." Areas of particular concern include resource efficiency, energy efficiency, pollution control, waste minimization, and indoor air quality.

	Calculated STC Ratings for Clay Masonry Walls[§]							
	Hollow Units		Grout Filled		Sand Filled		Solid Units	
Nominal Wall Thickness (in.)	Weight	STC	Weight	STC	Weight	STC	Weight	STC
3	—	—	—	—	—	—	30	45
4	20	44	38	47	32	46	35	46
6	32	46	63	51	50	49	55	50
8	42	48	86	55	68	52	75	53
10	53	50	109	60	86	55	95	57
12	62	51	132	64	104	59	115	61

[§] Based on unit dimension at smaller of specified less manufacturing tolerance; clay density of 120 lb/ft³; grout density of 144 lb/ft³; and sand density of 100 lb/ft³. STC values for grout filled and sand filled units assume the materials completely fill all void areas in and around the units. STC values for solid units are based on bed and head joints solidly filled with mortar.

Figure 5-12 STC ratings for clay masonry walls. (From The Masonry Society's *Standard Method for Determining the Sound Transmission Class Rating for Masonry Walls*, TMS 0302-00.)

Calculated STC Ratings for Lightweight Concrete Masonry Walls[§]

Nominal Wall Thickness (in.)	Density	Hollow Units	Grout Filled	Sand Filled	Solid Units
4	80	43	45	45	45
	85	43	46	45	45
	90	44	46	45	45
	95	44	46	45	45
	100	44	46	45	46
6	80	44	49	47	47
	85	44	49	47	47
	90	44	50	48	48
	95	44	50	48	48
	100	45	50	48	49
8	80	45	53	50	50
	85	45	53	50	50
	90	45	53	50	51
	95	46	53	51	51
	100	46	54	51	52
10	80	46	56	52	52
	85	46	56	53	53
	90	47	57	53	53
	95	47	57	53	54
	100	47	57	54	55
12	80	47	60	55	55
	85	47	60	55	55
	90	48	60	56	56
	95	48	61	56	57
	100	48	61	57	58

[§] Based on grout density of 140 lb/ft^3 and sand density of 90 lb/ft^3. Percent solid thickness of units based on mold manufacturer's literature for typical units as follows: 4 in. = 73.8% solid, 6 in. = 55% solid, 8 in. = 53% solid, 10 in. = 51.7% solid, and 12 in. = 48.7% solid. STC values for grout-filled and sand-filled units assume the materials completely fill all void areas in and around the units. STC values for solid units are based on bed and head joints solidly filled with mortar.

FIGURE 5-13 STC ratings for lightweight concrete masonry walls. (From The Masonry Society's *Standard Method for Determining the Sound Transmission Class Rating for Masonry Walls*, TMS 0302-00.)

The green building movement seeks to identify building materials that minimize environmental impacts in their creation and use and minimize health risks to building occupants. But there is no such thing as an environmentally perfect material. Product selection for green buildings is therefore a process of evaluation and compromise, seeking the best overall solution for a given program and budget. For example, steel may have more embodied energy than wood, but steel framing is more efficient and can produce smaller structural members and longer spans. Ceramic tile is more energy intensive than hardwood for flooring but requires no finish coatings and no chemical cleaners for maintenance. By the same token, masonry products may require more energy to produce than some other building materials, but their performance characteristics, durability, and chemical stability usually justify similar trade-offs.

Calculated STC Ratings for Medium Weight Concrete Masonry Walls§					
Nominal Wall Thickness (in.)	Density	Hollow Units	Grout Filled	Sand Filled	Solid Units
4	105	44	46	46	46
	110	44	46	46	46
	115	44	47	46	46
	120	45	47	46	47
6	105	45	50	48	49
	110	45	50	49	49
	115	45	51	49	50
	120	45	51	49	50
8	105	46	54	51	52
	110	47	54	52	53
	115	47	55	52	53
	120	47	55	52	54
10	105	48	58	54	55
	110	48	58	54	56
	115	48	58	55	57
	120	49	59	55	57
12	105	49	62	57	59
	110	49	62	57	59
	115	49	62	58	60
	120	50	63	58	61

§ Based on grout density of 140 lb/ft³ and sand density of 90 lb/ft³. Percent solid thickness of units based on mold manufacturer's literature for typical units as follows: 4 in. = 73.8% solid, 6 in. = 55% solid, 8 in. = 53% solid, 10 in. = 51.7% solid, and 12 in. = 48.7% solid. STC values for grout-filled and sand-filled units assume the materials completely fill all void areas in and around the units. STC values for solid units are based on bed and head joints solidly filled with mortar.

FIGURE 5-14 STC ratings for medium-weight concrete masonry walls. (From The Masonry Society's *Standard Method for Determining the Sound Transmission Class Rating for Masonry Walls*, TMS 0302-00.)

Masonry's multifunctional properties have always made it an attractive choice as a building material. From an environmental standpoint, this ability to serve more than one purpose is a particular bonus. Coatings are generally not required because most types of masonry already have a finished surface. Sound batts are not required because the masonry has inherent sound-damping capacity. Fireproofing is not required because masonry is noncombustible. And structural framing is eliminated in buildings where loadbearing systems can be used. The thermal mass of masonry can reduce the amount of insulation material required in some climates. It can also, when properly integrated with passive solar design techniques, reduce total energy consumption and reduce utility service demand through off-peak loading. Such multifunctional applications, as well as the long service life and low maintenance traditionally associated with masonry buildings, mean that the energy embodied in the materials goes further and delivers more than that embodied in many other materials. The rehabilitation of historic buildings, many of which are masonry, conserves the embodied energy already invested in such structures.

Masonry Properties

Nominal Wall Thickness (in.)	Density	Hollow Units	Grout Filled	Sand Filled	Solid Units
\multicolumn{6}{l}{Calculated STC Ratings for Normal Weight Concrete Masonry Walls§}
4	125	45	47	46	47
	130	45	47	46	47
	135	45	47	47	47
	140	45	49	47	48
	145	46	48	47	48
	150	46	49	47	48
6	125	46	51	49	51
	130	46	51	49	51
	135	46	52	50	51
	140	46	52	50	52
	145	47	52	50	52
	150	47	52	50	53
8	125	47	55	52	54
	130	48	55	53	55
	135	48	56	53	55
	140	48	56	53	56
	145	49	56	54	56
	150	49	57	54	57
10	125	49	59	56	58
	130	49	60	56	59
	135	50	60	56	59
	140	50	60	57	60
	145	50	61	57	61
	150	51	61	57	61
12	125	50	63	59	62
	130	51	63	59	63
	135	51	64	59	63
	140	51	64	60	64
	145	52	65	60	65
	150	52	65	61	66

§ Based on grout density of 140 lb/ft^3 and sand density of 90 lb/ft^3. Percent solid thickness of units based on mold manufacturer's literature for typical units as follows: 4 in. = 73.8% solid, 6 in. = 55% solid, 8 in. = 53% solid, 10 in. = 51.7% solid, and 12 in. = 48.7% solid. STC values for grout-filled and sand-filled units assume the materials completely fill all void areas in and around the units. STC values for solid units are based on bed and head joints solidly filled with mortar.

FIGURE 5-15   STC ratings for normal-weight concrete masonry walls. (From The Masonry Society's *Standard Method for Determining the Sound Transmission Class Rating for Masonry Walls*, TMS 0302-00.)

Generally, a building is evaluated throughout its life cycle, from construction through operation and demolition. The amount of energy consumed and the amount of waste generated at each phase, as well as the building's internal environment and its relationship to the external global environment, should enter into site considerations, design decisions, and product selections. Products and systems must demonstrate reduced life-cycle energy consumption, increased recycled content, and minimal waste products in manufacture, construction, use, and demolition. Such requirements may result in the introduction of mortarless interlocking masonry systems, a renewed interest in "bio-bricks," or the successful reintroduction of autoclaved cellular concrete block from Europe.

5.6.1 Resource Management, Recycled Content, and Embodied Energy

The raw materials for making clay brick are an abundant resource that is easily acquired and produces little waste. Clay-mining operations are regulated by the Environmental Protection Agency, and dormant pits have been reclaimed as lakes, landfills, and nature preserves. Recycled materials are not often used in the manufacture of clay brick, but additives such as oxidized sewage sludge, incinerator ash, fly ash, waste glass, paper-making sludge, and metallurgical wastes have been incorporated with varying degrees of success. The waste materials are either burned to complete combustion at the high kiln temperatures needed to bake the brick or encapsulated within the clay body where they cannot leach out. The primary energy cost associated with brick manufacturing is the fuel burned in the firing process.

Most brick kilns now use natural gas instead of coal. This has reduced sulfur dioxide emissions and also allows more precise control of fuel consumption. Waste heat from the firing kilns is also ducted and reused to dry unfired units. When the costs of transporting brick to job sites is factored in, the embodied energy is estimated at approximately 4000 Btu per pound of brick. Brick costs about 70% less to manufacture than it did in 1970.

The primary ingredients in CMU are the sand and aggregates, which account for as much as 90% of a unit's composition. These materials are abundant, easily extracted, and widely distributed geographically. Recycled materials such as crushed concrete or block and by-products such as blast furnace slag, cinders, and mill scale can be used for some of the aggregate. The portland cement used as the binder in concrete masonry is energy intensive in its production, but it accounts for only about 9 to 13% of the unit. Energy consumption for cement production has decreased 25% during the past 20 years, mostly as a result of more efficient equipment and production methods. The proportion of portland cement in CMU can be reduced by substituting fly ash, which is a by-product of coal-fired power plants.

Natural stone uses less energy in its production and fabrication than that of other masonry materials, but its transportation costs can be significantly higher. It is not unusual for a stone to be quarried on one continent, shipped to another for fabrication, and to yet another for installation. The use of local or regional building stones greatly reduces transportation and embodied energy costs.

5.6.2 Construction Site Operations

Masonry construction is generally less hazardous to the environment than some other building systems because most of the materials used are chemically inert. Mortar-mixing and stone-cutting operations can generate airborne particulate wastes such as silica dust. Keeping aggregate piles covered and using water-cooled saws can reduce this hazard. Modular dimensioning of masonry can reduce job-site waste by limiting construction to the use of only whole and half-size units.

Cleaning compounds, mortar admixtures, coatings, and the chemicals used to clean and maintain equipment may include potentially hazardous materials. Precautions should be taken in the disposal of such products, and runoff should be controlled to prevent the migration of chemicals into natural waterways and municipal storm sewer systems. On small cleaning projects, this may be a simple matter of temporary flashings and catch basins, but on large projects this may become a complex task. The rinse material should be tested after cleaning a sample wall area to make sure it is safe to dispose of in the public storm sewer system.

5.6.3 Indoor Air Quality and Building Ecology

When the cost of energy went up dramatically in the 1970s, building standards began to change. Construction was tightened up to reduce or eliminate air leakage and the heat loss or heat gain associated with it. Ventilation standards also changed, reducing the number of air changes per hour that the mechanical systems delivered to increase the efficiency of heating and cooling systems. Unfortunately, these changes also led to increased concentrations of chemical air pollutants in buildings. Many building products contain substances that are known to pose health risks through continued exposure.

Masonry products are generally inert and do not contribute to indoor air quality problems. They contain no toxins or volatile organic compounds, do not emit any chemical pollutants as they age, and will not support mold growth, and none of the natural stone that is typically used in building is known to emit radon.

5.6.4 Masonry and LEED

Masonry can contribute a possible 24 points to the LEED categories, including "Sustainable Sites," "Energy & Atmosphere," "Materials & Resources," and "Innovation in Design" (*see Fig. 5-16*). Most specifically, masonry is strong in the "Energy & Atmosphere" and "Materials & Resources" categories.

5.7 Masonry Costs

Exterior envelope materials are usually selected on the basis of both cost and aesthetics. An architect or building owner may begin with a mental image of the project that is related to its context, its corporate identity, and its budget. Masonry is very cost-competitive as an envelope material, but the decision to use masonry of one type or another is usually an aesthetic one. Material selections are based on color, texture, and scale.

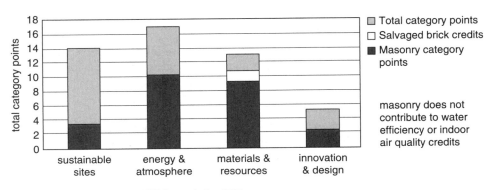

From The Masonry Society, TMS Responds, June 2004.

FIGURE 5-16 Potential contribution of masonry to LEED points.

The relative cost of different types of brick or different types of architectural block is related primarily to unit size and labor production. Within a selected size, however, aesthetic preference should govern unit selections because the cost of materials has only a small effect on the cost of the completed envelope. There are a number of other design and specification decisions that affect masonry cost, and these can be used to minimize budget limitations.

5.7.1 Factors Affecting Cost

Careful detailing and thoughtful design can enhance the cost economy of any building system. Conscientious planning and material selection, attention to detail, thorough specifications, and on-site field observation and inspection can all contribute to lower masonry costs. In masonry construction, unit size, unit weight, and modular dimensions have as much or more influence on mason productivity (and therefore on cost) as that of any other factors.

- Larger-face-size units increase the area of wall completed each day, even though the mason may lay fewer units because of greater weight. This option is simple and cost-effective. The higher price of larger units can be offset by lower labor costs and by earlier completion of the work. For some designs, larger masonry units may actually give better proportional scale with the size of the building as well.
- All other factors being equal, mason productivity decreases as unit weight increases. Selection of CMU weight (normal weight, medium weight, or lightweight) should be matched to project requirements for thermal resistance, fire resistance, water-penetration resistance, and loadbearing capacity.
- Running bond patterns generally increase mason productivity, whereas decorative patterns and even stack bond patterns decrease productivity.
- Colored mortar costs more than ordinary gray mortar.
- Proper planning with modular dimensions increases productivity and reduces cost.
- Analytically designed brick or CMU curtain walls can eliminate the need for shelf angles on buildings up to 100 ft or more in height.
- Mechanical and electrical lines and conduit are less expensive to place in double-wythe cavity walls than in single-wythe walls unless special concrete block units are used.
- Openings spanned with masonry arches or reinforced masonry lintels eliminate the need for steel angle lintels and the associated maintenance costs they include.

5.7.2 Value Engineering

In estimating the total cost of a building system or product, future as well as present costs must be considered. Value engineering and life-cycle costing methods evaluate expenses throughout the life of a building. For example, the fire resistance of masonry structures means lower insurance rates and lower repair costs if interior spaces do sustain damage from fire. Masonry thermal characteristics reduce energy consumption for heating and air conditioning, and the durability and finish of the surfaces also minimize maintenance costs.

The "value engineering" process seeks to provide better value for the owner on the overall structure. Value engineering usually means reducing costs, sometimes at the expense of quality. One consistent item subject to cost-reduction thinking is masonry flashing. This is not, however, a good place to save money. The quality of the masonry flashing is directly related to length of service and the ability of the walls to drain water. Building owners and designers should resist arguments made in favor of "value engineering" masonry flashing. It is a false savings in terms of building function.

CHAPTER 6
Expansion and Contraction

All building materials expand and contract to some degree with changes in temperature. Others also expand and contract with variations in moisture content. The thermal movement characteristics of most materials are known, and a standard coefficient can be used to calculate the expected expansion or contraction of a material for a given set of conditions. Concrete, stucco, and concrete masonry products shrink permanently, clay products expand permanently, and metals expand and contract reversibly. Such movement is accommodated through flexible anchorage and the installation of control joints in concrete masonry and expansion joints in brick.

6.1 Movement Characteristics

Masonry materials are relatively stable in thermal expansion and contraction compared with metals and plastics. In addition to thermal movement, however, brick experiences moisture expansion, and concrete masonry experiences moisture shrinkage. Some alternating shrinking and swelling occurs through normal wetting and drying cycles, but more important are the *permanent moisture expansion* of brick and the *permanent moisture shrinkage* of concrete masonry.

One of the principal causes of cracking in masonry walls is differential movement. All materials expand and contract with temperature changes, but at very different rates. All materials change dimension because of stress, and some develop permanent deformations when subjected to sustained loads. Masonry walls are much stronger than in the past because of high-strength units and portland cement mortars, but strength has come at the expense of flexibility. Masonry walls are relatively brittle and include thousands of linear feet of mortar joints along which cracks and bond line separations can occur. Cracking in the masonry can result from restraining the natural expansion or contraction of the materials themselves or from failure to allow for differential movement of adjoining or connected materials. Using masonry as we do today with steel and concrete structural frames requires careful consideration of the movement characteristics of each material.

6.1.1 Brick Expansion and Contraction

Fired bricks are at their smallest dimension when leaving the kiln. The natural moisture in the clay and the water added for forming and extrusion are evaporated during the firing process. After firing, clay products begin to rehydrate by absorbing atmospheric moisture. As the moisture content increases, the bricks expand permanently and irreversibly until the units reach equilibrium moisture content with their

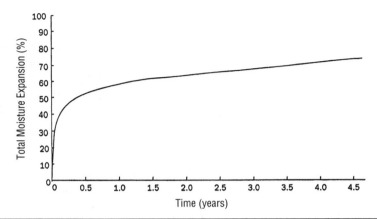

Figure 6-1 Graph showing permanent moisture expansion of clay brick over time. (*From BIA, Technical Note 18 Rev.*)

surrounding environment. For any given clay, the higher the firing temperature, the drier the brick and the greater the potential for moisture expansion. Some very minor expansion and contraction occurs with subsequent changes in moisture content, but it is the initial permanent and irreversible moisture expansion that governs brick masonry expansion joint design. The majority of brick moisture expansion occurs during the first year after manufacture, but it continues at a much slower rate for years (*see Fig. 6-1*).

Temperature changes cause minor dimensional variations in brick, but these changes are quite small and fully reversible, and the units return to their original size after being heated and cooled through the same temperature range.

6.1.2 Concrete Masonry Expansion and Contraction

Concrete masonry products are moist cured to hydrate the portland cement in the mix. Once the curing is complete, residual moisture evaporates, causing the units to shrink permanently and irreversibly until they reach an equilibrium moisture content with their surrounding environment. Temperature changes cause minor dimensional variations in concrete masonry units (CMU), but these changes are small and fully reversible, and the units return to their original size after being heated and cooled through the same temperature range. Coefficients of thermal expansion vary with different aggregates. The initial, permanent, and irreversible CMU moisture shrinkage significantly exceeds any subsequent thermal or moisture volume changes that might occur, so it is always this initial shrinkage that governs CMU control joint design.

Steel reinforcement increases the resistance of concrete masonry to the tensile stress of shrinkage. The most common method of shrinkage crack control is the use of horizontal joint reinforcement, which distributes the stress more evenly through the wall to minimize cracking. Control joints localize cracking to predetermined locations so that elastomeric sealants can be installed in the joint to prevent moisture penetration. Even though they are not exposed to the weather, interior CMU walls and partitions also need horizontal joint reinforcement and control joints to accommodate the permanent shrinkage the units experience as they lose their residual moisture. CMU walls that are reinforced to resist structural loads are also more resistant to shrinkage cracking.

6.1.3 Differential Movement

When clay and concrete masonry are combined or when masonry is combined with or attached to nonmasonry materials, cracking can be caused by the differential movement of the various materials or components. To avoid cracking and the increased moisture penetration it invites, differential movement must be accommodated by isolation, by flexible anchorage, and by expansion and control joints. Both restrained brick expansion and uncontrolled CMU shrinkage can cause cracking in masonry construction.

Cracks can occur at the corners of brick veneer walls or block/brick cavity walls. The brick walls expand in length more than the stud frame or CMU backing walls. Less so, brick/brick cavity walls can crack at the external corners because of the wider range of thermal and moisture exposure of the outer versus the inner wythe, especially if there is insulation in the open cavity.

Masonry walls built on concrete foundations that extend above grade can develop similar problems. The concrete contracts with moisture loss, and the brick expands with moisture absorption, so the masonry can slide off or break the foundation corners (see *Fig. 6-2*). Flashing at the brick lug serves as a bond break between the masonry and the foundation and allows independent movement without cracking. Roof and floor slabs poured directly on brick bearing walls can curl from the same opposing concrete shrinkage and brick expansion. If the slab warps, it can rupture the masonry at the building corners and cause horizontal cracks just below the slabs. To permit the differential movement without causing damage, a strip of flashing can be used to create a horizontal slip plane between the concrete slab and masonry. The slip

FIGURE 6-2 Opposing moisture expansion of brick and curing shrinkage of concrete slab can cause cracking or slippage at building corners if differential movement is not accommodated.

FIGURE 6-3 Parapets can expand and contract at a different rate than that of the building walls below.

plane should extend 12 to 15 ft back from the corners and terminate at a movement joint.

Brick parapet walls can develop similar problems. With both sides of the wall exposed, parapets are subject to temperature and moisture extremes. Differential expansion or contraction from the building wall below can cause parapets to bow, to crack horizontally at the roof line, and to overhang corners. The lateral stress is highest at the corners, where a brick facing wythe can literally push itself off the edge of the building (see *Fig. 6-3*). Flashing at the uppermost shelf angle at the roof line provides a convenient slip plane so that the parapet and the wall below can move independently of each other in the horizontal direction. This is good because the brick in a parapet will always expand a little more than the brick in the building wall. Additional expansion joints in the brick can accommodate differential movement without excessive sliding at the flashing plane and without bowing or cracking in the parapet itself. The additional joints should be added near the corners of the parapet. Locating expansion joints at the spacing recommended later in this chapter will absorb this movement and prevent the unintended cantilever.

Concrete structural frames can shorten over time from shrinkage or creep, and both concrete and steel frames can deflect. This shortening can transfer excessive stress to masonry cladding or in-fill walls. Failures are characterized by bowing, by horizontal cracks at shelf angles, by vertical cracks near corners, and by spalling of masonry units at window heads, shelf angles, and other points where stress is concentrated. Horizontal soft joints must be provided between the structural frame and the masonry to allow the frame to shorten without damage to the masonry.

Where interior or exterior steel columns are wrapped with masonry, the greater expansion and contraction of the column can be inadvertently transmitted to the masonry and cause cracking. To prevent this, a bond break material such as board insulation or roofing felt should be used to isolate the masonry from the steel, and flexible anchors should be used to accommodate the differential movement.

Long walls constructed without expansion joints can also develop shearing stresses in areas of minimum cross section, so diagonal cracks often occur between window and door openings, usually extending from the head or sill corners at openings. Vertical soil or foundation movements can also cause masonry cracking (*see Fig. 6-4*).

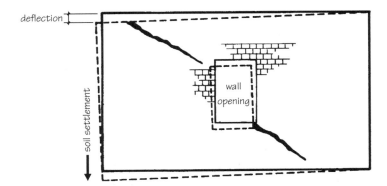

FIGURE 6-4 Diagonal, tapered cracking caused by foundation settlement at end of wall. (*From Clayford T. Grimm, "Masonry Cracks: A Review of the Literature,"* Masonry: Materials, Design, Construction, and Maintenance, *ASTM STP 992, H. A. Harris, Ed., ASTM International, West Conshohocken, PA.*)

6.2 Flexible Anchorage

Jointing details must provide flexible anchorage to accommodate differential movement. When masonry walls are connected to steel or concrete frame buildings, differential movement must be accommodated in the anchorage of one material to the other. Even if the exterior masonry veneer carries its own weight to the foundation without shelf angles or ledges, the columns or floors provide the lateral support that is required by code. Flexible connections should allow relative vertical movement without inducing stresses that could cause damage (i.e., they should resist the lateral tension and compression of wind loads but allow differential vertical movements). Several types of two-piece adjustable anchors were discussed in Chapter 4, and *Figs. 6-5 and 6-6* show ways in which they are used.

6.3 Movement Joints

In addition to the flexible anchorage of masonry backing and facing materials, control joints and expansion joints are used to alleviate the potential stresses caused by differential movement between materials and by thermal and moisture movement in the masonry itself. The terms *control joint* and *expansion joint* are not interchangeable. The two types of joints are different in both function and configuration (*see Fig. 6-7*).

Expansion joints in brick masonry cannot contain mortar or hard filler material because they must allow for the permanent moisture expansion of the brick as its increasing volume fills up the joint space. Expansion joints are open (void) joints that close as the brick expands. Expansion joints (or soft joints as they are often called) must be wide enough to accommodate the permanent, irreversible moisture expansion of the brick plus the alternating thermal expansion and contraction that occur throughout the life of the building.

104 Chapter Six

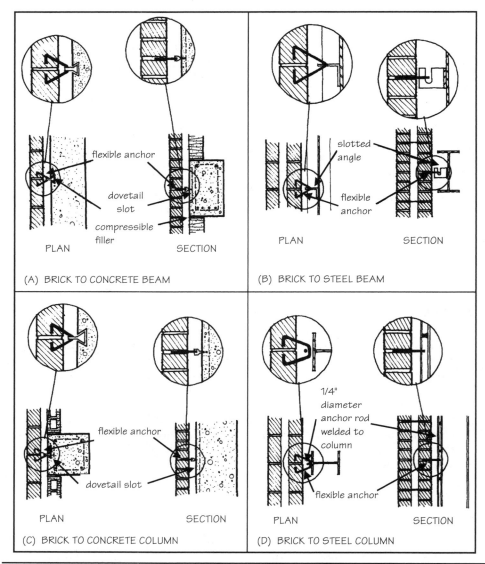

FIGURE 6-5 Flexible anchorage of brick masonry. (*From BIA*, Technical Note 18.)

Control joints in CMU construction do not have to be void of mortar or other hard fillers. Some types of CMU construction require that a wall be "keyed" across control joints, so it is often necessary to use a raked mortar fill or hard rubber shear key in CMU control joints. The permanent moisture shrinkage of a CMU exceeds its potential for subsequent thermal expansion, so once the CMU shrinks, the joint is opened up enough to accommodate the alternating thermal expansion and contraction that will occur

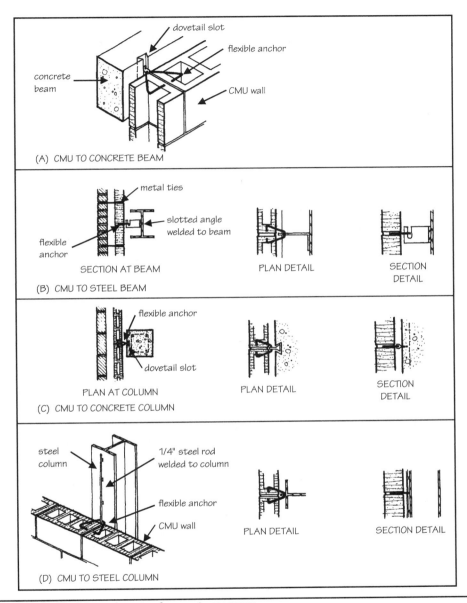

FIGURE 6-6 Flexible anchorage of concrete masonry.

throughout the life of the building. If it is not necessary to "key" the two adjacent sections of CMU wall, then an expansion joint will provide all the movement and crack control function that is necessary.

Isolation joints are used to separate masonry construction from adjacent or surrounding materials.

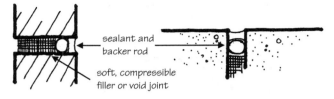

EXPANSION JOINTS IN MATERIALS, IN WALL SYSTEMS, OR IN BUILDINGS ALLOW FOR BOTH EXPANSION AND CONTRACTION

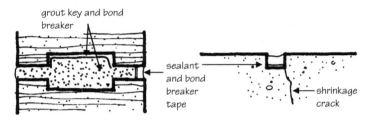

CONTROL JOINTS IN CONCRETE, STUCCO AND CONCRETE MASONRY CONTROL THE LOCATION OF SHRINKAGE CRACKS
(in concrete work, control joints are sometimes called contraction joints)

FIGURE 6-7 Expansion joints and control joints are different, and the terms should not be used interchangeably.

6.3.1 Joint Design

Control joints are continuous, weakened joints designed to accommodate the permanent shrinkage of portland cement–based products such as concrete, stucco, and concrete masonry. When stress development is sufficient to cause cracks, the cracking will occur at these weakened joints rather than at random locations. Horizontal joint reinforcement can be used to limit the size of shrinkage cracks, but strategically located control joints prevent the occurrence of random cracks and the increased moisture penetration they invite. Cracking is not as likely to occur in fully reinforced construction, as the reinforcing steel absorbs much more tensile stress than that absorbed by wire joint reinforcement.

In single-wythe CMU walls and CMU backing walls, control joints must also provide lateral stability between adjacent wall sections. *Figure 6-8* shows several common types of control joints, all of which provide a shear key for this purpose. Exterior control joints are sealed against moisture penetration with a backer rod or bond breaker tape and a high-quality silicone or urethane sealant at the exterior face of the wall.

Brick expansion joints cannot contain mortar or other hard materials because they must allow room for the brick to expand into the joint space (*see Fig. 6-9*). Compressible fillers may be used to keep mortar out of the joints during construction, as even small mortar bridges can cause localized spalling of the unit faces wherever movement is restricted. Filler materials and sealant backer rods should be at least as compressible as

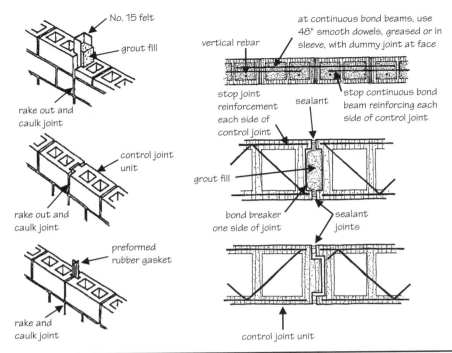

FIGURE 6-8 Vertical control joints in CMU for shrinkage crack control.

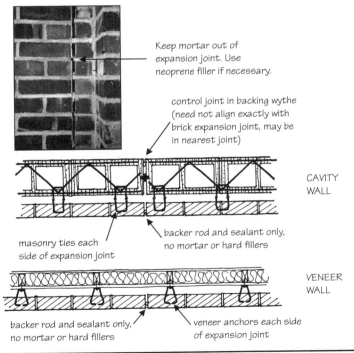

FIGURE 6-9 Expansion joints in brick masonry must be void of mortar or other hard materials that could restrict movement and cause localized spalling.

107

the joint sealant that will be used, and the compressibility of the sealant must be considered in calculating joint width.

Shelf angles are used in masonry veneer to support the dead load of a veneer at each floor or every other floor of a building. Shelf angles must be installed with a "soft joint" between the bottom of the angle and the top course of masonry below to accommodate differential vertical movement. This permits differential expansion and contraction of the veneer and structure to occur, as well as deflection and frame shortening, without the angle bearing on the veneer below (*see Fig. 6-10*). Shelf angles must provide continuous support around building corners (*see Fig. 6-11*) and should be bolted rather than welded in place to permit field adjustments (*see Fig. 6-12*). A bolted connection allows level adjustment of the angle and shimming outward if necessary. Bolted shelf angles are recommended over fixed welded connections. Flashing and weep holes must be installed above every shelf angle to collect moisture and drain it to the outside (*see* Chapter 7). *Figure 6-13* shows several types of soft joints for interior non-loadbearing CMU (or brick) partitions.

Vertical expansion and contraction in masonry parapets can cause distress in the coping. A cast stone coping may rotate and a metal coping may cup because of the differential movement between brick and CMU (*see Fig. 6-14*). The brick veneer is increasing in both height and length with permanent moisture expansion, and at the same time the concrete masonry is experiencing permanent moisture shrinkage. To minimize vertical movement in the parapet facing wythe, place a shelf angle at the roof level so that the brick above is limited to a few courses with less expansion potential.

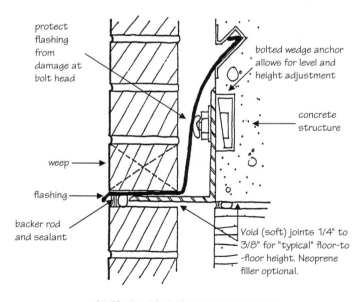

FIGURE 6-10 Soft joint below shelf angle.

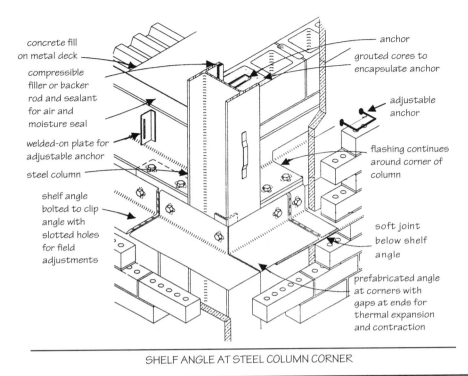

SHELF ANGLE AT STEEL COLUMN CORNER

FIGURE 6-11 Masonry shelf angles must provide continuous support at building corners. (*From W. Laska,* Masonry and Steel Detailing Handbook, *Aberdeen Group, 1993.*)

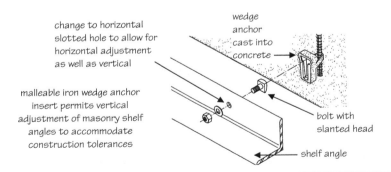

FIGURE 6-12 Adjustable shelf angle connection to concrete structural frame.

The required width for control joints and expansion joints can be determined by standard formulas to account for thermal movement, moisture movement, and the movement capability of the sealant itself in both extension and compression. ASTM C1472, *Standard Guide for Calculating Movement and Other Effects When*

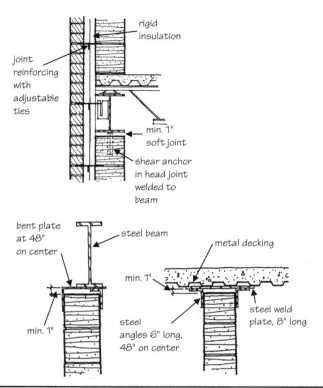

FIGURE 6-13 Horizontal soft joints to accommodate beam or floor deflection above.

Establishing Sealant Joint Width, provides formulas for calculating joint width that consider all of these factors as well as material fabrication and erection tolerances. When unanticipated construction tolerances result in joints that are wider or narrower than intended, sealant performance is affected. Narrow joints especially are a frequent cause of sealant joint failure. In order for the sealant to function properly, sealant industry sources recommend that for butt joints up to ½ in. wide, joint depth should be less than or equal to the width, with 2:1 a preferred ratio. Sealant depth should be constant along the length of the joint and should never be less than ¼ in. The calculations for joint width and spacing apply to continuous walls with constant height and thickness. Joint locations may be adjusted or additional joints may be required for other conditions.

ASTM C1472 provides a complete method and formulas for calculating required sealant joint widths. If the calculated joint is too wide for aesthetic considerations, the assumed spacing or panel length can be decreased and the width recalculated. Control and expansion joints for unit masonry construction are typically the same width as a 3/8-in. mortar joint, so calculations are used to determine joint spacing.

6.3.2 Joint Locations

The *Masonry Standards Joint Committee* (MSJC) and *International Building Codes* (IBC) require that the architect or engineer show the location of masonry control and expansion joints

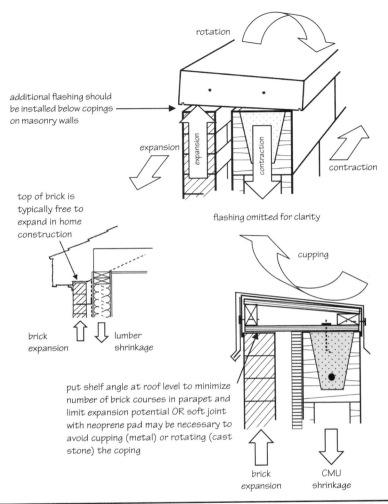

Figure 6-14 Differential movement in parapet walls can cause distress and possible water penetration through the coping.

on the drawings. For most buildings, the exact location of control and expansion joints will be affected by design features that create points of weakness or high stress concentration. Rule-of-thumb vertical movement joint locations for both brick and block construction include:

- Changes in wall height
- Changes in wall thickness (such as pilasters)
- Offsets in parallel walls (*see Fig. 6-15*)
- One side of openings 6 ft or less in width (*see Fig. 6-16*)
- Both sides of openings more than 6 ft wide

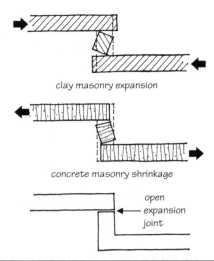

Figure 6-15 Movement at offsets in parallel walls requires placement of open expansion joint to prevent cracking and displacement.

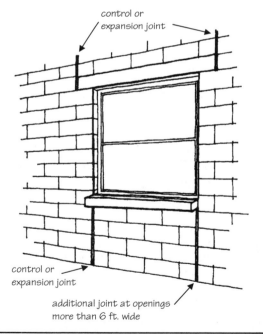

Figure 6-16 Movement joints at openings.

Masonry control and expansion joints should also be located coincidentally with expansion joints in floors, roofs, foundations, or backing walls. In brick/block multi-wythe walls, expansion and control joint alignment may be slightly offset in locations where head joints do not align at whole or half unit lengths (*see Fig. 6-17*). At long, straight sections of walls without openings or other architectural features, maximum expansion joint spacing should be calculated.

In brick walls, expansion joints should also be located near the external corners of buildings (*see Fig. 6-18*). The opposing push of the intersecting veneer wythes can crack through both the brick and mortar (*see Fig. 6-19*).

The joint reinforcement used to limit shrinkage cracking in concrete masonry walls affects the required location of control joints. The National Concrete Masonry Association (NCMA) provides tables to determine control joint spacing based on joint reinforcement size and spacing (*see Fig. 6-20*). Joint reinforcement should not continue across control joints (*see Fig. 6-21*). The details in *Fig. 6-22* show how to combine brick and CMU in the same wall with minimal risk of cracking.

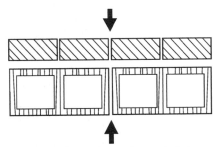

align expansion and control joints in facing and backing wythes if head joints align at whole and half unit lengths

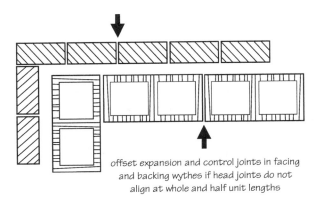

offset expansion and control joints in facing and backing wythes if head joints do not align at whole and half unit lengths

FIGURE 6-17 In brick/block multi-wythe walls, it is acceptable to offset expansion and control joint alignment in locations where head joints in facing and backing wythes do not align at whole or half unit lengths.

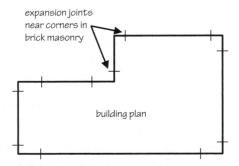

FIGURE 6-18 Locate expansion joints near corners in brick masonry.

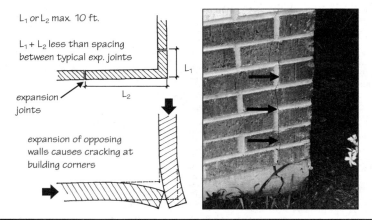

FIGURE 6-19 Locate expansion joints near corners in brick masonry construction to prevent cracking caused by expansion of opposing walls. (*From BIA*, Technical Note 18A.)

6.3.3 Accommodating Movement Joints in Design

Requirements for the location of movement joints in masonry are dictated by the expansion and contraction characteristics of the materials, but designers can also exercise some control over joint location and the aesthetic impact of the joints themselves.

The objective of movement joint placement is to divide a wall into smaller panels of masonry that can expand and contract independently of one another. The smaller the panels, the lower the cumulative stress will be, and the less likely it is that cracking will occur. Wall panels that are more square than rectangular also have less stress buildup. Movement joints will be less noticeable in the appearance of a building if the exterior elevations are designed with joint locations in mind instead of placing them simply by rule of thumb after the design is completed. Just as the joint pattern in a stucco façade is part of the overall design, so too should masonry joints be a design element in masonry buildings. Joints can even be articulated with special-shape units

Maximum Spacing of Horizontal Reinforcement to Meet Criteria of 0.025 in² per foot of Wall Height	
Reinforcement Size	Maximum Vertical Spacing (in.)
W1.7 (9 gauge) two wire§	16
W2.8 (3/16 in.) two wire§	24
W1.7 (9 gauge) four wire†	32
W2.8 (3/16 in.) four wire†	48
No. 3 bars	48
No. 4 bars	96
No. 5 bars or larger	144

§ Two-wire joint reinforcement = one wire per face shell.
† Four-wire joint reinforcement = two wires per face shell.

Recommended Control Joint Spacing For Above Grade Exposed Concrete Masonry Walls§±*	
Length to Height Ratio	Spacing Between Joints (ft)
1-1/2	25

§ Table based on horizontal reinforcement with equivalent area of at least 0.025 in² per foot of wall height to keep random cracks closed. See table above.
± Spacing based on experience over wide geographical area. Adjust spacing where local experience justifies, but not to exceed 25 ft on center.
* Applies to all concrete masonry units.

FIGURE 6-20 Control joint spacing for concrete masonry construction. (*From NCMA*, TEK Bulletin 10-2B.)

to make their visual impact stronger. Alternatively, the joints can be hidden in the shadow of a protruding pilaster, while the series of pilasters articulates the panelized sections of the wall.

The location of window and door openings often governs control and expansion joint placement because of the frequency of their occurrence. In general, joints should be located at one side of openings less than 6 ft wide and at both sides of openings wider than 6 ft. When the masonry above an opening is supported by a cast stone or reinforced CMU lintel, the adjacent movement joint must be located at the ends of the lintel as shown in *Fig. 6-16*. This creates an odd-looking pattern that is not very attractive. As an alternative, movement joints can be located at the midpoint between windows. If the spacing is relatively wide (or simply as an added measure of safety), joint

continuous joint reinforcement interferes with proper functioning of control joints

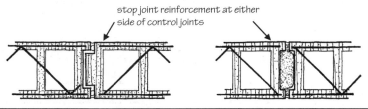

stop joint reinforcement at either side of control joints

FIGURE 6-21 Joint reinforcement should *not* be continuous across control joints.

reinforcement can be added in the courses immediately above and below the openings to strengthen the panel (*see Fig. 6-23*).

When the masonry is supported on a loose steel lintel that simply spans between the masonry on each side of a window or door opening, special detailing can be used to avoid offsetting the joint to the end of the lintel. A piece of flashing placed under the lintel-bearing area creates a slip plane so that the end of the lintel can move with the masonry over the window. With this detailing, the movement joint can then be placed adjacent to the window and run in a continuous vertical line (*see Fig. 6-24*). When the masonry above an opening is supported on shelf angles that are attached to the structure, a control or expansion joint can be located immediately adjacent to the opening and continue straight up the wall past the horizontal support. Joints that are aligned with the window jamb should extend onto the end of the sill itself to avoid the cracking that commonly occurs at this location (*see Fig. 6-25*).

Joint reinforcement can also be used to group closely spaced windows into larger panels so that the movement joints can be spread farther apart. The joints on either side of such a grouping must be sized large enough to accommodate the movement of the larger panel. In the elevation shown in *Fig. 6-26*, the two bed joints immediately above and below the groups of windows are reinforced with two-wire, truss-type joint reinforcement. Oversized movement joints can then be placed at either end of the window groupings. Because the joints are large, an offset or pilaster can be created in the wall at the joint locations to make them less noticeable. Calculating the expected movement in

Expansion and Contraction 117

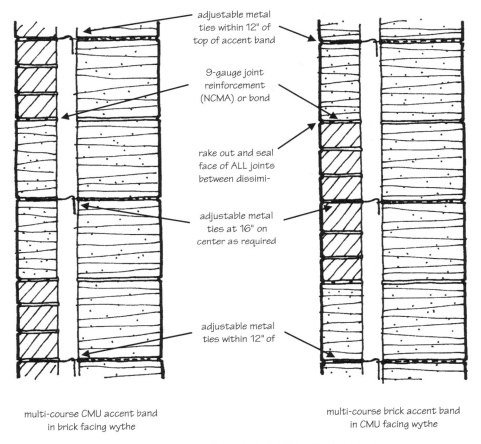

FIGURE 6-22 Differential movement between brick and CMU must be accommodated when combining units in the same wythe to create accent bands. (*Based on recommendations in NCMA TEK Bulletin 5-2A and BIA Technical Note 18A.*)

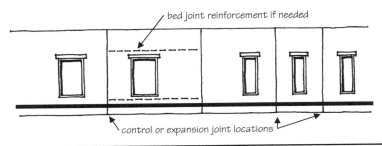

FIGURE 6-23 With proper planning, expansion and control joints can be placed between rather than immediately adjacent to window and door openings.

Chapter Six

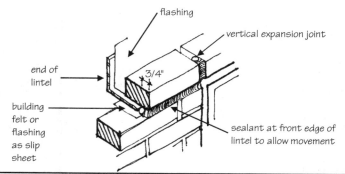

FIGURE 6-24 Expansion joint at window jamb.

FIGURE 6-25 Sealant joints at window jambs should extend onto the top of the sill to avoid cracking.

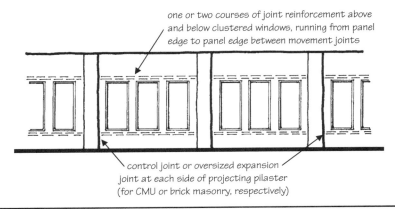

FIGURE 6-26　Grouping windows between movement joints.

the masonry panels in a situation like this is very important to be sure the joints are wide enough to *generously* accommodate the expected movement. With expansion and control joints, it is always more prudent to have too many than just enough.

Sealant color can also affect the appearance of expansion and control joints. Sealant may be selected to blend with either the units or the mortar color but should be slightly darker rather than lighter. Some architects use different sealant colors to blend with different bands of brick or block colors, alternating through the height of the façade to minimize the visual impact (*see Fig. 6-27*).

FIGURE 6-27　Alternating sealant colors can help minimize the visual impact of movement joints.

CHAPTER 7
Moisture and Air Management

The *International Building Code* (IBC) sets performance requirements for the weather resistance of exterior building walls. Specifically, the code mandates:

> *Exterior walls shall provide the building with a weather-resistant exterior wall envelope. The exterior wall envelope shall include flashing.... The exterior wall envelope shall be designed and constructed in such a manner as to prevent the accumulation of water within the wall assembly by providing a water resistive barrier behind the exterior veneer...and a means for draining water that enters the assembly to the exterior of the veneer....*

Responsibility for the building's weather resistance is assigned to both the designer and the contractor. In jurisdictions where the IBC has been adopted, the designer is obligated by law to provide both an overall design and a series of details that are capable, if properly constructed, of providing the weather resistance required by the code.

7.1 Definitions

Although it would seem obvious, the terminology for different types of water movement has been standardized in ASTM E2128, *Evaluating Water Leakage of Building Walls*.

- *Water penetration*: a process in which water gains access into a material or system by passing through the surface exposed to the water source.
- *Water absorption*: a process in which a material takes in water through its pores and interstices and retains it wholly without transmission.
- *Water permeation*: a process in which water enters, flows and spreads within, and discharges from a material.
- *Water infiltration*: a process in which water passes through a material or between materials in a system and reaches a space that is not directly or intentionally exposed to the water source.
- *Incidental water*: unplanned water infiltration that penetrates beyond the primary barrier and the flashing or secondary barrier system, of such limited volume that it can escape or evaporate without causing adverse consequences.

- *Water leakage*: water that is uncontrolled, exceeds the resistance, retention, or discharge capacity of the system or causes subsequent damage or premature deterioration.

7.2 Moisture Management

Masonry walls are not waterproof. Masonry materials are porous and absorptive. Rain can penetrate masonry walls through cracks or separations at the mortar bond line and through defects at copings, sealant joints, windows, parapets, and so on. Masonry does not support mold growth, but the sheathing and framing behind veneers can be a food source for mold spores. For successful performance, it is important to limit rain penetration and expedite moisture removal. There are several physical forces that move water into and through building envelopes:

- Gravity pulls moisture downward.
- Kinetic energy (wind-driven rain) pushes rainwater in through large openings.
- Surface tension causes water to cling to a surface as it moves from a vertical plane to a horizontal plane.
- Capillary action draws moisture into porous materials and allows it to diffuse through the material.
- Negative air pressure sucks moisture in through large and small openings (even hairline cracks) in the outer face of the wall.
- Moisture follows the path of least resistance.

The best and most effective approach to water management in cladding systems in general and masonry walls in particular is threefold:

- *Deflection*: limit rainwater penetration by deflecting or diverting water away from the building envelope and neutralizing the forces that transport water.
- *Drainage*: prevent moisture accumulation by draining water from walls with flashing and weeps and by controlling condensation.
- *Drying*: expedite drying of wall materials with airflow and breathable coatings.

Most contemporary masonry wall systems are designed as drainage walls in which penetrated moisture is collected on flashing membranes and drained through a series of weep holes. Design, workmanship, and materials are all important to the performance of masonry walls. Basic requisites include the following:

- Mortar head and bed joints must be full.
- Mortar must be compatible with and well bonded to the units.
- Mortar must cover the entire head and bed surfaces of the units to ensure maximum extent of bond.
- Appropriate flashing materials must be selected for the expected service life of the building.
- Flashing details must provide protection for all conditions.

- Flashing must be properly installed.
- Weep holes must be properly sized and spaced.
- Mortar droppings must not obstruct water flow to weep holes.
- Weep holes must provide rapid drainage of infiltrated moisture.

With adequate provision for moisture management, masonry wall systems provide long-term performance with little required maintenance.

7.3 Deflection: Limit Rain Penetration

We cannot keep rain, snow, and groundwater away from buildings completely, but we can design building envelopes that minimize the amount of water that passes through the building skin. The first approach in protecting masonry from water infiltration is to deflect or divert water away from the building and neutralize the forces that transport water.

7.3.1 Wall System Concepts

There are three different wall design concepts with respect to the manner in which they manage moisture. All above-grade and below-grade building envelope systems can be classified as one of these types or a combination of types (*see Fig. 7-1*).

Barrier wall systems fall into two categories. Face-sealed barriers must entirely exclude rain penetration at the exterior wall surface because there is no accommodation for moisture drainage. In *face-sealed barrier walls*, impervious materials shed rainwater with no absorption. Face-sealed barrier walls also rely on sealant joints to provide the first and only line of defense at vulnerable joints—there is no redundant protection. Precast concrete cladding, thin stone curtain walls, concrete tilt wall, and some metal panel systems use face-sealed barrier wall strategies as do below-grade walls. Barrier walls must prevent air movement and moisture infiltration. *Mass barrier walls* are sometimes called collection and retention walls or reservoir cladding systems because they absorb and store rainwater that is later evaporated back into the atmosphere before penetrating the full thickness of the wall. Historic multi-wythe masonry walls functioned as reservoir cladding systems.

Drainage wall systems (walls with drainage cavities or drainage planes) are more forgiving than barrier walls because they do not require the total exclusion of moisture. Drainage wall systems include masonry cavity walls and anchored veneers, stucco and exterior insulation and finish systems (EIFS) with a carefully designed and installed drainage plane, and some metal claddings. Drainage walls can tolerate some water penetration because their drainage capability prevents moisture accumulation and damage to materials. Masonry cavity walls and anchored veneer drainage walls provide superior resistance to water damage, but effective moisture drainage is critical to their successful performance.

Rain screen wall systems are a sophisticated variation on the drainage wall concept with the added feature of rapidly equalizing the air pressure between the cavity and the outside atmosphere. Blowing winds during a rain cause a low-pressure condition in the drainage cavity. In seeking a natural state of equilibrium, air moves from the high-pressure zone outside to the low-pressure zone in the cavity. With air infiltration, rainwater is carried through the wall face through minute cracks that may exist at the

- **Barrier wall** systems fall into two categories. Face-sealed barriers must entirely exclude rain penetration at the exterior wall surface because there is no accommodation for moisture drainage. Face-sealed barrier walls rely on sealant joints to provide the first and only line of defense - there is no redundant protection. Precast concrete cladding, thin stone curtain walls, concrete tilt wall, and some metal panel systems use face-sealed barrier wall strategies. Mass barrier walls are sometimes called collection and retention walls or reservoir cladding systems because they absorb and store rain water that is later evaporated back into the atmosphere.

- **Drainage wall** systems (walls with drainage cavities or drainage planes) are more forgiving than barrier walls because they do not require the total exclusion of moisture. Drainage walls can tolerate some water infiltration because their drainage capability prevents moisture accumulation and damage to materials. A continuous clear and open drainage cavity is required behind the exterior wythe. Flashing membranes and weep holes must be properly detailed, sized and spaced to control moisture flow and facilitate drying. Masonry cavity and veneer walls are water-managed walls with a drainage cavity.

- **Rain screen wall** systems are a sophisticated variation on the drainage wall concept with the added feature of rapidly equalizing the air pressure between the cavity and the outside atmosphere. Blowing winds during a rain cause a low-pressure condition in the cavity. In seeking a natural state of equilibrium, air moves from the high-pressure zone outside to the low-pressure zone in the cavity. With air infiltration, rainwater is carried through the wall face through any minute cracks which may exist at the mortar-to-unit interface. Under such a pressure differential, rainwater which would normally run down the face of the wall is literally sucked into the wall. Venting and compartmenting the cavity equalizes the pressure to eliminate the force which pulls moisture through the wall.

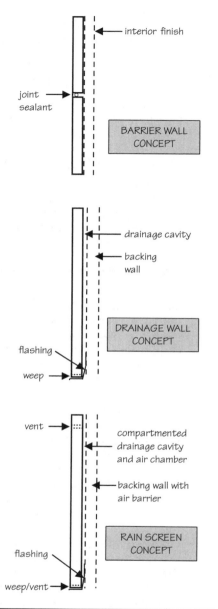

FIGURE 7-1 Exterior wall systems may incorporate a number of moisture protection strategies but can generally be divided into three basic wall types: barrier walls, drainage walls, and rain screen walls. (*From C. Beall and R. Jaffe,* Concrete and Masonry Databook, *McGraw-Hill, 2003.*)

mortar bond line. Under such a pressure differential, rainwater that would normally run down the face of the wall is literally sucked into the wall. Venting and compartmenting the drainage cavity equalizes the pressure to eliminate the force that pulls moisture through the outer wall face.

7.3.2 Masonry Wall Types

Masonry walls can take many forms and have been given many names based on their structural function (loadbearing, non-loadbearing, shear walls, and so on), their method of load resistance (composite or noncomposite), their composition (single-wythe, multi-wythe, or veneer), and their components (brick, CMU, or stone). This book focuses primarily on three types of masonry walls:

- *Single-wythe walls*, which are most commonly constructed of concrete block that is a minimum of 8 in. thick.
- *Cavity walls*, which consist of facing and backing wythes of masonry separated by an open drainage space.
- *Veneer walls*, which consist of a masonry facing that is anchored or adhered to a nonmasonry backing wall.

Single-wythe walls are usually constructed of concrete block. Grouted single-wythe masonry walls are sometimes referred to as "barrier" walls, but this is a misnomer. Single-wythe masonry walls are not capable of excluding all moisture at the exterior wall face as true barrier walls must be. Solidly grouted CMU walls are somewhat more resistant to water infiltration than ungrouted walls. Both grouted and ungrouted single-wythe walls should incorporate flashing and weeps.

Ungrouted single-wythe CMU walls are the most vulnerable to water infiltration and require the greatest care in material selection, design, and workmanship. The quality and extent of mortar bond is critical in the exclusion of rain penetration at the outer wall surface. The CMU must include integral water-repellent admixtures and should be supplemented by field-applied, penetrating water repellents to allow the units to shed water at the outer face. If the CMU cores are ungrouted and the units are water repellent, some infiltrated water will drain through the cores. Without the water repellent, moisture can soak through the full thickness of the block.

Any insulation used in the block cores must be moisture tolerant and must not interfere with moisture drainage through the cores. Perlite and vermiculite pellets meet this criteria as do polystyrene insulation inserts. Open-cell expanding foam insulation should be evaluated with caution because some can act as a sponge, soaking up whatever water enters the cores and prolonging the wall drying process. Single-wythe walls must also incorporate flashing and weeps to redirect penetrated moisture out of the wall. Some proprietary weep systems designed especially for single-wythe CMU walls are illustrated in Chapter 4.

A single-wythe CMU wall that is combined with interior furring strips or studs, batt or board insulation, gypsum board, paint or wall coverings is no longer a single-wythe wall. It is a wall *system* with a single wythe of masonry and will perform much differently than a traditional single-wythe wall.

Cavity walls have an outside facing wythe and an inside backing wythe of brick or CMU or a combination of the two. A drainage cavity separates the two wythes. One or both wythes may be loadbearing. Most commonly, cavity walls are constructed of concrete block backing wythes with a CMU, brick, or stone facing wythe. This is a "drainage" wall design and is the most resistant to water damage because it is all masonry and does not include materials susceptible to rot, mold, or physical deterioration.

A continuous clear and open cavity is required behind the exterior wythe of a cavity wall. Flashing membranes and weep holes must be properly detailed, sized, and spaced to control moisture flow and facilitate drying. Masonry cavity walls are considered to be "water-managed" systems because of their drainage capabilities.

Cavity walls permit moisture that enters the wall or condenses within the cavity to be collected on flashing membranes and drained through weep holes. At the base of the wall, and at any point where the cavity is interrupted, such as shelf angles, floors, or openings, a layer of flashing must be installed, and with it a row of weep holes. Building codes require a minimum drainage cavity width of only 1 in. The cavity in a drainage wall, however, should be at least 2 in. wide (exclusive of insulation) because narrower cavities are impossible to keep clear of mortar bridges across the narrow gap. Even walls with 2-in. air gaps can be difficult to keep clear of mortar bridges and mortar droppings and require diligent quality control measures during construction. Air spaces can be as wide as 4½ in. without engineering analysis if rigid wire ties are used. Even with careful workmanship during construction, every masonry drainage wall will have some mortar droppings.

> *As brick is laid, some mortar will protrude into the air space; however, the air space should not be clogged to the extent that it inhibits drainage.* (From BIA, *Brick Builder Notes*, Issue 1, 2007)
>
> *Some mortar droppings and protrusions are to be expected, but the key point is the performance of the cavity: Is it draining the water to the flashing?...If it is draining, then the cavity is clean enough.* (From International Masonry Institute [IMI] *Technology Brief* 2.5.2, June 2004)

To expect a perfectly clean drainage cavity is unrealistic. The most important thing is to minimize mortar droppings so that they do not block drainage to the weep holes.

To function properly as a pressure-equalized rain screen, a cavity wall must include an air barrier in the backing wythe, and the cavity must be divided into smaller compartments. The cavity must be blocked both horizontally and vertically to prevent wind tunnel and stack effects. Without an air barrier and compartmenting, the horizontal flow of air around building corners and through the backing wall prevents pressure equalization in the wall cavity. Shelf angles in conventional masonry cavity wall and veneer construction provide compartmental barriers to the vertical flow of air, but corners require special detailing (*see Fig. 7-2*). Each "compartment" must be properly vented so that the pressure change occurs as rapidly as possible. Rain screen vents should be added near the top of the wall or panel section and constructed in the same manner as open-head joint weep holes. Weep hole accessories inhibit airflow, so open-head joints are best. Aesthetically, rain screen vents look best if they are located at the same spacing as the weep holes in the lower course of the wall. Canada uses rain screen masonry walls almost exclusively, but they are not yet as widely accepted in the United States.

Veneer walls consist of an exterior masonry wythe and a nonmasonry backing wall. The most common form of masonry veneer is made up of a metal or wood stud backing wall with sheathing, a water-resistive barrier, a drainage cavity, and a single, nonloadbearing veneer wythe of brick, stone, or CMU. Masonry veneer walls are also "drainage" walls, but the sheathing and studs can deteriorate when exposed to enough moisture over a sufficient length of time. This makes them more vulnerable to water damage than cavity walls.

Like masonry cavity walls, properly constructed masonry veneer walls are "water-managed" walls. Water that enters the wall or condenses within the cavity is collected on flashing membranes and drained through weep holes. At the base of the wall and at

Moisture and Air Management

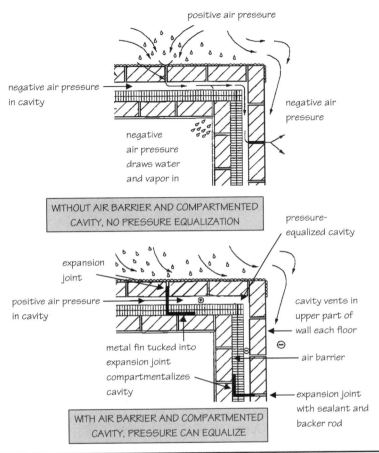

Figure 7-2 Compartmenting a masonry cavity wall at the corners. (*From Quiroutte*, Rain Penetration Control, Ottawa, Ontario. National Research Council of Canada 1985.)

any point where the cavity is interrupted, such as shelf angles, floors, or openings, a layer of flashing must be installed, and with it a row of weep holes. The masonry industry recommends a minimum cavity width of 2 in. rather than the code-required minimum of only 1 in. because a wider cavity is easier to keep clean. Residential-type corrugated metal anchors, however, cannot span more than a 1-in. cavity, so stronger commercial-type anchors must be used.

7.3.3 Masonry Units and Mortar

Proper selection of masonry units and mortar for expected weathering conditions is an important factor in water infiltration resistance. Clay brick for exterior use should be ASTM C216, *Standard Specification for Facing Brick*, Grade SW (*see* Chapter 2). High-suction brick usually produces walls with poor bond. High-suction brick and porous concrete block can absorb excessive water from the mortar and prevent complete cement hydration at the unit surface. Mortar generally bonds best to clay masonry units with moderate initial rates of absorption between 5 and 25 g·min^{-1}·30 sq in.$^{-1}$. Brick with initial rates of absorption (IRA) higher than 25 or 30 g/min should be thoroughly wetted and then allowed to surface-dry before laying. This produces better bond and more water-resistant joints. CMU should not be prewetted. CMU must be kept dry at the job site or the potential for shrinkage cracking in the wall will increase.

Mortars made from portland cement and lime generally produce higher flexural bond strengths and are therefore more resistant to water penetration than proprietary masonry cement mortar. Type N mortar has a lower cement content and higher lime content than Type S and therefore experiences less shrinkage cracking and bond line separations. Type N mortar is recommended for above-grade work with normal exposure. Types M and S should be used only for special conditions (*see* Chapter 3 for mortar-type recommendations).

Another way of limiting water penetration into masonry walls is properly constructed mortar joints. Cracking or bond line separations between mortar and masonry units can allow significant water penetration, especially under negative pressure conditions. Full mortar joints are essential in minimizing water penetration through masonry walls. The extent of bond is as important as bond strength. With only minor installation tolerances, all head joint and bed joint surfaces should be fully covered with mortar. Mortar joints that are not completely filled offer less resistance to rain penetration (*see Fig. 7-3*). Tooling, unit texture, mortar workability, and water retention are also important in obtaining water-resistant joints, but workmanship affects the water penetration resistance of mortar joints more than anything else. Good mortar bond must be achieved at all contact surfaces.

Concave and V-tooled joints are the most effective in excluding moisture at the wall surface. Steel jointing tools compress the mortar against the unit, forming a tight bond at the unit-mortar interface. Mortar joints that are not properly tooled may allow water to penetrate freely (*see Figs. 7-4 and 7-5*). Extra care in tooling should be given to the exposed bed joint at offset courses (*see Fig. 7-6*). If available, use a solid brick at the projection. If not, pack the core holes in the brick with mortar. Maximum projection should be no more than ½ in. In high-wind areas, place flashing in the bed joint just above the projection. Make sure it is sealed to the top of the brick before the joint is mortared, and install a row of weeps just above the flashing. Also make sure the flashing comes all the way out to the face of the wall or bring it over the top of the projection and turn down to form a drip. You will still have a shadow line. Recommended flashing details and profiles are discussed later in this chapter. Different flashing materials are compared in Chapter 4.

FIGURE 7-3 Mortar joints that are not completely filled offer less resistance to rain penetration.

FIGURE 7-4 Good workmanship produces a joint with a tight, compacted surface and good bond line adhesion. Poor workmanship produces rough joints with voids at the unit-mortar interface.

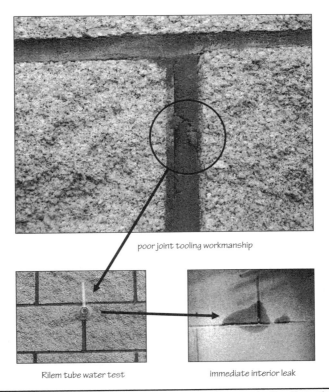

FIGURE 7-5 Poor workmanship in mortar joints increases water penetration through masonry walls.

Figure 7-6 The exposed bed joint at offset courses merits special care in joint tooling. If available, use a solid brick at the projection. If not, fill core holes in brick with mortar. Do not expose core holes. Maximum ½-in. projection.

Mortar must be mixed with the *maximum* amount of water conducive to good workability, bond, and complete hydration of the cement. This is just the opposite of concrete technology, which seeks the *minimum* amount of water necessary. Optimum water content is affected by the initial rate of absorption of the units and by weather conditions, so the mason should be allowed to judge the necessary amount of mixing water based on workability. In hot, dry, or windy conditions, moist curing of the masonry *after* construction (for both clay and concrete units) can enhance bond and weather resistance by ensuring complete cement hydration (*see* Chapter 15).

7.3.4 Parapets

Much of the water that gets into masonry walls enters at the top. A roof overhang is the best protection for the tops of masonry walls, but not all architectural styles lend themselves to overhangs. When parapets are used, they are arguably the most important detail in deflecting and diverting water away from masonry walls. Parapets must be carefully detailed to allow expansion, contraction, and differential movement and to prevent water from penetrating the top of the wall. Parapets have more extreme weather exposure than the walls below because the tops and corners of buildings get wet every time it rains, whereas the lower walls usually stay drier depending on wind direction and duration of the rain event (*see Fig. 7-7*).

The details in *Figs. 7-8 and 7-9* illustrate both metal and cast stone copings used with brick and concrete masonry parapet structures. Because the joints in a masonry coping are vulnerable to moisture penetration as well as expansion and contraction, brick rowlock copings are a bad choice (*see Fig. 7-10*). The cross-joints and face joints in cast stone copings should *always* be sealed with a durable elastomeric sealant and backer rod. Mortared joints will allow water infiltrations when they inevitably crack, and a thin "smear" of sealant across the top of mortared cross-joints will fail (*see Fig. 7-11*).

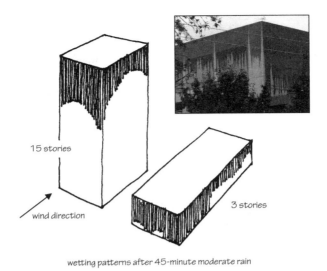

wetting patterns after 45-minute moderate rain

Figure 7-7 Wind and wetting patterns at the tops of buildings subject parapets to extreme weather exposure.

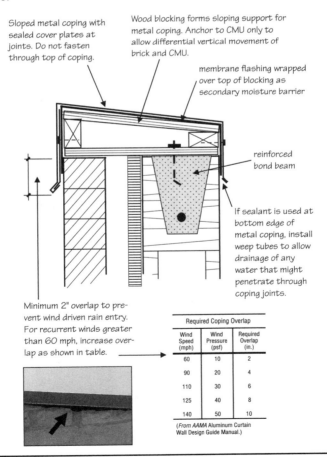

Figure 7-8 Metal parapet coping.

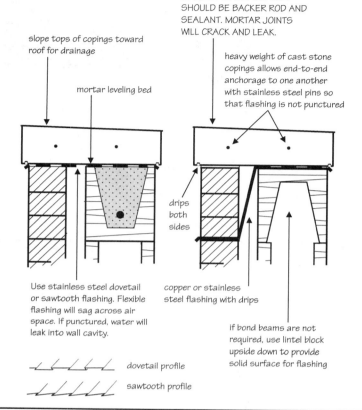

Figure 7-9 Cast stone parapet copings.

Figure 7-10 Brick rowlock copings have a high probability of cracking.

Moisture and Air Management 133

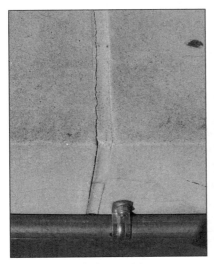
cracked mortar joint

thin, delaminated sealant smeared over mortar joint

FIGURE 7-11 Rake out mortar joints in masonry copings and use backer rod and elastomeric sealant to prevent water penetration.

The tops of parapet copings must be sloped to drain water and melting snow. They should slope toward the roof so that water does not streak down the face of the masonry wall. Secondary flashing should *always* be installed underneath parapet copings to provide redundant protection against water infiltration. All penetrations through this secondary flashing must be sealed. Anchoring cast stone copings end to end avoids the necessity of penetrating the flashing and is an accepted method of installation (*see Fig. 7-12*).

FIGURE 7-12 Coping stones anchored end to end with stainless steel pins.

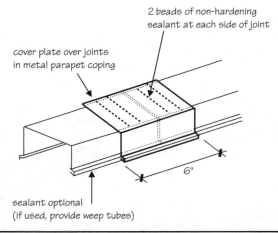

FIGURE 7-13 Cover plate over joints in metal parapet coping.

Metal copings provide the best protection for masonry parapets. The metal is impervious to moisture and can be installed in lengths requiring a minimum number of joints. Because every joint that occurs on a horizontal surface is an opportunity for a leak, the fewer joints there are, the greater the probability of keeping the wall dry. Metal copings should have cover plates at the joints between sections, with a double bead of nonhardening sealant on each side of the joint (*see Fig. 7-13*). If metal copings are not adequately sloped to drain water, they will eventually cup in the middle and hold water. The sealant under the joint cover plates is then the only defense against this slight hydrostatic force. The bottom edge of a metal coping can be caulked, but if it is, weep tubes should be installed in the sealant joint to permit drainage of any moisture that may infiltrate the top surface. If the vertical leg of the metal coping laps over the brick far enough, this sealant is not necessary (*see Fig. 7-14*). Leaving the coping too short does not protect the wall and leaves it open to wind-driven rain penetration.

Masonry copings should overhang the parapet and have drips to break the surface tension and keep water from flowing across and back into the joint between the coping

Required Height of Coping Overlap		
Wind Speed (mph)	Wind Pressure (psf)	Required Overlap (in.)
60	10	2
90	20	4
110	30	6
125	40	8
140	50	10

(*From AAMA* Aluminum Curtain Wall Design Guide Manual.)

FIGURE 7-14 Minimum overlap of metal coping over top edge of masonry.

and the wall below. This bed joint is very vulnerable to water penetration because expansion and contraction along its length usually cause the mortar to crack along the bond line. Where a parapet intersects a higher wall, a saddle flashing must be used to prevent moisture infiltration. This is particularly critical for veneer construction, where the backing wall may include components that are easily damaged by moisture or that support mold growth (*see* Chapters 8, 9, and 10).

Where the parapet meets the roof membrane, two-piece reglets facilitate the best interface between the roofing and masonry work. Several counterflashing and reglet details are shown in *Figs. 7-15 and 7-16*. Make sure that the horizontal leg of the reglet

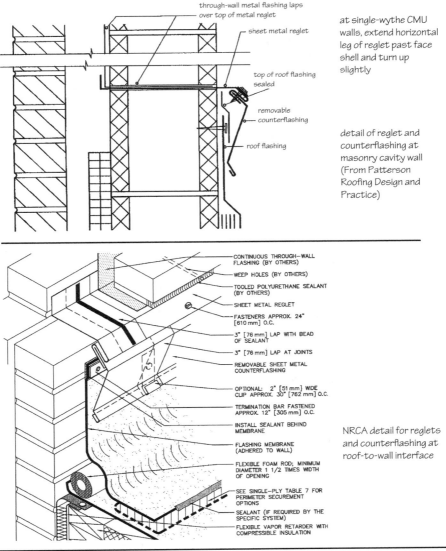

FIGURE 7-15 Reglets and counterflashing at roof-wall intersection.

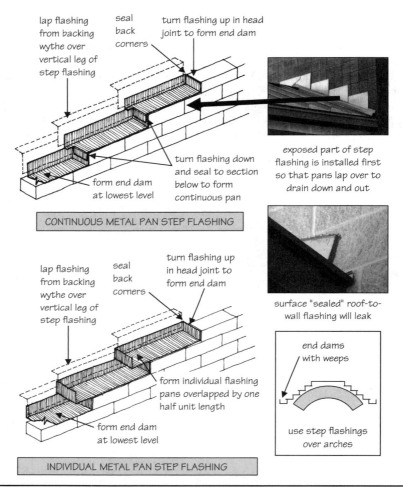

FIGURE 7-16 Two different methods of forming step flashing at roof-to-wall intersection.

or counterflashing extends completely through the brick wythe or the CMU face shell and is turned up or folded to prevent water from reaching the interior of the wall. Surface-mounted reglets with a sealant trough are not recommended for masonry construction because rain that penetrates the wall can bypass the reglet and flow in behind the roofing membrane. Surface-mounted reglets are intended for concrete or tilt-up parapets because concrete is both monolithic and more dense than masonry.

One final recommendation for masonry parapets is not to seal the back of the parapet with any type of coating or covering that is not breathable. A wet parapet will dry out more quickly if it can evaporate moisture from both sides. This is particularly important in cold climates where saturated masonry can freeze and spall the face of the units. In warmer climates, it can cause efflorescence. Where the parapet is only tall enough to accommodate the required 8-in. height of the roof flashing, there is no alternative to the masonry being covered by the roof membrane. On taller walls, however, a simple solution is to use metal roof panels as a rain screen that allows the masonry to breathe. Coatings require maintenance and periodic recoating, but metal panels provide a solution that will last the life of the building.

7.3.5 Coatings and Water Repellents

Concrete block is much more porous than brick. Stucco, cement-based paints, block fillers, and acrylic paints were once typically used to limit the absorption of rainwater at the exterior face of block walls. After "architectural" block came on the market, the preferred coating became elastomeric paint and eventually clear water repellents. Elastomeric paints no longer enjoy widespread use in new construction except for the backs of buildings where plain gray block is often used and for some utilitarian exterior walls around loading docks and trash dumpsters. A single coat of elastomeric paint leaves pinholes on the block surface (*see Fig. 7-17*). A second coat will usually fill in these holes, but even then, the coating does not provide enough protection to be used

elastomeric paint on smooth block

elastomeric paint on split face block

FIGURE 7-17 A single coat of elastomeric paint leaves pinholes in the surface of CMU.

Figure 7-18 Paint and other coatings on CMU walls should be breathable to avoid trapping moisture, which could cause bubbling and delamination.

on horizontal or low-sloped surfaces. The pinholes do help the masonry to breathe so that the coating cannot trap moisture in the wall (*see Fig. 7-18*). A single coat of elastomeric paint is more breathable than two coats.

Integral water repellents are now the most widely used protective treatment for CMU, and most manufacturers now market their block with an integral water-repellent admixture. To achieve good bond, the mortar must also be treated with a compatible water-repellent admixture from the same manufacturer as the block water repellent. To prevent variations in field mixes, the mortar admixture is packaged in premeasured quantities appropriate for standard mortar batch sizes. No water repellent, regardless of its chemical composition, will solve the problems of poorly designed or constructed masonry walls, but water repellents can reduce the absorption of CMU (*see Fig. 7-19*). Water repellents are not a substitute for flashing and weep systems and they cannot turn a single-wythe CMU wall into a barrier wall no matter how much is applied. Water repellents should serve only as a single component of a CMU wall and should never be relied upon as the only line of defense.

Integral water repellents prevent concrete block from wicking moisture through to the interior face, but they cannot stop water from coming through cracks or voids that develop at the mortar-to-unit bond line. Manufacturers recommend that CMU walls built with block that contains an integral water repellent should be supplemented with a surface-applied water repellent after the walls are constructed. Surface-applied water repellents can bridge only the smallest hairline cracks or separations. This provides some protection at the unit-to-mortar bond line, but water can still penetrate through larger cracks.

Water repellents can reduce CMU absorption while still permitting the wall to breathe. Compatibility of each specific block with the appropriate surface-applied water repellents is an important consideration that affects performance. The complicated chemistry and ingredients of surface-applied water repellents and the variations in pore size of different concrete block make it difficult for an architect or mason contractor to determine the most appropriate product. Both the block manufacturer and

Moisture and Air Management

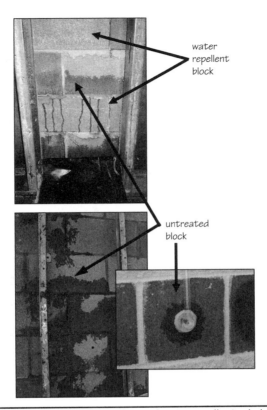

FIGURE 7-19 Concrete block with and without integral water-repellent admixture.

the water-repellent manufacturer should be consulted to ensure that the right product is used for any given block. Product specifications should require that CMU manufactured with an integral water repellent must comply with the performance criteria of NCMA TEK 19-7, *Characteristics of Concrete Masonry Units with Integral Water Repellent*.

Although water repellents reduce the absorption of CMU, they do not have a negative effect on grout bond. Integral water repellents are designed to resist a wind-driven rain that is equivalent to 2 in. of hydrostatic pressure. Wet grout in a CMU core exerts a much higher hydrostatic pressure, so the grout paste pushes itself into the unit pores. Absorption is then essentially the same as that with grout and an untreated block. Compressive strength and bond strength are comparable and provide adequate structural performance.

7.4 Drainage: Prevent Moisture Accumulation

All exterior building materials can tolerate getting wet, but staying wet for too long can cause damage or deterioration that might not otherwise occur. The ability of any cladding system to dry quickly after each rain prolongs the service life of its constituent materials. Masonry construction relies on the effective design, location, and construction of flashing and weeps and the drying characteristics of the wall.

7.4.1 Flashing

Although water penetration can be limited through good design and workmanship, it is impossible to entirely prevent moisture from entering a masonry wall. Flashing is used to collect and divert penetrated water, and weep holes expedite its removal. High-quality materials, appropriate design, and good workmanship are critical because flashing failure can lead to significant damage and replacement is expensive.

Flashing should be installed at all locations within a wall where the downward flow of water is interrupted. Flashing should be placed over all wall openings and at all window sills, spandrels, caps, copings, and parapet walls (*see Fig. 7-20*). Single-wythe walls can drain moisture through ungrouted cores and also require flashing at the same

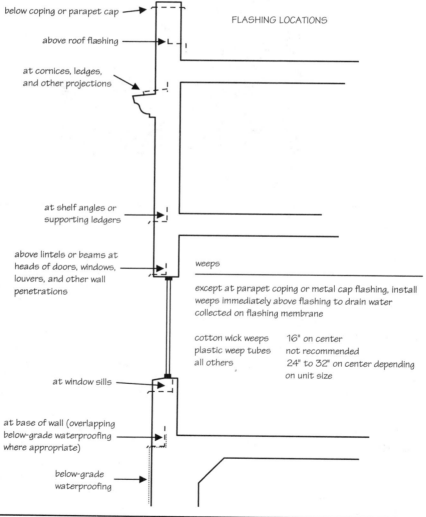

FIGURE 7-20 Masonry flashing and weep placement. (*From C. Beall and R. Jaffe,* Concrete and Masonry Databook, *McGraw-Hill, 2003.*)

coping, parapet, head, sill, and base locations. Masonry walls should be set on a recessed ledge at the slab so that flashing and weeps at the base of the wall are below the finish floor elevation and a minimum of 6 in. above finish grade.

Metal flashings have enough stiffness to span unsupported across a wall cavity. Flexible membrane flashings, however, must have continuous support to avoid sagging and holding of water and to avoid accidental punctures. The top of the vertical leg of any flashing has to be sealed or capped so that water cannot run behind it (*see Fig. 7-21*). Flashing should always extend to or beyond the face of the masonry and be sealed to the substrate so that moisture collected on the surface cannot flow around, underneath, and back into the wall. Without exception, industry sources agree that extending flashing to or beyond the exterior face of the wall is the *minimum standard of care*.

- "Flashing must extend to or beyond the exterior face of the masonry in order to be effective." (The Masonry Society, *Masonry Designer's Guide*)
- "It is imperative that flashing be extended at least to the face of the brickwork." (BIA, *Technical Note 7*)

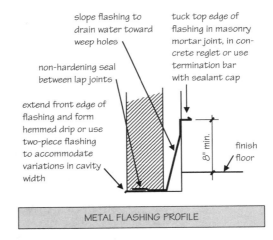

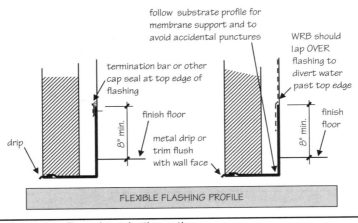

FIGURE 7-21 Flashing profile and termination options.

- "The flashing should continue beyond the exterior face of the masonry a minimum of ¼" and terminate with a sloped drip edge." (NCMA TEK 19-4A)
- "Flashing should . . . extend beyond the face of the wall." (Portland Cement Association, *Concrete Masonry Handbook*)
- "Extend flashing to outside face of wall and terminate as detailed on the drawings." (The Masonry Society, *TMS Annotated Guide to Masonry Specifications*, Article 3.03F)
- "Detail drawings should indicate amount and shape of flashing projection beyond the face of the masonry." (The Masonry Society, *TMS Annotated Guide to Masonry Specifications*, Annotation to Article 3.03F)
- "The Flashing should be extended beyond the exterior face of the wall." (ASTM C1400, *Reduction of Efflorescence Potential in New Masonry Walls*)
- "The Flashing should be extended beyond the exterior face of the wall." (ASTM E2266, *Design and Construction of Low-Rise Frame Building Wall Systems to Resist Water Intrusion*)

Flashing should never be held back from the face of the wall because of the risk of water flowing around the front edge and back into the wall, where it can pool in the cores of the brick or block below or drain into a door or window head below. Some flashing materials cannot be exposed to ultraviolet radiation because it can cause bleeding, emulsification, and staining. Flashings that cannot be exposed may be extended past the face of the wall and cut off flush after the units are laid and the mortar set (*see Fig. 7-22*). Flashings may also be installed with a separate metal drip (*see Fig. 7-23*).

Figure 7-22 Flashing can be extended beyond the wall face and trimmed flush after the units are laid and the mortar has set.

Moisture and Air Management 143

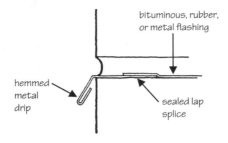

Figure 7-23 Flashing can be installed with a separate metal drip edge.

Water cannot flow underneath flashing that is adhered to the substrate. A self-adhering membrane or a metal flashing adhered with mastic or other adhesive can be used (*see Fig. 7-24*). An angled drip edge is supposed to make the water drip off and not flow around the flashing and back underneath. The horizontal leg of the drip should be adhered to the substrate.

Flashing should be continuous around corners (*see Fig. 7-25*), and at horizontal terminations the flashing should be turned up to form an end dam (*see Fig. 7-26*). Where masonry abuts door jambs, curtain walls, storefront systems, or other cladding materials, stop flashing in the first head joint adjacent to the interface and form an end dam. Where structural framing interrupts the backing wall, flashing should be continued across the face of the framing, and the gaps between the backing wall and columns or spandrel beams should be sealed against air and moisture penetration (*see Fig. 7-27*).

As important as it is to bring flashing all the way out to the face of the wall, it is equally important that adjacent pieces of flashing be lapped 4 to 6 in. and sealed to prevent water from getting behind or underneath. For metal flashing, a double bead of nonhardening butyl caulk or urethane sealant will seal the lap and still accommodate thermal expansion and contraction. Where metal flashing crosses a control or expansion joint, this type of lap splice works well. Flexible flashing materials usually have enough elasticity to span a movement joint without a splice at the precise joint location. Flexible flashing should always be fully supported and follow the profile of the substrate without any gaps behind or underneath. If the flashing is inadvertently

self-adhered rubberized asphalt flashing set
flush with face of wall

copper flashing set flush with face of wall and
adhered with mastic

FIGURE 7-24 Flashing adhered to horizontal substrate.

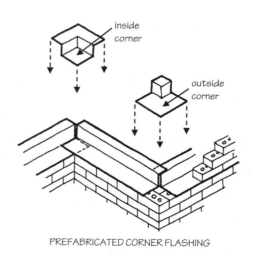

PREFABRICATED CORNER FLASHING

FIGURE 7-25 Flashing must be continuous at building corners.

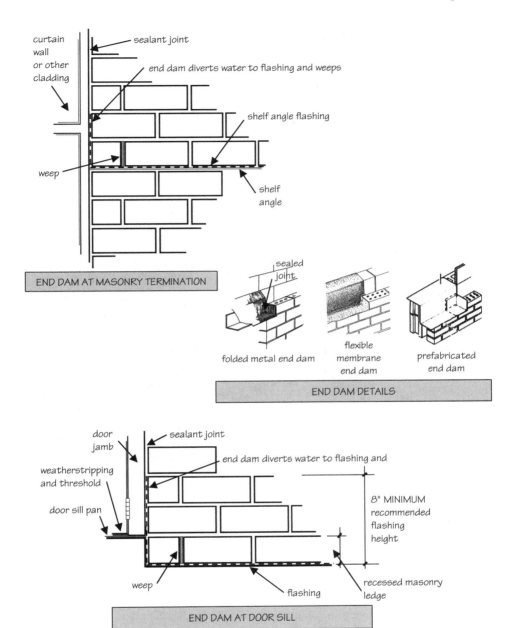

FIGURE 7-26 Form end dams wherever flashing terminates at windows, doors, and against adjacent construction.

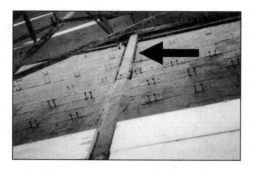

Figure 7-27 Seal gaps in backing wall against air and water penetration.

punctured during installation, these gaps can conduct water laterally behind the flashing.

7.4.2 Flashing Locations

It is not coincidence that roofs frequently leak at the intersection with masonry parapets, and masonry parapets often leak at the intersection with roofing. Where the work of two trades must interface to form a water-resistant barrier, the blame for failure can often go either way. In the case of masonry parapets and roofing, it is not so much a matter of poor workmanship on the part of either trade, but rather the manner in which the interface is designed and constructed. *Figures 7-15 and 7-16* illustrate two-piece flashing, counter-flashing, and reglet options that work well for parapet-roof intersections.

Shelf angles at intermediate floor lines interrupt the downward flow of water in the drainage cavity, so they must be detailed with flashing and weeps. The flashing course should be set below the finish elevation of the adjacent floor. A traditional shelf angle detail is shown in *Fig. 7-28*. The alternative details shown in *Fig. 7-29* are intended to

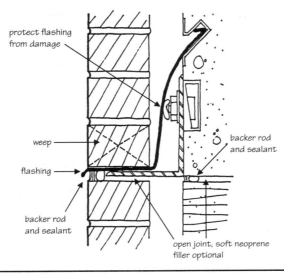

Figure 7-28 Traditional shelf angle detail.

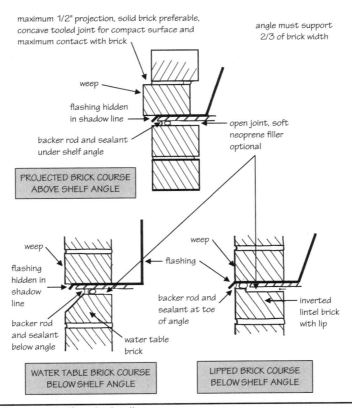

Figure 7-29 Alternate shelf angle details.

FIGURE 7-30 Strong horizontal color bands disguise wide soft joints at shelf angles.

reduce the visual impact of the wide joint at shelf angles or to articulate them using special-shape units or a simple projected course. This creates a strong shadow line in which the joint and flashing are hidden. The appearance of horizontal soft joints can also be minimized by changing the unit pattern or the unit color for a few courses above or below the shelf angle to create a strong horizontal band (*see Fig. 7-30*).

The rule of thumb that flashing extend all the way to the outside face of the wall can be interpreted slightly differently at shelf angles. If the shelf angle is recessed from the face of the wall and the backer rod and sealant are also recessed, the flashing must come out past the sealant joint. This ensures that the required collection and drainage of water in the wall will not be compromised.

Flashing at window penetrations is also critical in preventing water infiltration. Industry-standard window flashing details are thoroughly illustrated in ASTM E2112, *Standard Practice for Installation of Exterior Windows, Doors and Skylights*. Where backing walls are of stud frame and sheathing, the substrate must be wrapped and protected from moisture damage. Windows and doors that are recessed even slightly from the face of the wall will be better protected against leaks than those installed flush with the outer plane of the masonry.

Window openings in veneer walls are typically wrapped with flashing tape. Sheathing at veneer walls that is not inherently moisture resistant is required by code to be protected by a water-resistant barrier (WRB) such as building paper, felt, building wrap, or a spray or troweled-on membrane. The vapor permeance of water-resistant barriers should always be taken into consideration in masonry wall design. The same is true of the mastic coatings typically used on CMU backing walls in multi-wythe construction. CMU backing walls are typically coated with a damp-proofing mastic to shed water and reduce air penetration. They do not require a separate water-resistant barrier because the mastic coating serves the same purpose.

Window and door flashing must be integrated with the masonry flashing so that every component of the wall is protected. Where flashing is installed on loose steel lintels spanning individual punched windows, the *Brick Industry Association* (BIA) and *National Concrete Masonry Association* (NCMA) both recommend forming end dams to prevent collected water from running back into the drainage cavity. At window sills also, they recommend flashing end dams at each side of the opening. Weeps should be installed in the first course above the lintel and the first course above the sill flashing. The masonry above each door or window lintel should have a minimum of two weeps.

Flashing at the base of a masonry wall is simple. First, keep the flashing and weeps below the finish floor level. A recessed masonry lug of 4 in. provides a safety factor of sorts in case there are any problems with the flashing or weeps at this level. Flashing and weeps should also be above the adjacent grade or sidewalk. The finish grade (including mulch in landscape beds) should slope away from the building as it should for any type of building. Too often, the flashing and weep course are set above grade only to be buried by topsoil and mulch placed at the end of the job. The masonry, civil engineering, and landscape drawings as well as accessibility requirements should be coordinated to avoid this type of issue. Weeps that are below grade cannot drain a masonry wall no matter how good the flashing details are.

For masonry basement walls, the same type of waterproofing is used as for any other type of basement walls. At the top of the basement walls, the below-grade waterproofing membrane should be tied in to the above-grade wall flashing (see Fig. 7-31). Whenever the veneer continues below grade, keep the flashing and weeps above finish grade and grout solid below the flashing to provide support.

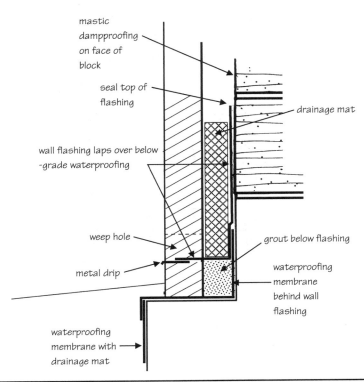

FIGURE 7-31 Transition from wall flashing to below-grade waterproofing.

7.4.3 Weeps

Weep holes are located in the masonry course immediately above flashing. Spacing of open-head joint weep holes should be no more than 24 in. on center for brick or 32 in. on center for concrete block. Cotton wicks should be spaced a maximum of 16 in. on center to compensate as much as possible for their slow wick drying. Small-diameter (3/8 in.) round weep tubes are too easily obstructed by mortar droppings, insects, or other debris (*see Fig. 7-32*). Tubes that include a rope wick are acceptable if they are

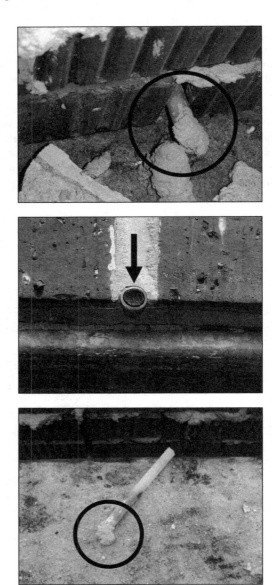

FIGURE 7-32 Small-diameter tube weeps are easily clogged with mortar or other debris.

FIGURE 7-33 Weeps that end up buried by the landscaping are as useless as those plugged by mortar.

spaced no more than 16 in. apart. Larger square plastic tubes can be spaced the same as for open-head joint weeps. The proprietary weep inserts described in Chapter 4 are typically designed to fit into an open-head joint in modular brick. They do not significantly restrict drainage, so weeps in which they are used can be spaced the same distance as for open weeps.

Weep hole drainage can also be obstructed on the outside of the wall. The most common error is setting the finish floor and/or masonry ledge elevation too close to grade. The civil engineering design has to be coordinated well enough that the finished grades allow enough height for topsoil, mulch, turf, and other landscape elements to be installed and still remain below the weeps. Weeps that end up buried by the landscaping are as useless as those plugged by mortar (*see Fig. 7-33*).

The drainage cavity must be kept as free of mortar droppings as possible so that there is free flow of water to the weeps. Maintaining an unobstructed path for water to reach the weeps is easier now with some of the proprietary products on the market (*see Fig. 7-34*). These "drainage mat" devices work better than the pea gravel that was once used for the same purpose, and they are easier to handle and install in the field. Along with the construction techniques described in Chapter 15, they enhance the ability of masonry walls to drain water quickly after a rain. Even drainage mats can become obstructed, however, if the mortar droppings are excessive. Using a drainage mat does not mean that good workmanship is not still a vital part of the success of masonry walls. Mortar droppings still need to be kept to a minimum.

7.5 Drying: Evaporation and Venting

The flashing and weeps discussed earlier remove the bulk of the rainwater that enters the masonry and moves through to the drainage cavity. There is still residual water that has been absorbed by the masonry units themselves. This is true of brick and stone more so than CMU because most CMU are now manufactured with integral water repellents that prevent absorption into the units. If CMU are not treated with an integral water repellent, then they will absorb water in the same way as the brick and

even drainage mats can become obstructed if the mortar droppings are excessive

FIGURE 7-34 Proprietary mortar deflection devices help maintain unobstructed flow to the weeps.

stone. Masonry cavity walls and anchored veneers dry residual absorbed water by evaporation.

Drying takes place toward the exterior and toward the cavity side of the walls. Absorbed water is evaporated at the exterior face of the wall, but it is also driven inward by radiant heat (*see Fig. 7-35*). This vapor transport mechanism makes it more important than ever to address the condensation potential of every wall. An inward vapor drive is typical of hot, humid climates, but solar-driven vapor in a masonry wall can occur, to one degree or another, in any climate.

Exterior evaporation occurs more quickly with both warm air and radiant heating. Cool temperatures and cloudy days will slow the process. The north sides of buildings dry slowly and hold moisture long enough to grow mold or mildew on the outside of the wall. Any wall that cannot dry quickly enough also runs the risk of efflorescence and prolonged inward vapor drive. Besides shade, irrigation sprinklers can also create problems, even on sunny walls. The constant wetting of masonry walls prevents them from drying completely before they are wet again (*see Fig. 7-36*).

One way of dissipating some of the water vapor directly from the drainage cavity is to vent the top of the wall. Vents are the same as weep holes, but they must be the

Moisture and Air Management   153

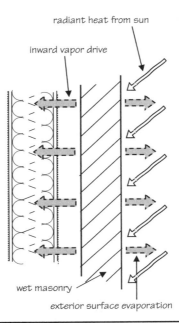

FIGURE 7-35   Radiant heat evaporates moisture at the exterior face of the wall and drives water vapor inward.

FIGURE 7-36   Irrigation sprinklers can keep a masonry wall from drying out.

open-head joint type to be truly effective. Vents at the top of the wall must be coupled with open-head joint weeps at the bottom of the wall to create effective airflow. The various types of weep inserts on the market have been found to reduce airflow significantly, and mortar droppings at the base of the wall will also impede air movement. Minimum requirements for any measurable airflow are shown in *Fig. 7-37*.

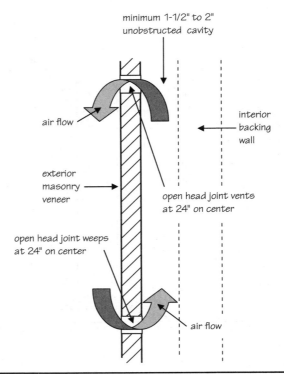

FIGURE 7-37 Ventilation drying of cavity moisture.

Even with weep inserts or with some mortar blockage, venting a masonry cavity wall or veneer wall still has some advantages. Moist air that is warmed by radiant heat rises to the top of the cavity by natural convection. Vents at the top of the cavity, even if they do not allow optimum air movement, provide a path for rising, vapor-laden warm air to escape.

7.6 Controlling Air and Vapor Movement

The first part of this chapter addressed the water permeance of masonry walls, but masonry is also vapor permeable and air permeable. A detailed discussion of vapor movement and related condensation is beyond the scope of this book, but the air and vapor characteristics of masonry walls will be described here.

Moisture vapor movement through building envelopes and building wall assemblies are controlled by:

- Water-resistive barriers
- Vapor retarders
- Air barriers

There are many different materials that can be used to perform these functions, and some that can perform more than one. Air barriers may be vapor resistant or vapor permeable, and water-resistive barriers are not necessarily vapor resistant. If the potential exists for condensation within the wall, careful thought should be given to material selection and location. Each wall design should be evaluated independently to determine the need for and optimum location of these various moisture control elements.

7.6.1 Water-Resistive Barriers

In masonry cavity walls, damp-proofing mastic or other liquid applied membranes are typically used on the cavity face of the backing wythe. These membranes are not the same as below-grade waterproofing because they cannot withstand hydrostatic pressure. They can shed water, and in a masonry cavity wall, they shed that water into the drainage cavity between the inner and outer wythes. The primary function that damp-proofing has traditionally served in masonry cavity walls is that of a WRB. In anchored and adhered masonry veneers, a sheet or membrane WRB is typically attached to or part of the nonmasonry backing wall.

7.6.2 Vapor Retarders

Vapor diffusion is the process by which water vapor migrates directly through a material driven by vapor pressure differentials. Vapor retarders are used to control vapor diffusion. Vapor diffusion and air movement are independent of each other. Vapor diffusion may be inward at the same time that air movement is outward. Both masonry units and mortar are vapor permeable. Water vapor diffuses easily from one side of a masonry wall to the other.

The traditional rule-of-thumb recommendation for vapor retarders was once very simplistic. For cold climates, vapor retarders were recommended on the inside of the insulation layer, and for hot and humid climates, vapor retarders were recommended on the outside of the insulation layer. By this measure, a damp-proof coating on the face of the backing wythe in a masonry cavity wall would be appropriate only in a warm climate. What we have learned in more recent years is that the incidence of condensation on or within walls and the placement of vapor retarders can be much more complex than rules of thumb.

Damp-proof coatings have a wide range of permeance ratings. Some are considered vapor permeable, and some are considered vapor resistant. Vapor-resistant damp-proof coatings will function as vapor retarders—whether you want them to or not. Vapor-permeable damp-proof coatings will resist the passage of liquid water but allow the diffusion of water vapor. Vapor-permeable damp-proof coatings will not function as vapor retarders—whether you want them to or not. Vapor-permeable damp-proof coatings should be used where no vapor retarder is required or where the vapor retarder function is provided by another material in a different location.

The permeance rating of any membrane or damp-proofing that is used on the backing wythe of a masonry cavity wall should be known so that it can be considered as one component of the overall wall design. Paint finishes for interior or exterior masonry surfaces should be selected with their permeance ratings in mind. Low-permeance coatings (or a buildup of multiple coating layers) may blister or delaminate under certain conditions. The characteristics of individual products should be investigated before they are specified.

Vinyl wall coverings function as vapor retarders. So do rubber bases, blackboards, lockers, and other items mounted on a building's exterior walls. The use of interior vinyl wall coverings, especially in combination with the solar-driven water vapor discussed earlier, can cause extensive condensation and mold growth. Anytime interior finishes include the use of vinyl wall coverings on exterior walls, and careful calculation and professional judgment are required to avoid condensation.

156 Chapter Seven

The drainage cavity in a masonry wall can be used to collect condensed water vapor. If the vapor drive is from the exterior to the interior, a low-permeance damp-proof coating on the backing wall can prevent water vapor from diffusing any further inward. If the temperature at the face of the backing wall is below the dew point of the water vapor, the vapor will condense and flow down the drainage cavity toward the weeps.

7.6.3 Air Barriers

Much larger quantities of water vapor move into and out of buildings by air transport than by diffusion (*see Fig. 7-38*). Wherever there is air movement, there is vapor movement. The majority of air leaks in buildings occur through seams and gaps between adjacent elements, at wall and roof penetrations, and at door and window openings. U.S. Department of Energy studies showed that less than 1% of vapor transport through residential building envelopes is caused by diffusion, and the rest is attributed to air movement. Because of the quality of construction, commercial buildings may be slightly more air tight, but airflow is still by far the primary transport mechanism for water vapor.

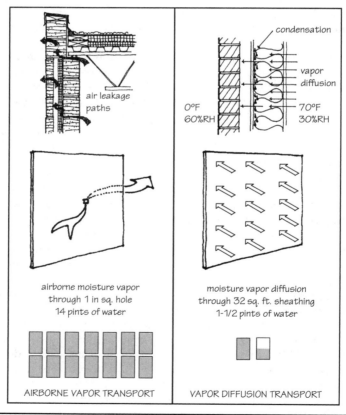

Figure 7-38 A much larger quantity of water vapor is transported through the building envelope by air leaks than by diffusion. (*From Quirouette*, The Difference Between a Vapor Barrier and an Air Barrier, *and Canadian Homebuilder's Association*, Builder's Manual.)

Airflow may be in the opposite direction of vapor diffusion. One of the primary influences on airflow is the heating and air conditioning system. Excessive air may be pushed outward or drawn inward depending on whether the HVAC system is operated at positive or negative air pressure. Air pressure differentials can change continuously based on exterior wind direction and speed. Masonry cavity walls and anchored veneers provide an air space immediately behind the outer wall surface, which can act as a buffer in minimizing air pressure differentials across the interior portion of the wall. No pressure equalization or reduction will occur, however, unless the backing wall includes an air barrier.

Air barriers are used to control the infiltration and exfiltration of airborne water vapor. The damp-proof coatings described earlier that can serve as WRBs and vapor retarders can also function as part of an air barrier system in masonry cavity walls. The barrier is incomplete, however, unless it includes sealing of all wall perimeters, penetrations, seams, and voids. An air barrier system must provide continuity from walls to roofs and from walls to windows and doors. Anywhere there is a breach in the air barrier, airflow will be concentrated, and localized wetting from condensation may occur.

Air can move readily through mortar bond line separations and through unsealed penetrations and wall openings in masonry veneers. The building wraps that are often used on stud backing walls behind masonry veneers are vapor permeable but function both as WRBs and air barriers. They must be integrated with window and door openings, penetrations, and wall-to-roof intersections.

Masonry walls are air permeable. Air can move through mortar bond line separations, but in single-wythe CMU walls, air may also move through units if they have an open pore structure (*see Fig. 7-39*). There are a number of different ways that a single-

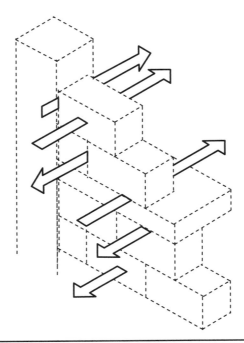

FIGURE 7-39   Air permeance of a CMU wall.

wythe CMU wall can be made to function as an air barrier. The NCMA has done extensive research and testing, and the following CMU wall assemblies are considered to meet air leakage rates of less than 0.04 cfm/sq ft at 75 Pa air pressure differential:

- Fully grouted concrete masonry
- Concrete masonry with a portland cement/sand parge coat
- Concrete masonry with a ½-in. stucco or plaster coating
- Concrete masonry walls coated with one application of block filler and two applications of a paint or sealer coating
- 12-in. CMU sealed with at least two coats of commercial-grade latex paint
- 8-in. CMU with a single coat of high-quality latex paint
- 8-in. CMU with a single coat of masonry block filler

Paint or block filler can be applied to either the interior or exterior side of the concrete masonry.

CHAPTER 8
Single-Wythe Wall Details

Single-wythe walls are most commonly constructed of concrete block but may also be built of hollow brick. Single-wythe walls may be loadbearing or non-loadbearing. Walls of hollow brick or block provide the options of grouting the cores for greater mass, stability, and water penetration resistance and adding steel reinforcement for flexural strength. The structural design of loadbearing single-wythe walls is discussed in Chapter 14. Chapters 6 and 7 describe in detail the basic concepts of movement control and water penetration resistance, but this chapter addresses architectural details that are specific to single-wythe concrete block walls.

8.1 Single-Wythe Concrete Masonry Unit Walls

Grouted, reinforced concrete block and hollow brick walls of a single unit thickness can be designed as structural loadbearing walls or as non-loadbearing curtain walls. Single-wythe concrete masonry unit (CMU) curtain walls are widely used for retail buildings such as the "big box" stores. They provide a durable and economical cladding typically used over prefabricated steel structures. Non-loadbearing masonry curtain walls can be designed by empirical height-to-thickness ratios or by engineering analysis as described in Chapter 14.

CMU curtain walls stand outside the structural frame and are self-supporting. They rely on the structure only for lateral support against overturning and the transfer of wind loads to the structure. Masonry curtain walls are designed to span either horizontally or vertically between lateral support connections. For vertically spanning curtain walls, fixed anchorage is provided at the foundation and flexible anchorage at the floor lines (*see Fig. 8-1*). Long walls are usually designed to span horizontally across the face of columns or cross-walls. Masonry curtain walls do not require a backing wall for additional anchorage or load transfer.

Single-wythe curtain walls must be stiff enough to resist wind loads between anchorage points without bending. To increase stiffness, many walls incorporate grout and reinforcing steel. Reinforced and unreinforced pilasters can also be used to stiffen unreinforced curtain walls and increase the heights to which they can be built (*see Fig. 8-2*). For horizontal spans, joint reinforcement and bond beams provide flexural resistance. Truss-type joint reinforcement is stiffer than ladder type, but its angular cross-wires can interfere with grout flow if the wall has grouted cells (*see Fig. 8-3*). Type N mortar is best for single-wythe wall construction, but type S may be used as an alternative when there are compelling, project-specific requirements (*see* Chapter 3 for mortar recommendations).

160 Chapter Eight

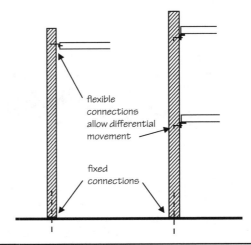

FIGURE 8-1 CMU curtain wall lateral connections.

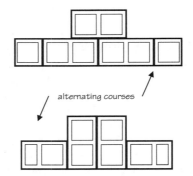

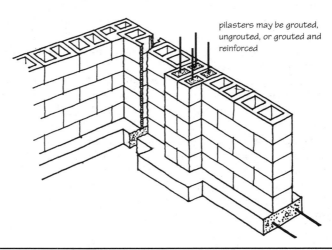

FIGURE 8-2 CMU pilasters increase wall stiffness.

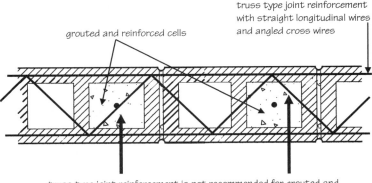

FIGURE 8-3 Ladder- and truss-type joint reinforcement for single-wythe CMU walls.

8.2 Insulation

Single-wythe CMU walls do not offer much thermal resistance on their own. Loose fill insulation such as perlite or vermiculite were once the only insulation products available for single-wythe walls. For walls that will be exposed on both the interior and exterior, loose fill insulation or rigid insulation inserts are the only choices. Loose fill insulation can settle over time, leaving an uninsulated thermal bridge at the top of the wall. Open-head joint weeps have to be screened to prevent the insulation from spilling out through the holes. Walls with loose fill insulation have thermal bridges at the cross-webs of the units and at the mortared head joints. In partially grouted walls, there are thermal bridges at the grouted cores as well. This makes loose fill much less effective than other insulation methods. Other methods of insulating single-wythe walls include interior rigid insulation and faced or unfaced batt insulation (*see Fig. 8-4*).

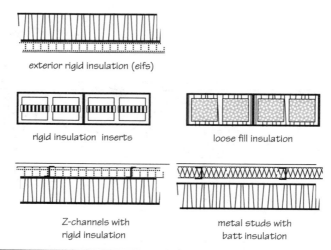

FIGURE 8-4 Methods of insulating single-wythe CMU walls. (*Adapted from C. Beall and R. Jaffe,* Concrete and Masonry Databook, *McGraw-Hill, 2003.*)

8.3 Interior Finishes

Depending on the application, single-wythe walls may remain exposed on the interior or an interior finish may be added later. Where studs and drywall are used as an interior finish, they are independent of the masonry curtain wall, and there are no anchors connecting the two. The interior finish may be supported by hat channels attached to the interior face of the wall or metal studs standing just inside, but not touching, the masonry wall (*see Fig. 8-5*).

Interior finishes can have a significant effect on the thermal and moisture performance of a single-wythe wall. Although there may be a small air space between the masonry and any interior studs, it is not a controlled drainage cavity. There is no sheathing or weather-resistive barrier, and the "air space" is not outfitted with flashing and weeps. Once metal studs, insulation, gypsum board, and paint or vinyl wall coverings have been added, it is no longer a true single-wythe wall. It is a wall "system" that includes a single wythe of masonry. Insulation type and location affect vapor transmission and the possibility of condensation because the location of the dew point changes. The drainage, drying, and vapor transmission characteristics of the wall are significantly different from those of a single-wythe wall without added components.

8.4 Water Penetration Resistance

Masonry walls are not waterproof. Single-wythe walls in particular are vulnerable to water penetration even when they are fully grouted and reinforced. Full mortar joints, good bond between units and mortar, and concave tooled joints are imperative and provide the first line of defense in limiting the amount of water entering the wall. Water-repellent admixtures in the block and mortar will reduce surface absorption and water migration but cannot stop the penetration of moisture through mortar joint bond line separations or other voids. Extra care is warranted in single-wythe walls because

Single-Wythe Wall Details

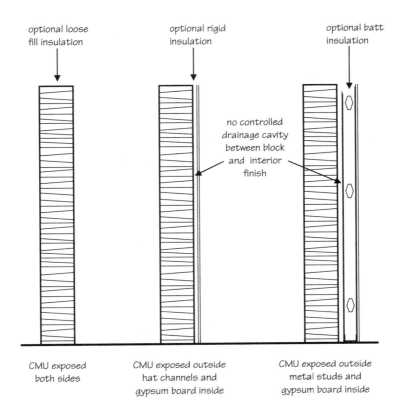

FIGURE 8-5 Interior CMU wall finishes.

they lack the redundant protection of a controlled drainage cavity with a water-resistive barrier. Single-wythe walls are less forgiving of minor workmanship defects that would otherwise be tolerated by a cavity wall or anchored veneer. Water-repellent admixtures and field-applied water repellents are imperative in single-wythe walls, but they cannot compensate for poor design or workmanship and should not be relied upon as the only method of controlling water penetration.

The typical method of flashing single-wythe walls is to step the flashing up between two half-thickness units at the flashing course (*see Fig. 8-6*). The flashing prevents mortar bond, so intermittent vertical reinforcing steel is used to maintain the structural integrity of the wall. The steel is designed to resist the entire lateral load, without any reliance on mortar bonding at the bed joints. Flashing penetrations at vertical reinforcing bars must then be sealed to maintain the integrity of the flashing (*see Fig. 8-7*). Through-wall flashing at the top and bottom of the wall, at window and door openings, and at other wall penetrations are necessary to collect penetrated moisture so that it can be drained out at the bottom of the wall (*see Fig. 8-8*).

164 Chapter Eight

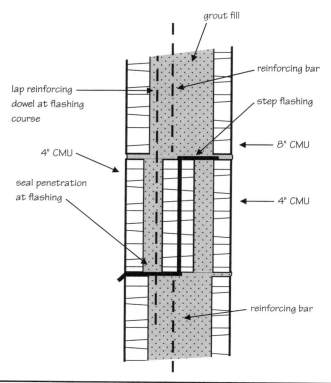

FIGURE 8-6 Step flashing through a single-wythe wall.

FIGURE 8-7 Unsealed flashing penetrations allow water infiltration.

Single-Wythe Wall Details   165

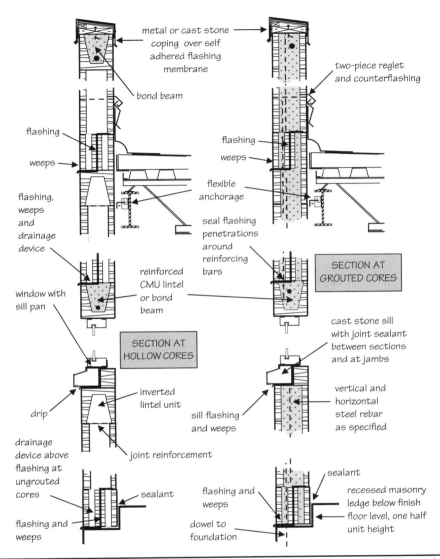

FIGURE 8-8 Flashing locations for a partially grouted and reinforced single-wythe CMU curtain wall with exposed block on interior.

8.5 Parapet Details

Parapets are a high priority for redundant protection against water penetration. Metal copings provide the best protection (*see Fig. 8-9*). They are impervious to moisture and can be installed in lengths requiring a minimum number of joints. Metal copings should have sealed splice plates at the joints to keep water out and to accommodate the differential movement between the masonry and the metal. Wood blocking is typically used to create a slope to support the coping and minimize cupping.

166 Chapter Eight

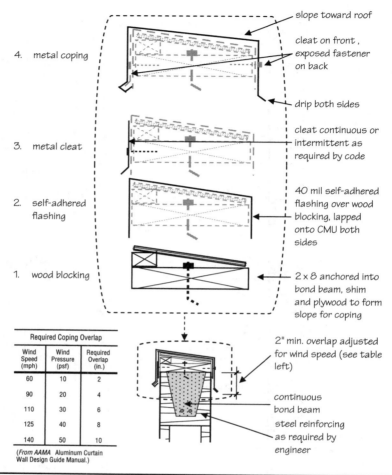

FIGURE 8-9 Metal coping detail for single-wythe walls.

Flashing (typically self-adhered membrane flashing) should be installed over the wood blocking and down onto the face of the masonry. This covers the gap between the blocking and the CMU and acts as a second line of defense if the metal coping itself leaks. The vertical legs of metal copings should extend at least 2 in. below the top of the masonry and turn out to form a drip.

If the wall is split-face block, the irregularities of the block surface make it difficult to install a sealant joint that is both functional and attractive. Some architects use a smooth face block at the top course of the wall to provide a joint that is more uniform (see Fig. 8-10). If you simply turn a split-face block around and use the smooth side, there will be no color variation. If the joint between the metal coping leg and the face of the masonry is sealed, insert weep tubes to drain any water that may penetrate the coping itself.

Single-Wythe Wall Details

FIGURE 8-10 Smooth course at top of wall is easier to seal to metal trim than split-face block.

A cast stone coping is shown in *Fig. 8-11*. Because mortar can shrink and crack, the joints in a cast stone coping are very vulnerable to moisture penetration. The head (top and face) joints in masonry copings should be raked out and filled with a backer rod or bond breaker and sealant to form a joint with the correct width-to-depth geometry.

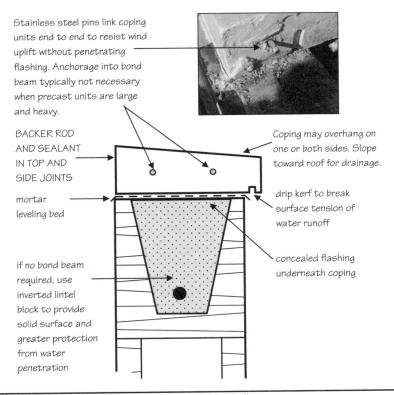

FIGURE 8-11 Cast stone coping detail for single-wythe walls.

The sealant joints can be made to look more like mortar if sand is sprinkled on the surface before the sealant sets. A metal or membrane flashing should be installed below the coping to provide redundant protection against water penetration.

8.6 Roof-to-Wall Details

This is an interface between two trades performing their work at different times, so the roof and wall systems need to be considered together. The perimeter flashing on a roof has to be a minimum of 8 in. above the roof surface. A parapet wall must be tall enough to accommodate the roof flashing height and the length of the vertical leg of a metal coping. When parapet walls are relatively short, the roof membrane is typically wrapped up the back and onto the top of the parapet and the coping or the flashing underneath the coping is lapped over the top edge of the roofing membrane. This creates a bad situation when it is time to replace the roof. If the coping and the roof membrane are tied together, the coping will have to be removed to do the roofing work. It is better if the parapet wall is higher and a metal counterflashing can cap the roof membrane separate from the coping.

A reglet is installed by the mason as the wall is laid up, and the wall flashing is lapped over it (*see Fig. 8-12*). The roofer inserts a metal counterflashing into the reglet when the roof work is complete. Surface-mounted reglets should not be used on masonry walls. Masonry is too porous, and water can saturate the area above and flow behind the reglet and into the wall. In a single-wythe wall, the lapped flashing and reglet provide a simple water-resistant detail.

Roof membranes should not cover the backs of masonry parapet walls. The backs of masonry parapet walls should be allowed to breathe so that they can dry from both sides. An elastomeric coating with a relatively high permeance will do, but it requires maintenance over the life of the building. Metal panels are simple and maintenance free. The coping width has to be extended to cover the top of the panels.

Where a joist or beam is structurally supported by the masonry wall, an 8-in. block unit is omitted at the bearing point and a 4-in. closure unit later added at the exterior face in the same spot. Joist and beam pockets often leak because the flashing course is interrupted. *Figure 8-13* shows a detail recommended by the *National Concrete Masonry Association* (NCMA) for this condition. The flashing must be formed of metal because it has to be able to take and hold its shape.

8.7 Window Head Details

Window head details in a single-wythe CMU wall are fairly straightforward. The openings are spanned by reinforced CMU lintels bearing on either side. The flashing course is above the lintel, stepped up through the wall thickness in the usual manner. It is best to have a drip above the window (*see Fig. 8-14*) to break the surface tension of the runoff water and prevent it from clinging to the surface all the way back to the window sealant joint. A drip will have to be saw cut into the blocks before they are placed in the wall.

8.8 Window Sill Details

Window sills can take a number of different shapes. Because concrete blocks are nominally 8 in. tall, some architects like to use a cast stone sill that is the same height. This

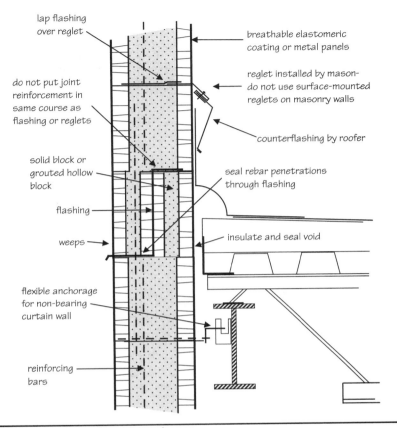

FIGURE 8-12 Flashing at roof-to-wall interface.

prevents any interruption in the coursing height of the wall. Others prefer to use a 4-in.-high sill for better scale, but it may require a 4-in. course just below the sill to stay on module with the 8-in. block (*see Fig. 8-15*). Window sills have high weather exposure and can be a significant source of rain penetration. Sills that are constructed of concrete block units are very porous and can absorb a lot of water. The number of units it takes to span the width of the window also means that there are numerous mortar joints that can shrink and crack. Cast stone sills are a better choice because they are larger with fewer joints, and they are much less porous than block. The sill should project past the face of the wall slightly and always have a drip on the bottom.

The window sill pan should turn up at the back and sides and extend at least to the front edge of the window framing, depending on the shape of the sill. The sill pan must be set in a full bed of sealant and care taken not to block the window weeps. All joints on the top of a CMU or cast stone window sill should have a backer rod and sealant rather than mortar, including the joint between the end of the sill and the jamb opening. The most important thing in a window sill is slope. The top of the sill should have a 15° positive slope away from the window.

170 Chapter Eight

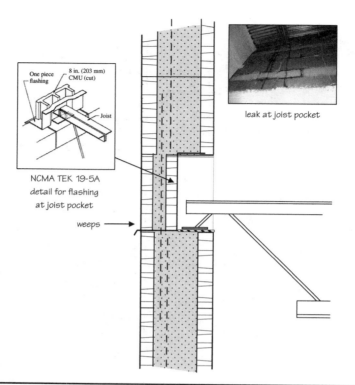

FIGURE 8-13 Flashing at joist or beam bearing pocket.

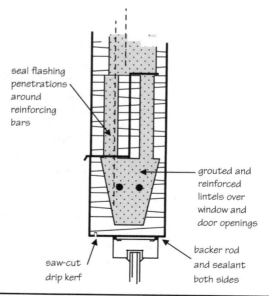

FIGURE 8-14 Flashing at window or door head.

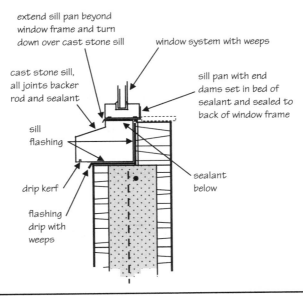

FIGURE 8-15 Window sill detail.

8.9 Base Flashing Details

Masonry walls and veneers are usually set onto a recessed ledge at the edge of the slab. Sometimes in loadbearing structures, engineers prefer to set the walls on top of and at the same elevation as the finish floor level. The manner in which the two scenarios are flashed is essentially the same. *Figure 8-16* shows a recessed ledge that is nominally 4 in. deep.

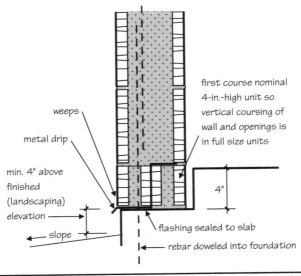

FIGURE 8-16 Single-wythe CMU wall with a masonry ledge.

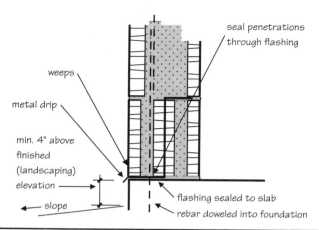

FIGURE 8-17 Single-wythe CMU wall without a masonry ledge.

A "half-high" unit at the base of the wall maintains the typical 8-in. module of vertical coursing so that window and door opening dimensions are not affected. Where there is no ledge (*see Fig. 8-17*), the entire wall consists of standard 8-in. units, but the flashing takes the same physical configuration as the wall with the ledge. In either instance, the flashing must be fully sealed to the concrete at the bottom.

8.10 Miscellaneous Details

It is absolutely imperative that the block used for single-wythe walls be manufactured with an integral water repellent and that the mortar be mixed on site with the companion water-repellent admixture. Without these, a single-wythe CMU wall will almost certainly leak. Water repellency is an important part of the wall's ability to drain water

FIGURE 8-18 This recess and raked joint is a leak waiting to happen in a single-wythe wall.

downward through ungrouted cores rather than wicking it through to the inside face of the block.

Recessed courses add vulnerability to the wall by diminishing the thickness of the narrow face shell bed joints. Recessed courses also create a small ledge that can funnel water into the wall if the joint is not well tooled and compacted against the units (*see Fig. 8-18*).

Service penetrations must be sealed at the outside face of the wall with a high-quality joint sealant material. Mortar should be held back away from junction boxes, hose bibbs, louvers, and other penetrations to allow for a backer rod and a properly sized and formed sealant joint.

CHAPTER 9
Multi-Wythe Wall Details

The structural design of multi-wythe loadbearing walls is discussed in Chapter 14. Chapters 6 and 7 describe in detail the basic concepts of movement control and water penetration resistance, but this chapter addresses architectural details that are specific to multi-wythe masonry walls.

9.1 Multi-Wythe Walls

Multi-wythe masonry walls include both composite walls and cavity walls. Composite walls are usually two wythes in thickness with a mortared or grouted collar joint in between and are bonded with *rigid* metal ties. Composite walls are typically loadbearing. In cavity walls, the backing and facing wythes are separated by an open cavity and connected with *adjustable* metal ties. Cavity walls may be either loadbearing or non-loadbearing. In loadbearing applications, the backing wythe typically supports axial loads from the floor and roof systems, and the facing wythe supports only lateral wind loads and its own weight.

When concrete masonry is used as the backing wythe in a cavity wall, the joint reinforcement required to control shrinkage cracking can be fitted with adjustable ties to connect to a facing wythe of brick or stone or concrete masonry. Three-wire joint reinforcement and joint reinforcement with fixed tab ties can also be used to connect the wythes of some types of cavity walls, but they do not provide flexibility for differential movement between wythes. Spacing requirements for different types of wire ties are covered in Chapter 14.

9.1.1 Composite Walls

Because of the differential moisture shrinkage of concrete masonry and moisture expansion of clay masonry, composite walls should have backing and facing wythes of the same material; that is, a concrete masonry backing with concrete masonry facing or a brick backing wall with a brick facing. Composite walls can be laid with the backing and facing wythes separated only by a ¾-in. collar joint, which is filled with mortar as the wall is built. Most composite walls, however, are constructed with a wider space between the wythes to accommodate reinforcing bars (*see Fig. 9-1*). These wider spaces can be grouted in low lifts as the wall is built or in high lifts after several courses or an entire story height is built.

Composite walls are less resistant to rain penetration than cavity walls. Composite walls resist rain penetration primarily by absorbing and storing moisture until it is evaporated back to the atmosphere. Composite walls are not "barrier" walls, but should

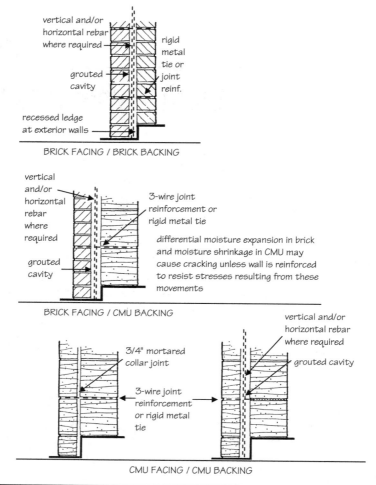

FIGURE 9-1 Composite masonry walls. (*From C. Beall and R. Jaffe*, Concrete and Masonry Databook, *McGraw-Hill, 2003.*)

instead be considered as "mass" walls or "reservoir" walls. Like single-wythe walls, they are relatively unforgiving of design and construction errors because they lack the redundant protection of a drainage cavity.

9.1.2 Cavity Walls

Cavity walls are among the strongest and most durable of exterior building wall systems and are usually the first choice for buildings that will have a long service life such as schools and government buildings. Cavity walls consist of two or more wythes of masonry units separated by an open cavity at least 2 in. wide. The wythes may be brick or concrete block anchored to one another with metal ties that span the cavity (*see Fig. 9-2*). Cavity walls are most often constructed with concrete block as the backing wythe and brick as the facing or with a CMU backing and facing.

Multi-Wythe Wall Details   177

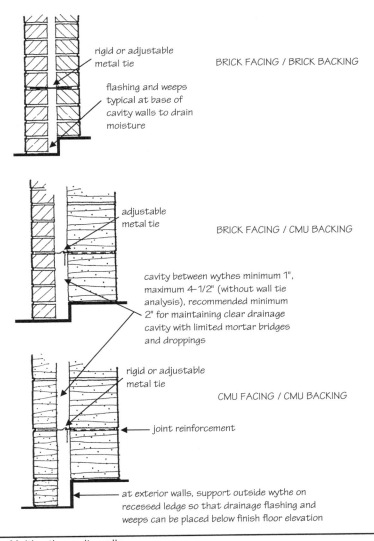

FIGURE 9-2 Multi-wythe cavity walls.

One of the major advantages of cavity wall construction is the increased resistance to rain penetration that results from the physical separation of the inner and outer wythes. This separation also increases thermal resistance by providing a dead air space and allows room for additional insulating materials if desired. The open cavity, when it is properly fitted with a system of flashing and weep holes, provides drainage for moisture that may penetrate the exterior or form as condensation within the cavity.

Periodic wetting and drying are not harmful to masonry or to the components that make up a cavity wall. Cavity walls are designed to collect and drain moisture efficiently so that there is no extended saturation that could cause efflorescence, freeze-thaw damage, or corrosion of metal ties. To maintain functionality, the cavity must be kept clear of mortar droppings, and the flow of moisture to the weeps must be unobstructed. Despite the fact that moisture is designed to move through a masonry cavity wall in a controlled manner, mold growth cannot occur because there is no food source to sustain the proliferation of mold spores. Cavity walls are, in fact, more durable in resisting moisture damage than almost any other type of wall.

Both wythes of a cavity wall must resist wind loads and other lateral forces. Metal ties transfer these loads from one wythe to the other in tension and compression and must be solidly bedded in the mortar joints to perform properly. Crimped ties with a water drip in the center should not be used because the weakened plane created can cause buckling of the tie and ineffective load transfer.

Two-piece adjustable ties permit differential thermal and moisture movements between the backing and facing wythes of a cavity wall. When constructed of dissimilar materials such as brick and CMU, this differential movement can be significant. A concrete masonry backing wall experiences irreversible moisture shrinkage as the latent moisture from the manufacturing process evaporates, and a brick facing experiences irreversible moisture expansion as the brick reabsorbs atmospheric moisture after it is fired (*see* Chapter 6). These opposing movements can be accentuated when cavity insulation increases the temperature differential between the inner and outer wythes. The *International Building Code* (IBC) and *Masonry Standards Joint Committee* (MSJC) *Building Code Requirements for Masonry Structures* both require that *adjustable* ties be spaced at a maximum of 16 in. on center vertically and horizontally with no more than 1.77 sq ft of wall area per tie (*see* Chapter 14).

It is imperative that exterior CMU facing wythes have an integral water repellent and that the accompanying mortar be treated with a compatible water-repellent admixture from the same manufacturer. Type N mortar is best for cavity wall construction. Type S should be used as an alternative only when there are compelling, project-specific requirements (*see* Chapter 3 for mortar recommendations). Cavity walls should be protected against moisture penetration in accordance with the principles outlined in Chapter 7, relying primarily on a system of flashing and weeps to collect and expel rain or condensate moisture. CMU backing walls can be constructed as in-fill panels, as curtain walls (*see* Chapter 8), or they can be constructed as load-bearing walls (*see* Chapter 14). The cavity face of the concrete block backing should be coated with a vapor-resistant or vapor-permeable damp-proofing mastic or other membrane or material to provide a water-resistive barrier. The damp-proof coating may also function as an air barrier if the perimeter joints and penetrations are also sealed.

9.2 Insulation

Insulation may be added in cavity walls, including vermiculite, perlite, or rigid boards. Chapter 5 discusses thermal properties of masonry walls and the types of insulation typically used in them. When rigid board insulation is to be installed in the cavity, the clear distance between the face of the insulation and the back of the exterior wythe should be 2 in. Codes limit the maximum distance between backing and facing to 4½ in. This limitation is based on the stiffness and load transfer capability of wire ties. With a 2-in. open cavity, this would permit a maximum insulation thickness of 2½ in.

9.3 Interior Finishes

Depending on the application, the interior wythe of a cavity wall may remain exposed on the interior or an interior finish may be added later. Where studs and drywall are used as an interior finish, they are independent of the cavity wall, and there are no anchors connecting the two. The interior finish may be supported by hat channels attached to the interior face of the wall or metal studs standing just inside, but not touching, the masonry wall (refer to Chapter 8).

Interior finishes can have a significant effect on the thermal and moisture performance of a cavity wall. Insulation type and location affect vapor transmission and the possibility of condensation because the location of the dew point changes. The drying and vapor transmission characteristics of a cavity wall are significantly different when they have interior finishes.

9.4 Water Penetration Resistance

Cavity walls offer the best and most reliable water penetration resistance of any type of masonry wall. Through-wall flashing at critical locations collects penetrated moisture so that it can be drained out of the wall through weeps (*see Fig. 9-3*). Water that penetrates at mortar bond separations, windows, and other locations cannot damage the wall or cause mold growth. Because neither the masonry units nor mortar are subject to deterioration from periodic wetting, water penetration from whatever source is little threat to the system. Facing wythes that stay wet for extended periods may develop efflorescence, which is an unsightly but merely cosmetic issue. Neither efflorescence nor cryptoflorescence (efflorescence that occurs below the surface) cause physical damage to units or mortar.

Drying the back of the facing wythe and the drainage cavity itself occurs more rapidly if the cavity is vented as described in Chapter 7. Basically, by providing open-head joint weeps at the base of the wall and open-head joint vents near the top of the wall, moisture vapor can escape out the top of the cavity. Weep inserts that fit into the head joints restrict airflow considerably and should not be used if venting is to be effective. The airflow itself is not of a high volume, but the vents at the top of the wall at least allow warm, moist air that naturally rises inside the cavity to escape to the exterior. Mortar droppings at the base of the wall must be kept to an absolute minimum if any airflow is to be established.

The surfaces can be spalled off of brick that repeatedly freezes and thaws *while it is saturated with moisture*, but cavity walls dry fairly rapidly, which limits the occurrence of this problem. Composite walls dry much more slowly than cavity walls because there is no air space behind the facing wythe. Freeze-thaw spalling is much more likely in composite walls. Masonry walls do not support mold growth because the materials that make up the units and mortar do not provide a food source for the mold spores. Only the dust or dirt on the face of a masonry wall will feed mold, and the amount is limited enough to be considered negligible.

Metal joint reinforcement and wall ties can corrode, but if they are hot-dip galvanized as required by code and held back from the face of the wall by the prescribed 5/8 in., there is little likelihood of this happening. In coastal climates where the air is heavily salt-laden, all metal accessories in all masonry walls should be upgraded from galvanized to stainless steel because of the severe environment.

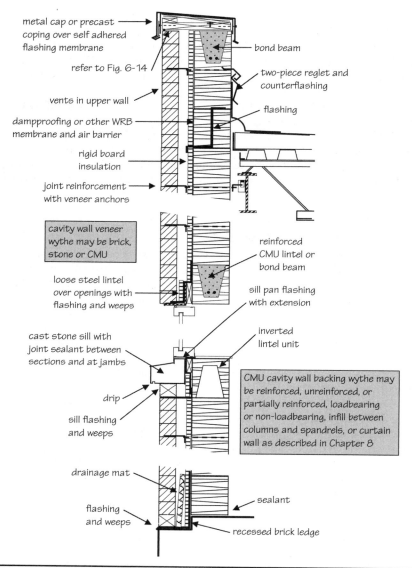

FIGURE 9-3 Basic brick and CMU cavity wall details.

9.5 Parapet Details

Both cast stone and metal copings are used on multi-wythe masonry walls. Metal copings provide the best protection (*see Fig. 9-4*). They can be installed in lengths requiring a minimum number of joints. The joints between sections should have sealed splice plates to prevent water penetration. Wood blocking is used to create a slope to support a metal coping and minimize cupping. Because cavity walls are wide, standing seams

Multi-Wythe Wall Details

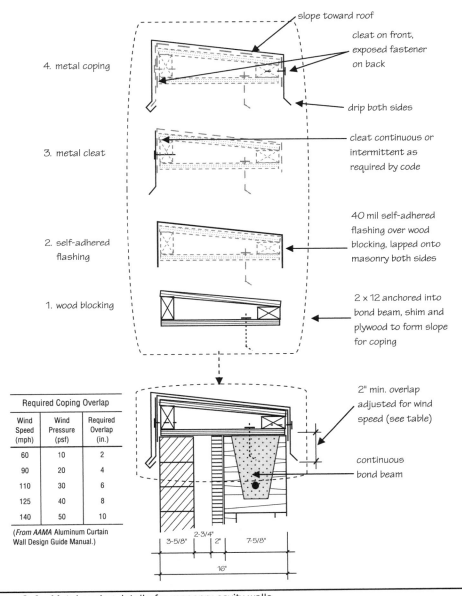

FIGURE 9-4 Metal coping details for masonry cavity walls.

are often used on metal copings to stiffen the section and provide additional protection against cupping. A metal coping that is cupped in the middle and has poorly sealed joints will allow water into the wall. Self-adhered flashing should be installed under the metal coping as a second line of defense if the coping itself leaks or is displaced in high winds. The vertical legs of metal copings should extend at least 2 in. below the top of the masonry and turn out to form a drip. The coping overlap and the drip act as a capillary break to prevent water from running back into the wall.

A cast stone coping is shown in *Fig. 9-5*. Because mortar can shrink and crack, the joints in a cast stone coping are very vulnerable to moisture penetration. The cross-joints in masonry copings (top and sides) should be raked out and filled with a backer rod or bond breaker and sealant to form a joint with the correct width-to-depth geometry. The sealant joints can be made to look more like mortar if sand is sprinkled on the surface before the sealant develops a skin. A metal flashing should be installed below the coping to provide redundant protection against water penetration. A membrane flashing may sag at the air space and collect water. If the flashing is punctured or improperly sealed, that water will penetrate into the wall cavity. The coping units should be linked together end-to-end rather than tie into the wall below so that the flashing is not penetrated.

Where a masonry cavity wall parapet intersects a higher wall, a metal or cast stone coping cannot simply be surface sealed against the face of the higher wall. The flashing membrane below the coping must be sealed against the backing wythe of the upper wall (*see Fig. 9-6*).

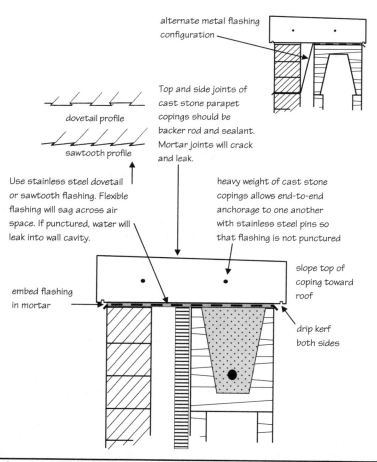

FIGURE 9-5 Cast stone coping for masonry cavity walls.

Multi-Wythe Wall Details

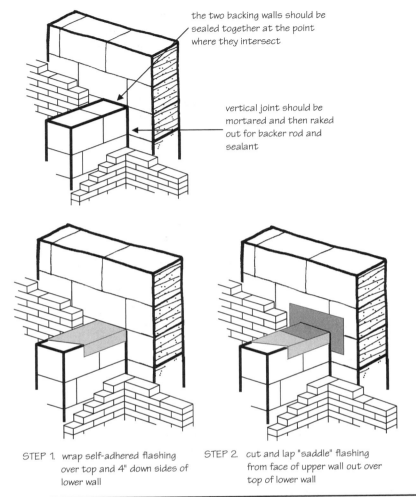

FIGURE 9-6 Flashing low cavity wall to higher cavity wall.

9.6 Roof-to-Wall Details

The work of two trades must interface to form a water-resistant barrier where the roof meets a masonry parapet wall. The systems used to form the interface must allow the two trades to perform their work in sequence without damaging the other. Roof flashing must be turned up onto the face of the parapet wall and terminated a minimum of 8 in. above the level of the roof surface. Where it terminates, metal through-wall flashing or counterflashing is used to cap the roof flashing (see Fig. 9-7). Through-wall flashing and two-piece reglets and counterflashing provide the best interface between a masonry parapet wall and the perimeter roof flashing. Reglets designed to be placed in the mortar joint are installed by the mason. The roofing contractor installs the counterflashing when the roof is installed. If through-wall flashing is also needed to block

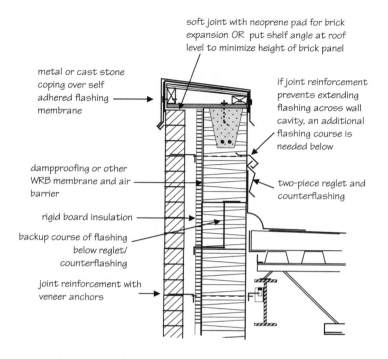

FIGURE 9-7 Roof-to-wall intersection at cavity walls.

moisture flow into the wall cavity below the roof level, a separate flashing should be located lower in the wall. There are additional roof-to-wall details in Chapter 7.

9.7 Shelf Angle Details

Shelf angles are used in cavity walls to support the dead load of the outer wythe at intermediate floors. At each shelf angle, flashing and weep holes must be installed to collect moisture and drain it to the outside. Metal flashing must be brought beyond the face of the wall and turned down to form a drip, and flexible flashing must be brought to the face of the wall and trimmed flush as described in Chapter 7. Flexible flashing membranes can also be installed with a separate metal drip. Flashing should never be stopped short of the face of the wall because this will permit moisture to flow around and underneath, where it can pool in the cores of the brick or block or into a window head below. Self-adhered flashing or metal flashings and drips bedded in sealant or mastic on the shelf angle will prevent this backflow.

The horizontal joints at shelf angles can be quite wide to accommodate the thickness of the angle plus the "soft" joint for vertical brick expansion. These wide joints are necessary, but their appearance can be minimized using an inverted lipped brick or articulated with special-shape units, bond pattern changes, color variations, and so on (*see Fig. 9-8*). Additional shelf angle details and discussion are included in Chapter 7.

Multi-Wythe Wall Details

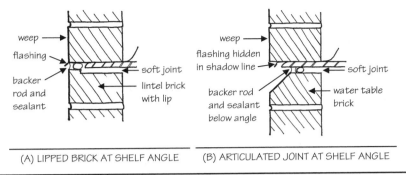

FIGURE 9-8 Alternate shelf angle details to minimize or articulate the sealant joint.

For architects who strongly object to the appearance of horizontal soft joints in a brick masonry façade, the best alternative is to design the veneer as a curtain wall without shelf angles. The veneer rests on the slab and is anchored to the backing wall in the usual way, but it extends upward multiple stories without intermediate shelf angle supports. Most building codes permit this type of construction to a height of at least 100 ft but may require the submittal of engineering calculations and special detailing to the building official. The compressive strength of the units is more than adequate to support the dead load of multiple stories of masonry above (*see Fig. 9-9*). The parapet cap and any terminations underneath balconies or other protruding or recessed elements must be carefully detailed to allow for differential vertical expansion and contraction (refer to Chapter 6). This wall system is permitted only in Seismic Design Categories A, B, and C. It works best with stacked windows or vertical curtain wall strips because the veneer must be able to expand vertically adjacent to the windows. Very careful detailing is required.

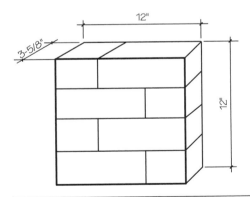

- Weight of brick veneer = 41 psf x 100 ft. veneer height = 4,100 psf = 28.5 psi.
- Gross cross sectional area of one sq.ft. of brick veneer = 12 in. x 3-5/8 in. = 43.5 sq.in.
- COMPRESSIVE LOAD = 28.5 psi x 43.5 sq.in. = 1,239.75 psi
- ALLOWABLE COMPRESSIVE LOAD for veneer of moderate strength brick (4,500 psi) and ASTM C270 proportion method Type N mortar combination = 1500 psi

FIGURE 9-9 Allowable vertical load on foundation-supported brick veneer without shelf angles.

9.8 Window Details

The flashing at the lintel above window and door openings collects water from the drainage cavity in the upper wall and prevents it from flowing into the window head itself. The window sill pan flashing and the flashing underneath the masonry sill protect the wall below from excessive water penetration from the window weep system and from the rain that impinges on the exposed surface of the masonry sill. Masonry sills must always be positively sloped away from the window. A minimum of 15° is usually recommended (*see Fig. 9-10*).

Masonry window sills are exposed to the same amount of rain as a roof. Brick sills and CMU sills are very vulnerable to water penetration because mortar shrinkage can cause cracks to form in the joints. Longer cast stone sills have fewer joints and therefore less vulnerability. The joints in masonry sills and at the jambs of masonry sills should be filled with a backer rod and sealant to minimize the risk of water penetration. The window jamb and the adjacent masonry must be integrated to ensure that there is continuity in the air barrier seal (*see Fig. 9-11*).

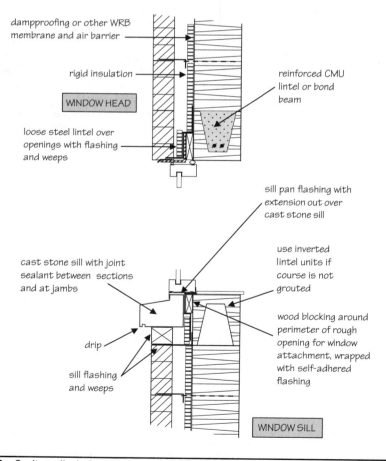

FIGURE 9-10 Cavity wall window details.

Multi-Wythe Wall Details   **187**

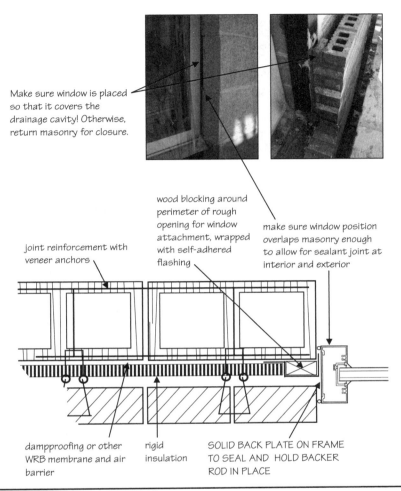

FIGURE 9-11 Window jamb in cavity wall.

9.9 Base Flashing Details

The base of a masonry cavity wall is a simple detail. The outer wythe of the wall is set on a recessed masonry ledge below the finish floor elevation. If it is a brick facing wythe, the ledge is typically one brick below the floor. If it is a CMU facing wythe, the ledge is typically one-half the height of a block. Flashing is installed under the bottom course of the outer wythe and turned up the face of the backing wall. The metal ties should begin at the second course of the CMU backing so they do not interfere with the flashing. The flashing on the backing wall should extend higher than the drainage mat in the cavity so that mortar droppings cannot collect above the flashing. If the metal ties begin in the second CMU joint, that is the maximum height of the flashing. Choose the drainage mat height that is appropriate for the unit dimensions, and choose the flashing width that will extend just above it (*see Fig. 9-12*).

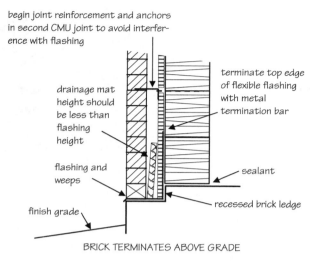

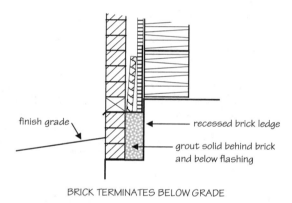

FIGURE 9-12 Base flashing for cavity walls.

Typically, masonry walls should terminate above grade. Some architects prefer the appearance of brick that extends slightly below grade. There are also conditions such as sloping grade where the brick may be intermittently below grade even though the foundation and the brick ledge are stepped downhill. If the base of the outer wythe of a cavity wall is below grade or has only minimal clearance above grade for any reason, the flashing and weeps must still be kept above grade to function properly. The cavity space below the flashing should be grouted solid and the flashing brought out to the face of the wall two to three courses above finish grade (including landscape fill). The same is true where a sidewalk or other flatwork will be poured against a masonry wall at a later date. This must be anticipated by grouting the cavity in the bottom one or two courses and bringing the flashing out above that level. A second layer of flashing should be installed on the ledge and extend up behind the upper flashing. This brings the weeps well above grade regardless of the situation. Masonry flashing and weeps should never be obstructed by anything inside or outside of the wall that would inhibit drainage.

Multi-Wythe Wall Details 189

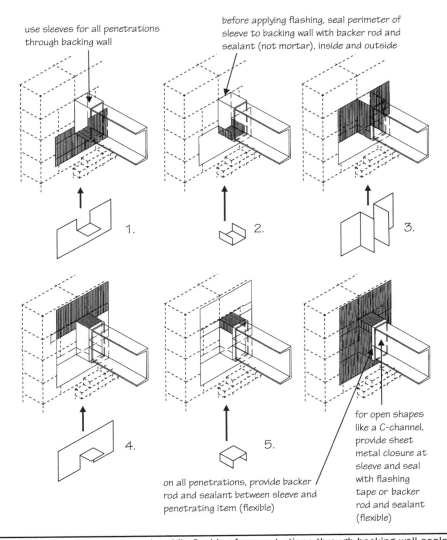

FIGURE 9-13 Provide sleeves and saddle flashing for penetrations through backing wall sealed with self-adhered flashing membrane, backer rod, and sealant.

9.10 Miscellaneous Details

Where large elements such as beams penetrate a cavity wall, they should be sealed at the backing wall with saddle flashing similar to that used at the intersection of parapet copings and high walls (*see Fig. 9-13*). Smaller service penetrations through a cavity wall should also be sealed at the backing wall with flexible flashing or a backer rod and high-quality joint sealant. Sealing penetrations at the backing wall provides the best protection against water penetration because the sealing material is not directly exposed to rain or ultraviolet deterioration.

CHAPTER 10
Anchored Veneer Details

Chapters 6 and 7 describe in detail the basic concepts of movement control and water penetration resistance. This chapter addresses architectural details that are specific to anchored masonry veneer walls.

10.1   Brick and Concrete Masonry Unit Veneer

Masonry cladding over a nonmasonry backing wall is called a veneer. Brick veneer is most commonly used over wood stud walls in residential buildings and over metal stud backing in commercial buildings. In most parts of the country, concrete masonry unit (CMU) veneer over stud wall backing is used almost exclusively for commercial construction. An open cavity between the backing wall and the masonry veneer allows drainage of moisture that penetrates the wall or condensate that forms within it. Stud backing walls are vulnerable to corrosion and decay, and some sheathing materials support mold growth, so moisture penetration control is critical to success (*see Fig. 10-1*).

Codes regulate the design of anchored veneers by prescriptive requirements based on empirical data. The veneer chapter of the *Masonry Standards Joint Committee* (MSJC) *Building Code Requirements for Masonry Structures* limits use of the prescriptive design method to walls subject to design wind pressures of 25 psf or less. Higher wind pressures require analytical design. The MSJC Code has prescriptive requirements for every aspect of veneer design and construction (*see Fig. 10-2*) and special requirements for seismic areas (*see Fig. 10-3*).

Connectors used to attach masonry facing wythes to masonry backing walls are called ties. Connectors used to attach masonry veneers to nonmasonry backing walls are called anchors. Code requirements for spacing of veneer anchors are shown in *Fig. 10-4*. Additional anchors should be located within 12 in. of openings larger than 16 in. in either dimension at a spacing not to exceed 36 in. on center.

For securing masonry veneer to single-family wood frame construction, corrugated sheet metal anchors are typically used. These should be 22-gauge galvanized steel, at least 7/8 in. wide × 6 in. long. Corrosion-resistant nails should penetrate the stud a minimum of 1½ in. exclusive of sheathing. The free end of the anchor should be embedded at least 2 in. into the mortar bed. Corrugated anchors are weak in compression and provide load transfer only if the horizontal leg is properly aligned in plane with the mortar bed joint and the nail is positioned within ½ in. of the 90° bend (*see Fig. 10-5*). Anchors randomly attached to the backing wall and bent out of plane to align with bed joints cannot transfer loads. Corrugated anchors may be used only in low-rise construction and only if the cavity width does not exceed 1 in.

192 Chapter Ten

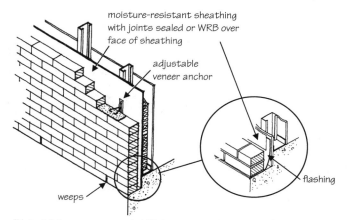

*Note: Minimum recommended 2" air space requires wire anchors. Maximum cavity width for corrugated sheet metal anchors is 1".

FIGURE 10-1 Brick veneer walls.

| MSJC Prescriptive Requirements for Anchored Masonry Veneer ||
Item	Minimum Requirements
Applicability	prescriptive requirements may not be used where basic wind speed exceeds 110 mph (25 psf)
Support over openings	unless the veneer is self-supporting (e.g., masonry arches), veneer above openings must be supported on non-combustible steel, concrete, or masonry lintels with minimum 4 in. bearing on each side and deflection limited to 1/600
Maximum height above non-combustible foundation	30 ft., with an additional 8 ft. permitted for gable ends, *except* if veneer with cold-formed steel stud backing exceeds this height, it shall be supported by non-combustible construction for each story above the height limit (unless designed by engineering methods)
Anchors	• corrosion resistant wire anchors not less than 9 gauge, or corrugated sheet metal not less than 7/8 in. wide, 22 gauge • embedded in mortar joint at least 1-1/2 in. with at least 5/8 in. mortar or grout cover to the exterior face • maximum 1 in. between veneer and sheathing with corrugated anchors
Anchor spacing	• maximum 32 in. on center horizontal x 18 in. on center vertical • adjustable two-piece anchors of W1.7 or 22 gauge corrugated sheet metal maximum 2.67 sq.ft. of wall area per anchor • additional anchors around all openings larger than 16 in. in either dimension, spaced 3 ft. on center and within 12 in. of opening
Air space	minimum 1 in. clear air space
Flashing	designed and detailed to resist water penetration into the building interior, with backing system designed and detailed to resist water penetration
Weepholes	minimum 3/16 in. diameter, maximum spacing 33 in., located immediately above flashing
Differential movement	design and detail veneer to accommodate differential movement

FIGURE 10-2 Code requirements for masonry veneer. (*Based on MSJC Building Code Requirements for Masonry Structures, ACI 530/ASCE 5/TMS 402.*)

Anchored Veneer Details

MSJC Prescriptive Seismic Requirements for Anchored Masonry Veneer	
Seismic Risk	Minimum Requirements
Seismic Design Categories A and B	Basic code requirements, no special provisions
Seismic Design Category C	Basic code requirements plus the following special provisions • Isolate sides and top of anchored veneer from structure so that vertical and lateral seismic forces resisted by the structure are not imparted to the veneer.
Seismic Design Category D	Same as Category C plus the following special provisions • Support the weight of anchored veneer for each story independent of the other stories. • Reduce the maximum wall area supported by each anchor to 75% of that normally required (maximum horizontal and vertical spacings are unchanged). • Provide continuous, single-wire joint reinforcement of minimum W1.7 wire at a maximum spacing of 18 in. on center vertically.
Seismic Design Categories E and F	Same as Category D plus the following special provisions • Provide vertical expansion joints at all returns and corners. • Mechanically attach anchors with clips or hooks to joint reinforcement required above.

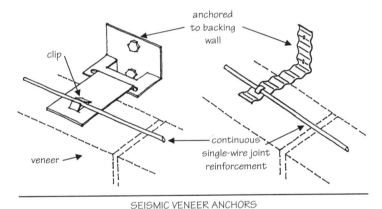

FIGURE 10-3 Seismic requirements for masonry veneer. (*Based on MSJC* Building Code Requirements for Masonry Structures, *ACI 530/ASCE 5/TMS 402.*)

Masonry veneer is anchored to metal stud frames with 9-gauge corrosion-resistant wire hooked through a slotted connector or looped eye for flexibility. Anchors are attached through the sheathing and into the studs with corrosion-resistant, self-tapping screws. Stainless steel screws with a rubber washer provide better performance than ordinary galvanized screws and should always be used in commercial construction.

Masonry veneers are rigid compared with the flexibility of stud walls in resisting lateral loads. The *Brick Industry Association* (BIA) recommends a deflection limit of $L/600$ to provide adequate stiffness in the studs. Lateral bracing or stiffeners in the stud wall may also be required for adequate rigidity to prevent veneer cracking and subsequent moisture intrusion. Sheathing or gypsum board must be attached on both sides of the stud wall to add stiffness, and-studs should never be less than 16 gauge. Stud spacing should not exceed 16 in., and galvanizing should be by hot-dip process, in accordance with ASTM A525, G60 or G90 coating.

194 Chapter Ten

Veneer Anchor Spacing	
Maximum Spacing, Horizontal x Vertical (in. x in.)	Maximum Wall Area Per Anchor (sq.ft.)
32 x 18	adjustable two-piece anchors of W1.7 (9 gauge) wire and 22 gauge corrugated sheet metal anchors, 2.67 all others, 3.5

Recommended Corrosion Protection for Veneer Anchors and Joint Reinforcement	
Application	Corrosion Protection
Interior	mill galvanized ASTM A653, Class G60
Exterior walls and interior walls exposed to mean relative humidity of 75% or more	hot-dip galvanized ASTM A153

FIGURE 10-4 Requirements for masonry veneer anchors.

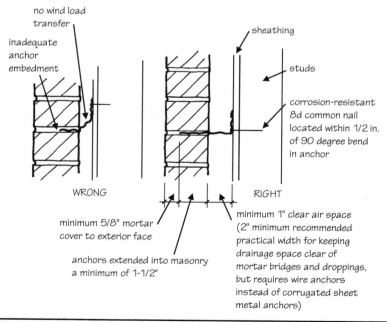

FIGURE 10-5 Requirements for corrugated anchors.

Anchored Veneer Details   195

Type N mortar is best for veneer wall construction. Type S should be used only if there are compelling, project-specific requirements (*see* Chapter 3 for mortar recommendations). Veneer walls should be protected against moisture penetration in accordance with the principles outlined in Chapter 7, relying primarily on a system of flashing and weeps to collect and expel rain or condensate moisture. Basic commercial and residential veneer details are shown in *Figs. 10-6 and 10-7*.

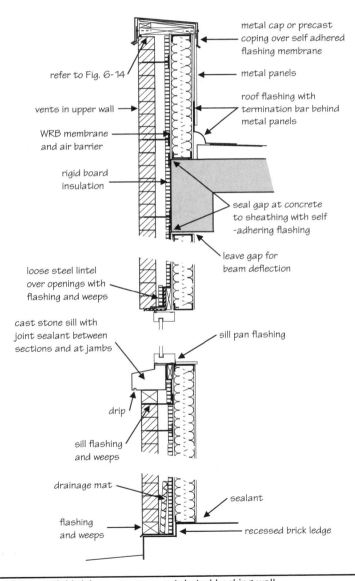

FIGURE 10-6 Commercial brick veneer over metal stud backing wall.

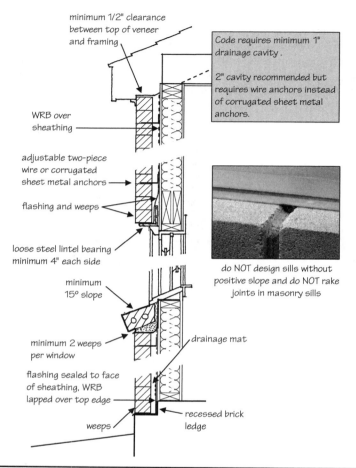

FIGURE 10-7 Residential brick veneer over wood stud backing wall.

10.2 Insulation

Veneer walls have traditionally been insulated with fiberglass batts fitted between the studs. The vapor permeance ratings of foil and kraft facings on batt insulation should be taken into consideration in the wall design as well as thermal bridging at the studs. These are two big issues that have contributed to condensation, water damage, and mold growth in veneer walls in the past. Rigid insulation in the drainage cavity instead of batt insulation eliminates the thermal bridges at the studs and moves the dew point further outward in the wall.

When rigid board insulation is to be installed in the cavity, the clear distance between the face of the insulation and the back of the veneer should be 2 in. Codes limit the maximum distance between backing and facing to 4½ in. This limitation is based on the stiffness and load transfer capability of wire anchors. With a 2-in. open cavity, this would permit a maximum insulation thickness of 2½ in.

10.3 Water Penetration Resistance

The drainage capacity of masonry veneer walls makes them a good second choice after cavity walls for effective rain penetration resistance. Unlike cavity walls, the backing wall in veneer systems is susceptible to water damage. Codes require a water-resistive barrier (WRB) over sheathing that is not inherently water resistant itself. Flashing at critical locations in masonry veneer walls collects penetrated moisture so that it can be drained out of the wall through weeps.

Drying the back of the veneer and the drainage cavity itself occurs more rapidly if the cavity is vented as described in Chapter 7. Basically, by providing open-head joint weeps at the base of the wall and open-head joint vents near the top of the wall, moisture vapor can escape out of the top of the cavity. Weep inserts that fit into the head joints restrict airflow considerably and should not be used if venting is to be effective. The airflow itself is not of a high volume, but the vents at the top of the wall at least allow warm, moist air that naturally rises inside the cavity to escape to the exterior. Mortar droppings at the base of the wall must be kept to an absolute minimum if any airflow is to be established.

10.4 Parapet Details

Both cast stone and metal copings are used on commercial masonry veneer walls. Metal copings provide the best protection because they have the fewest joints (*see Fig. 10-8*). Wood blocking is used to create a slope to support a metal coping and minimize cupping. The joints between sections should have sealed cover plates to prevent water

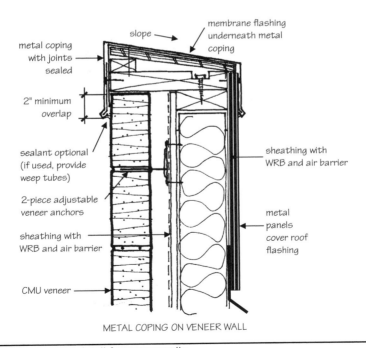

Figure 10-8 Metal coping detail for veneer walls.

penetration. A metal coping that is cupped in the middle and has poorly sealed joints will allow water into the wall. Self-adhered flashing should be installed under the metal coping as a second line of defense if the coping itself leaks or is displaced in high winds. The vertical legs of metal copings should extend at least 2 in. below the top of the masonry and turn out to form a drip. The coping overlap and the drip act as a capillary break to prevent water from running back into the wall.

When cast stone copings are used, the mortar joints are very vulnerable to water penetration. The cross-joints in cast stone copings (top and sides) should be raked out and filled with a backer rod or bond breaker and sealant to form a joint with the correct width-to-depth geometry. A metal or membrane flashing layer should be installed across the top of the wall below the coping to provide redundant protection against water penetration. The sealant joints can be made to look more like mortar if sand is sprinkled on the surface before the sealant develops a skin. The coping units should be linked together end-to-end rather than tie into the wall below so that the flashing is not penetrated.

Where a masonry veneer wall parapet intersects a higher wall, the metal coping cannot simply be surface sealed against the face of the higher wall. The flashing membrane below the coping must be sealed against the backing wall of the upper veneer (*see Fig. 10-9*).

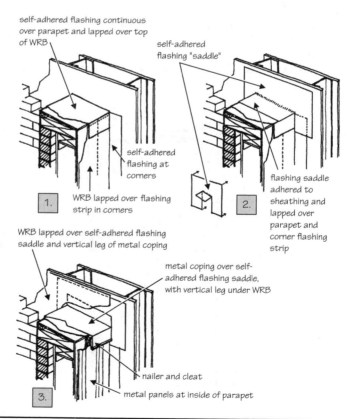

Figure 10-9 Saddle flashing at parapet to intersecting high wall in brick veneer.

10.5 Roof-to-Wall Details

The work of two trades must interface to form a water-resistant barrier where the roof meets a parapet wall. Roof flashing must be turned up onto the face of the parapet wall and terminated a minimum of 8 in. above the level of the roof deck. The inside of a masonry veneer wall parapet is usually constructed of metal studs. The simplest way to finish off the inside of the parapet wall and tie in to the roof perimeter flashing is with metal panels (*see Fig. 10-10*). The top edge of the roof flashing can be sealed with a termination bar, and the metal panels protect it from direct rain exposure.

10.6 Shelf Angle Details

Shelf angles are used in veneer walls, as they are in cavity walls, to support the dead load of the veneer at intermediate floors. At each shelf angle, flashing and weep holes must be installed to collect moisture and drain it to the outside (*see Fig. 10-11*). Metal flashing must be brought beyond the face of the wall and turned down to form a drip, and flexible flashing can be installed with a separate metal drip. The horizontal expansion joints ("soft" joints) below shelf angles can be quite wide to accommodate vertical brick expansion plus the thickness of the steel angle itself. Wide joints are necessary, but their appearance can be altered using a variety of techniques discussed in Chapter 7.

For architects who strongly object to the appearance of horizontal soft joints in a brick masonry façade, the best alternative is to design the veneer as a curtain wall without shelf angles. The veneer rests on the slab and is anchored to the backing wall in the usual way, but it extends upward multiple stories without intermediate shelf angle supports. Most building codes permit this type of construction to a height of at least 100 ft but may require the submittal of engineering calculations and special detailing to the building official. The compressive strength of the units is more than adequate to support the dead load of multiple stories of masonry above (*see* Chapter 9). The parapet cap and any terminations underneath balconies or other protruding or recessed

FIGURE 10-10 Metal panels inside masonry veneer parapet wall.

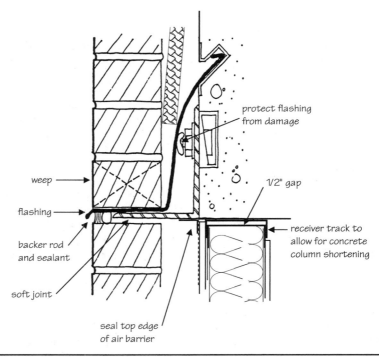

Figure 10-11 Shelf angle flashing.

elements must be carefully detailed to allow for vertical expansion of brick or shortening of CMU veneers. When the structural frame is concrete, column shortening due to the concrete shrinkage must also be considered. This wall system is permitted only in Seismic Design Categories A, B, and C. It works best with stacked windows or vertical curtain wall strips because the veneer must be able to expand vertically adjacent to the windows. Very careful detailing is required.

10.7 Window Details

The flashing at the lintel above window and door openings collects water from the drainage cavity above and prevents it from flowing into the window frame head. The window sill pan flashing and the flashing underneath the masonry sill protects the wall below from excessive water penetration from the window weep system and from the rain that impinges on the exposed surface of the masonry sill. *Figures 10-12 and 10-13* show flashing details for commercial aluminum "punched" windows and for residential windows with nailing fins.

The most important thing in a window sill is slope. The top of the sill should have a 15° positive slope away from the window. Masonry window sills are exposed to the same amount of rain as a roof. Brick sills and CMU sills are very vulnerable to water penetration because mortar shrinkage can cause cracks to form in the joints. Longer cast stone sills have fewer joints and therefore less vulnerability. The joints in masonry sills and at the jambs of masonry sills should be filled with a backer rod and sealant to minimize the risk of water penetration.

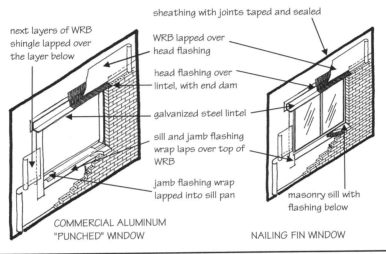

FIGURE 10-12 Commercial and residential window flashing. (Refer to ASTM E2112, *Standard Practice for Installation of Exterior Windows, Doors and Skylights*, for industry-standard window flashing details.)

10.8 Base Flashing Details

The base of a veneer wall at grade should be the "last catch" at collecting and removing water that has penetrated the wall or condensed within it. As simple as the detail may seem (*see Fig. 10-14*), the problem most frequently encountered (other than excessive mortar droppings) relates to height above grade and slope.

Accessibility requirements have led many architects to set the finish floor elevation of the building too close to grade. If the site is relatively flat and has poor drainage, this can become a big issue. Flashing and weeps at the base of the wall that are not able to drain effectively are useless. Worse than that, if water puddles against the wall because of poor site drainage, the water can actually run back into the wall cavity through the weeps. Unlike cavity walls, the materials in a stud backing wall can be water damaged. The worst-case scenario is water getting into the building at the interior floor line. If there is absolutely no wiggle room to elevate the finish floor sufficiently and still have the space required for ADA entrance slopes, it is critical that the flashing extends well above the interior floor elevation, that the flashing is not punctured or damaged during installation, and that mortar droppings are held to an absolute minimum.

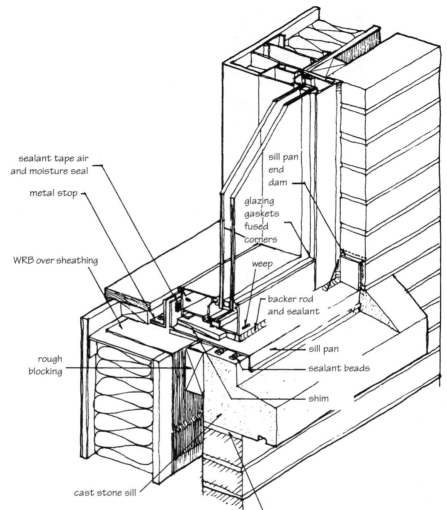

With window sill pan lapped over top of large precast or cast stone sill, flashing under masonry sill is usually not necessary. With brick rowlock or CMU or porous stone sill, water can penetrate through mortar joints or through porous units, so through-wall flashing and weeps should be installed under the masonry sill.

Figure 10-13 Commercial window sill flashing. (*Adapted from Nashed,* Timesaver Standards for Exterior Wall Design, *McGraw-Hill, 1996.*)

Masonry veneers should be set on a recessed ledge below the finish floor elevation. If it is a brick veneer, the ledge is typically one brick below the floor. If it is a CMU veneer, the ledge is typically one-half the height of a block. Flashing is installed under the bottom course of the masonry and turned up the face of the backing wall. The anchors should begin above the flashing so the fasteners do not puncture it. The flashing on the backing wall should extend higher than the drainage mat in the cavity so that

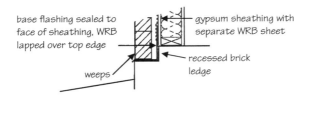

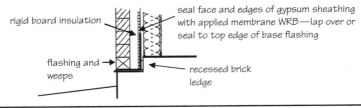

Figure 10-14 Base flashing for brick veneer walls.

mortar droppings cannot collect above the flashing. Choose the drainage mat height that is appropriate for the unit dimensions, and choose the flashing width that will extend just above it.

Typically, masonry walls should terminate above grade. Some architects prefer the appearance of brick that extends slightly below grade. There are also conditions such as sloping grade where the brick may be intermittently below grade even though the foundation and the brick ledge are stepped downhill. If the base of the veneer is below grade or has only minimal clearance above grade for any reason, the flashing and weeps must still be kept above grade to function properly. The cavity space below the flashing should be grouted solid and the flashing brought out to the face of the wall two to three courses above finish grade (including landscape fill). The same is true where a sidewalk or other flatwork will be poured against a masonry veneer at a later date. This must be anticipated by grouting the cavity in the bottom one or two courses and bringing the flashing out above that level. A second layer of flashing should be installed on the ledge and extend up behind the upper flashing. This brings the weeps well above grade regardless of the situation. Masonry flashing and weeps should never be obstructed by anything inside or outside of the wall that would inhibit drainage.

Two levels of flashing are necessary if the clearance above grade is minimal. One layer of flashing is installed on the masonry ledge, the first one or two courses of the veneer are grouted solid, and another layer of flashing installed above that. This brings the weeps well above grade even though the ledge has only minimal clearance.

CHAPTER 11
Adhered Veneer Details

Chapters 6 and 7 describe in detail the basic concepts of movement control and water penetration resistance, but this chapter addresses architectural details that are specific to adhered veneer.

11.1 Adhered Masonry Veneer

A veneer is defined as "a nonstructural facing attached to a backing for the purpose of ornamentation, protection, or insulation, but not bonded to the backing so as to exert a common reaction under load." There are two basic methods of attaching masonry veneer. The previous chapter described anchored veneer, which is the more common of the two methods. This chapter describes adhered veneer, in which the masonry units are secured by adhesion with a bonding material to a solid backing. Adhered veneer does not support its own weight.

Codes limit the weight of adhered veneer to 15 lb/sq ft. Individual units are limited to a 36-in. maximum face dimension and 5 sq ft in area. In some seismic areas, weight is limited to 10 lb/sq ft. Units weighing less than 3 lb/sq ft are not limited in dimension or area. The bond of an adhered veneer to the supporting element must be designed to withstand a shearing stress of 50 psi. Differential thermal and moisture movement characteristics should be considered in selecting backing and facing materials. An expanding clay masonry facing (thin brick) and a contracting concrete or concrete masonry backup are not compatible when relying exclusively on an adhesive bond. Code requirements do not limit the length or height of adhered veneer except as necessary to control expansion and contraction. Any movement joints that occur in the backing or the frame must be carried through the bedding mortar and the veneer as well. Type S mortar is recommended for adhered veneer applications.

11.2 Thin Brick Veneer

Thin brick units are made from clay or shale and are kiln-fired much like typical full-size brick (*see Fig. 11-1*). Thin brick is ½ to 1 in. thick with face sizes the same as those of conventional brick. When completed, a wall faced with thin brick gives the appearance of a conventional brick masonry wall. ASTM C1088, *Thin Veneer Brick Units Made From Clay or Shale*, includes Exterior and Interior Grades and appearance Types TBS, TBX, and TBA. The appearance Types are similar to those of full-size facing brick. There is no minimum compressive strength for thin brick.

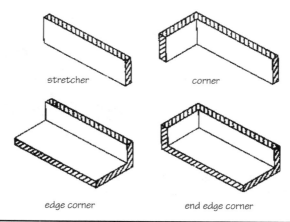

FIGURE 11-1 Thin brick veneer units.

The most common face size of thin brick is the same as standard modular brick with nominal dimensions of 2-2/3 × 8 in. The actual face dimensions are 3/8 to ½ in. less than the nominal dimensions to allow for the thickness of a mortar joint.

11.3 Adhered Manufactured Stone Masonry Veneer (AMSMV)

Manufactured stone is a type of concrete masonry unit manufactured as a thin, lightweight, non-loadbearing masonry veneer product. The units come in random sizes and shapes that simulate the look of natural stone. They are made from cement, aggregates, pigments, water, and other additives. Even though they can be made to look like brick, this is a cementitious product distinctly different from the thin brick described earlier.

11.4 Installation Methods

Adhered veneer is installed in a completely different manner than mortar-bedded unit masonry and stone. Adhered veneers may be applied over concrete or masonry substrates or over wood-framed or metal stud–framed walls (*see Fig. 11-2*). Deflection limitations for backing walls are recommended to be L/1000 to avoid cracking the veneer. Over stud and sheathing backing walls, a water-resistive barrier (WRB), drainage mat, and paper-backed lath are mechanically attached, and a scratch coat of mortar is applied and allowed to set for at least 48 hours. Some consider the drainage mat optional, but research studies have found that an air gap such as that provided by a thin (¼ in.) drainage mat significantly improves the drying time of adhered veneers and reduces the potential for water damage.

Adhered veneers may be applied directly to concrete and masonry substrates without a WRB. The surfaces must be clean and all form release oils and other contaminants removed by sandblasting, pressure washing, or other effective means of providing a good bonding surface. If there is any doubt about the bonding capability of the backing wall, metal lath should be installed the same as for stud wall construction.

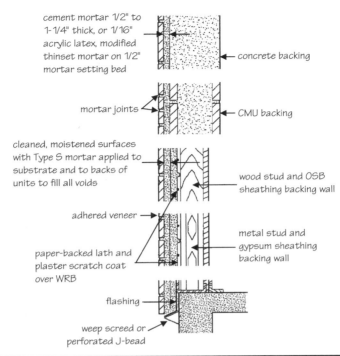

Figure 11-2 Adhered veneer may be installed on concrete, masonry, or stud-framed backing walls. (*From J. Chrysler et al.*, Masonry Design Manual.)

Before beginning installation, both the scratch coat or setting bed and the brick or stone units themselves are dampened evenly and allowed to surface dry. Mortar is then spread on the backs of the units about ½ in. thick (or more to compensate for uneven substrates) and covering the surface completely (*see Fig. 11-3*). When the unit is placed,

Figure 11-3 Inadequate mortar coverage will not produce good bond.

it must be wiggled or rotated slightly to spread the mortar and get rid of air pockets. The mortar should squeeze out beyond the edges of the units on all sides, even in "dry stack" stone patterns. Stone that is to have grouted joints is installed from the top of the wall downward to keep the stone clean during installation. Stone that will be installed without grouted joints in a "dry stack" or "tight-fitted" design is installed from the bottom of the wall upward.

The biggest challenge adhered veneer has faced is adhesion to cast-in-place, precast, and tilt-up concrete. Polymer-modified acrylic latex or thinset mortar additives and bond enhancers are usually recommended for application to concrete backing walls. Concrete masonry unit (CMU) backing walls typically have enough surface texture that the bond enhancers are not needed. CMU backing walls should not be constructed of units with an integral water-repellent admixture because the water repellent will inhibit mortar bond. Whenever there is any doubt about the surface characteristics of the backing wall, it is recommended that lath be applied to the wall with a scratch coat of mortar as it is for application over studs and sheathing. An *American Society for Testing and Materials* (ASTM) standard for minimum installation requirements for AMSMV is currently under development to assist in the specification of these systems.

Differential movement between the adhered veneer and the substrate or building frame should be accommodated by expansion joints. Keeping the panel sizes small minimizes the buildup of movement stresses. Horizontal expansion joints should be placed at every story height or change in backing material. Vertical expansion joints should be at the spacing recommended for the concrete or masonry backing material. In framed walls and masonry substrates as well, expansion joint locations should follow the rules of thumb for brick masonry (windows, changes in wall height, etc.).

11.5 Water Penetration Resistance

Adhered veneers, because they are constructed of thin units, are even more susceptible to rain penetration than standard masonry unit construction. Substrate walls of studs and sheathing are required by code to include a two-layer WRB. Concrete masonry walls may require a damp-proof coating to stop water penetration in exterior applications. Where this is the case, metal lath will be required because mortar will not bond to the damp-proofing. Precast and tilt-up concrete backing walls are generally considered to be resistant to water penetration as long as the panel joints are properly sealed, so they do not typically require any additional damp-proofing treatment.

Full mortar coverage is essential to the performance of adhered veneers. Even small voids or air pockets behind the units can collect and hold water. Tight-fitted stone patterns (narrow joints only partially filled with mortar) are not recommended in severe winter climates because of the possibility of freeze-thaw expansion breaking the bond between the units and the backing wall.

Flashing must be installed at copings, windows, doors, the base of the wall, and penetrations. The flashing must be integrated with the WRB and lap over the window nailing flange and over the vertical leg of a weep screed at the base of the wall. Detailing is basically the same as for stucco cladding.

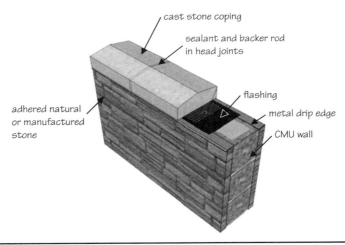

FIGURE 11-4 Natural or adhered manufactured stone masonry veneer over CMU with cast stone parapet. (*From Rocky Mountain Masonry Institute*, Adhered Natural Stone Veneer Installation Guide.)

11.6 Parapet Details

Adhered veneers can be installed with masonry or metal copings as detailed previously in Chapters 9 and 10 for multi-wythe and anchored veneer walls, respectively (*see Fig. 11-4*). As always, metal or membrane flashing must be installed below any coping over a masonry wall, and the head joints in masonry copings must be raked out and finished with a backer rod or bond breaker tape and sealant.

11.7 Roof-to-Wall Details

In residential applications where a sloped roof meets a higher wall with adhered veneer, the WRB and wall flashing must be integrated with and lap over the top of the roof edge flashing. A metal "kick-out" flashing at the edge of the roof keeps water from getting behind the veneer (*see Fig. 11-5*).

11.8 Window Details

Window head and sill details are the same as for a stucco or EIFS cladding, with the WRB integrated with the membrane flashing at the window nailing fins (*see Fig. 11-6*).

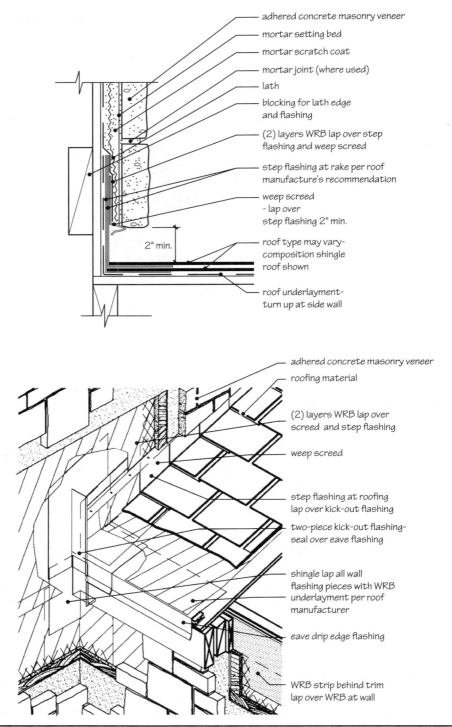

FIGURE 11-5 Roof-to-wall detail with metal kick-out flashing. (*From Masonry Veneer Manufacturer's Association*, Installation Guide for Adhered Concrete Masonry Veneer.)

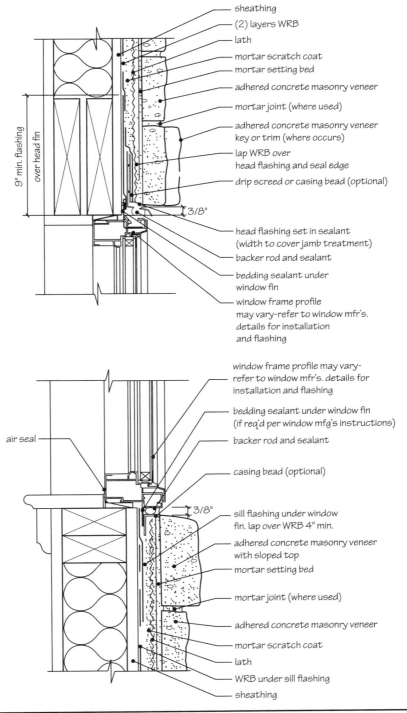

FIGURE 11-6 Nailing flange window head and sill details in adhered veneer over stud frame and sheathing. (*From Masonry Veneer Manufacturer's Association,* Installation Guide for Adhered Concrete Masonry Veneer.)

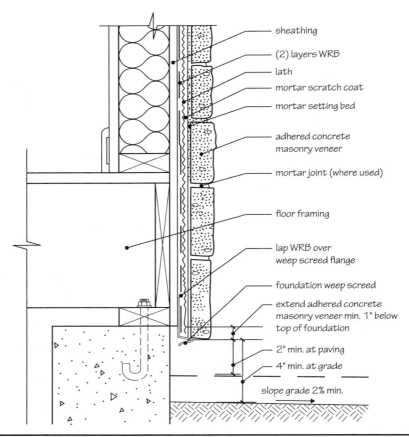

FIGURE 11-7 Flashing and weep screed at base of wall. (*From Masonry Veneer Manufacturer's Association*, Installation Guide for Adhered Concrete Masonry Veneer.)

11.9 Wall Base Details

At the base of the wall, the veneer should extend below the finish floor elevation but be clear of the exterior finish grade by a minimum of 2 in. (*see Fig. 11-7*). The WRB should lap over the vertical leg of the weep screed (or perforated J-bead) to direct moisture out of the wall.

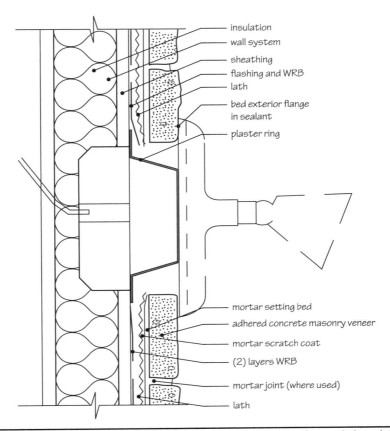

Figure 11-8 Flashing at wall penetration. (*From Masonry Veneer Manufacturer's Association, Installation Guide for Adhered Concrete Masonry Veneer.*)

11.10 Miscellaneous Details

At wall penetrations, flashing should lap the penetrating item where it passes through the backing wall (*see Fig. 11-8*). Self-adhered rubberized asphalt flashing tape is flexible enough to follow the contour of conduit, sleeves, or other cylindrical items. Wrap the bottom first and then lap the top flashing over the bottom.

CHAPTER 12

Special Wall Types

Special masonry wall types include non-loadbearing interior partitions, screen walls and fences, and glass block panels. Of the three, interior partitions are the most frequently used. In commercial construction, nonbearing partitions are used as fire and smoke partitions and for area separation. Screen walls and fences are used primarily in residential construction to give privacy between homes and along the perimeter of subdivisions. Glass block panels may be used to control daylighting, to provide security glazing, or to serve a strictly decorative function.

12.1 Interior Partitions

Partitions are interior, non-loadbearing walls one story or less in height and support no vertical load other than their own weight. They may be separation elements between spaces, as well as fire, smoke, or sound barriers. Non-loadbearing interior partitions are self-supporting, but they do not carry the vertical compressive load of the structure.

Code requirements for interior non-loadbearing partitions are based on empirical lateral support requirements expressed as length- *or* height-to-thickness (h/t) ratios. The *Masonry Standards Joint Committee* (MSJC) *Building Code Requirements for Masonry Structures*, ACI 530/ASCE 6/TMS 402, and the *International Building Code* (IBC) both prescribe a maximum h/t of 36 for interior nonbearing partitions (*see Fig. 12-1*).

Lateral support can be provided by cross-walls, columns, pilasters, or buttresses, where the limiting span is measured horizontally, or by floors, roofs, beams, clips, angles, or anchors, where the limiting span is measured vertically. Lateral support members must have sufficient strength and stability to transfer all overturning moments to adjacent structural members or to the foundation. Arbitrary span limitations, of course, do not apply if the walls are designed by engineering analysis.

Based on an h/t ratio of 36 as prescribed in the IBC and MSJC codes, a single-wythe, 4-in. brick partition without reinforcing steel is limited to a 12-ft span, whereas a 6-in. brick partition can span 18 ft between supports, and an 8-in. hollow brick partition can span 24 ft between supports. If the partition is securely anchored against lateral movement at the floor and ceiling, and if the height does not exceed these limits, there is no requirement for intermediate walls, piers, or pilasters along the length of the partition. If additional height is required, the 8-in. hollow brick can be reinforced every 24 ft or pilasters can be added at 12- or 18-ft intervals for the 4-in. and 6-in. walls, respectively. Lateral support is required in only one direction and can be either floor or ceiling anchorage (*see Fig. 12-2*) or cross-walls, piers, or pilasters, but need not be both.

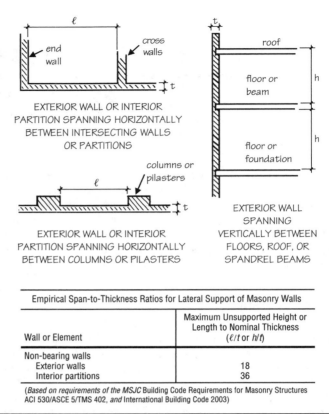

FIGURE 12-1 Lateral support requirements for empirically designed non-loadbearing masonry walls and partitions.

Concrete block partitions are widely used as interior fire, smoke, and sound barriers. Decorative units can be left exposed, but standard utility block is usually painted, textured, plastered, or covered with gypsum board. Wood or metal furring strips can be attached, or sheet materials may sometimes be laminated directly to the block surface.

Hollow concrete block partitions can be internally reinforced to provide the required lateral support in lieu of cross-walls or projecting pilasters. A continuous vertical core at the required interval is reinforced with steel reinforcing bars and then grouted to form an in-wall column.

12.2 Screen Walls and Fences

Perforated masonry screen walls may be built with specially designed concrete block or clay tile units, with standard concrete blocks laid with cores oriented horizontally, with brick or block laid in an open pattern, or with combinations of these units (*see Fig. 12-3*). As sun screens, the walls are often built along the outside face of a building to provide shading for windows. Screen walls are also used to provide privacy without blocking

Special Wall Types 217

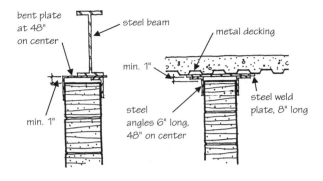

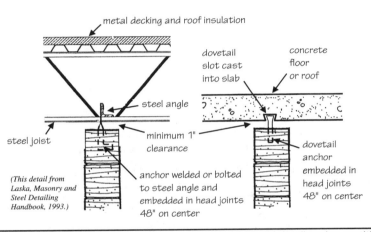

FIGURE 12-2 Examples of lateral support at the tops of interior, non-loadbearing partitions.

airflow and to form interior and exterior area separations. The function of the wall influences finished appearance, from strong and heavy to light and delicate. Dark colors absorb more heat and reflect less light into interior spaces. Relatively solid wall patterns block more wind, and open patterns allow more ventilation.

Screen walls are governed by the same h/t ratio for lateral support requirements as empirically designed masonry walls and partitions, but those with interrupted bed joints should be designed more conservatively because of reduced flexural strength and lateral load resistance.

Concrete masonry screen wall units should meet the minimum requirements of ASTM C129, *Nonloadbearing Concrete Masonry Units*. Brick should be ASTM C216, *Facing Brick* Grade SW, and clay tile units should be ASTM C530, *Structural Clay Non-Loadbearing Screen Tile* Grade NB. Mortar for exterior screen walls should be Type N.

Solid, uncored brick is used to build what some call "pierced" walls by omitting the mortar from head joints and separating the units to form voids. The walls may be laid up in single- or double-wythe construction. In double-wythe walls, separate header or

FIGURE 12-3 Examples of masonry screen wall units and bonding patterns.

rowlock courses alternate with stretcher courses to form different patterns. Double-wythe walls are more stable than single-wythe designs because of the increased weight, wider footprint, and through-wall bonding patterns. Piers may be either flush with the wall or projecting on one or both sides. The coursing of the screen panels must overlap the coursing of the piers to provide adequate structural connection. Regardless of exact design, however, the pattern of units in a pierced wall must provide continuous vertical paths for load transfer to the foundation.

Concrete screen block and clay screen tile are made with a decorative pattern of holes in the units, so it is not necessary to separate them with open-head joints. Most unit types are designed to be laid with continuous vertical and horizontal mortar joints in stack bond patterns. The larger area of mortar bedding increases the lateral load

Special Wall Types

resistance of the wall. The continuous bed joints accommodate the installation of horizontal joint reinforcement, and bond beam courses can be added at the top and bottom of the wall for even greater strength.

The *National Concrete Masonry Association* (NCMA) has done considerable research on concrete masonry screen walls, and as a result, more is known about this type of unit strength and wall performance than any other type of screen wall. Units should have a minimum compressive strength of 1000 psi (gross) when tested with the cores oriented vertically, and face shells and webs should be at least ¾ in. thick. Truss-type joint reinforcement should be spaced 16 in. on center vertically.

Lateral support for *concrete masonry fences* is usually provided by reinforced pilasters or by internal vertical reinforcement (*see Fig. 12-4*). Foundations should be placed

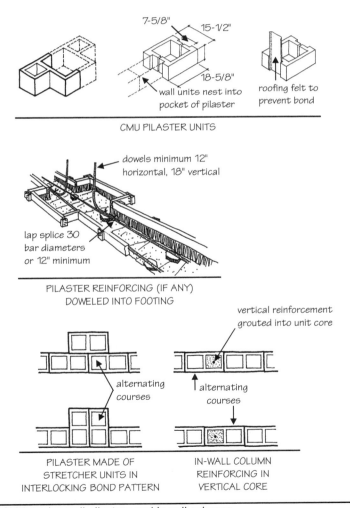

FIGURE 12-4 CMU garden wall pilasters and in-wall columns.

in undisturbed soil below the frost line. For stable soil conditions where frost heave is not a problem, a shallow continuous footing or pad footing provides adequate stability. Where it is necessary to go deeper to find solid bearing material, where location in relation to property lines restricts footing widths, or where the ground is steeply sloping, a deep pier foundation provides better support. In each instance, the supporting pilaster is tied to the foundation by reinforcing dowels. A vertical control joint should be provided on one side of each pilaster support. Joint reinforcement in the panel sections should stop on either side of the control joint. The designs shown in *Figs. 12-5 and 12-6* are based on wind loading conditions but are not intended to resist lateral earth pressure as retaining walls. Concrete masonry fences require joint reinforcement and control joints for shrinkage crack control the same as for other CMU walls. Stucco may be applied directly to concrete masonry unit fences, with control joints in the same locations as the CMU control joints (*see Fig. 12-7*).

Brick fences may take a number of different forms. A straight wall without pilasters must be designed with sufficient thickness to provide lateral stability against wind and

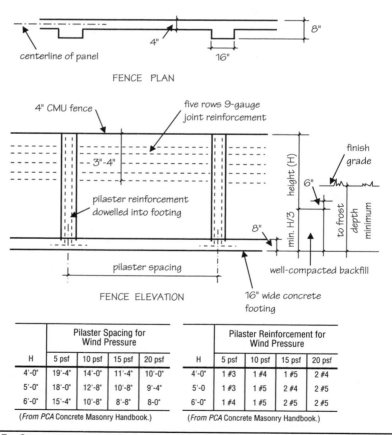

FIGURE 12-5 Concrete masonry pier and panel fences. (*From Randall and Panarese,* Concrete Masonry Handbook, *Portland Cement Association.*)

Special Wall Types

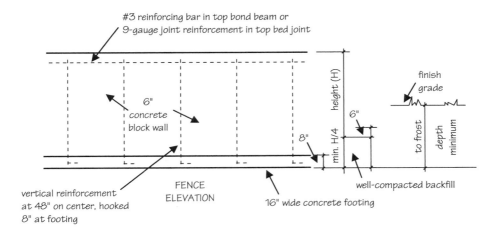

	Vertical Reinforcement for Wind Pressure			
H	5 psf	10 psf	15 psf	20 psf
4'-0"	1 #3	1 #3	1 #4	1 #4
5'-0"	1 #3	1 #4	1 #5	1 #5
6'-0"	1 #3	1 #4	1 #5	2 #4

(*From PCA* Concrete Masonry Handbook.)

Control Joint Spacing for CMU Fences		
Joint Reinforcement Spacing (in. on center vertically)	Control Joint Spacing Expressed as Ratio of CMU Panel Length to Panel Height, L/H	Maximum Control Joint spacing (ft.)
None	2	40
8	2-1/2	45
16	3	50
32	4	60

(*From NCMA* TEK Bulletin 10-1.)

FIGURE 12-6 Reinforced concrete masonry fences without pilasters. (*From Randall and Panarese,* Concrete Masonry Handbook, *Portland Cement Association.*)

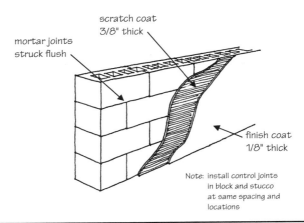

FIGURE 12-7 Direct two-coat plaster on concrete masonry.

impact loads. A rule of thumb is that for a 10-lb/sq ft wind load, the height above grade should not exceed three-fourths of the square of the wall thickness ($h \leq 0.75t^2$). If lateral loads exceed 10 lb/sq ft, the wall should be designed with reinforcing steel. Traditional brick fences are multi-wythe and bonded with brick headers laid in a variety of patterns (*see Fig. 12-8*). Fences laid in running bond pattern more commonly use metal ties to

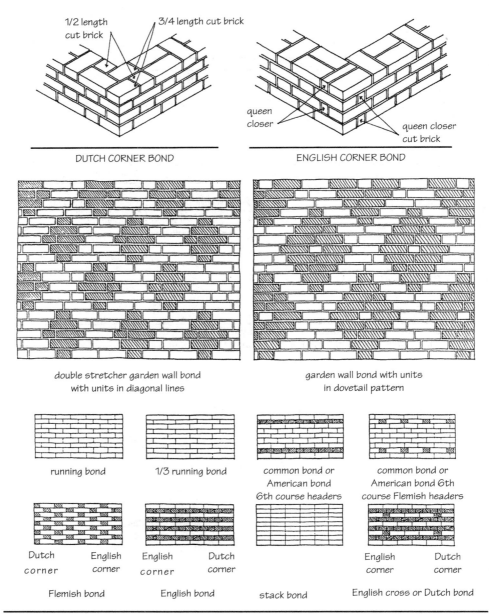

FIGURE 12-8 Traditional multi-wythe, masonry bonded brick fences.

Special Wall Types 223

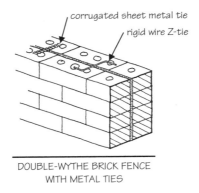

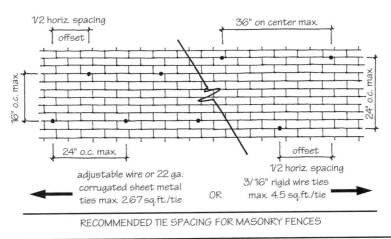

FIGURE 12-9 Multi-wythe masonry fences bonded with metal ties. (*Based on veneer anchor spacing from* International Residential Code for One- and Two-Family Dwellings, *2003.*)

connect the two wythes (*see Fig. 12-9*). Heavier ties can be spaced farther apart than light-gauge wire or corrugated sheet metal ties. Several sizes and shapes of masonry-bonded and metal-tied pilasters are shown in *Fig. 12-10*.

Brick "pier-and-panel" fences are composed of a series of thin panels (nominal 4 in.) braced intermittently by reinforced masonry piers (*see Fig. 12-11*). Reinforcing steel and foundation requirements are given in the tables in *Fig. 12-12*. Foundation diameter and embedment are based on a minimum soil-bearing pressure of 3000 lb/sq ft. Reinforcing steel requirements vary with wind load, wall height, and span. Horizontal steel may be individual bars or wires or may be prefabricated joint reinforcement, but it must be continuous through the length of the wall with splices lapped 16 in.

Because the panel section is not supported on a continuous footing, it actually spans the clear distance between foundation supports, functioning as a deep wall beam (*see* Chapter 14). Masons build the sections on temporary 2 × 4 wood footings that can be removed after the wall has cured for at least 7 days.

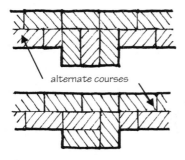
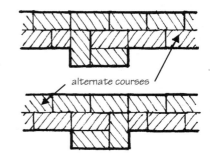

MASONRY-BONDED PILASTERS FOR TRADITIONAL MASONRY-BONDED BRICK FENCES

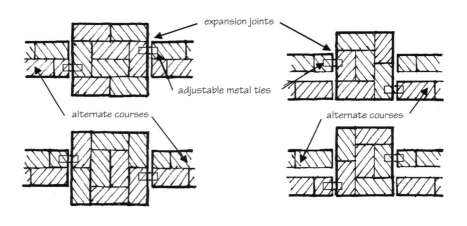

METAL-TIED PILASTERS FOR MULTI-WYTHE, METAL-TIED BRICK FENCES

Figure 12-10 Masonry-bonded and metal-tied pilasters for brick masonry fences. (*Adapted from Harry C. Plummer,* Brick and Tile Engineering, *BIA, 1962.*)

Serpentine walls and "folded plate" designs are laterally stable because of their shape. This permits the use of very thin sections without the need for reinforcing steel or other lateral support. For non-loadbearing walls of relatively low height, rule-of-thumb design based on empirically derived geometric relationships is used.

Because the wall depends on its shape for lateral strength, it is important that the degree of curvature be sufficient. Recommendations for brick and concrete block walls are illustrated in *Fig. 12-13*. The brick wall is based on a radius of curvature not exceeding twice the height of the wall above finished grade and a depth

Special Wall Types

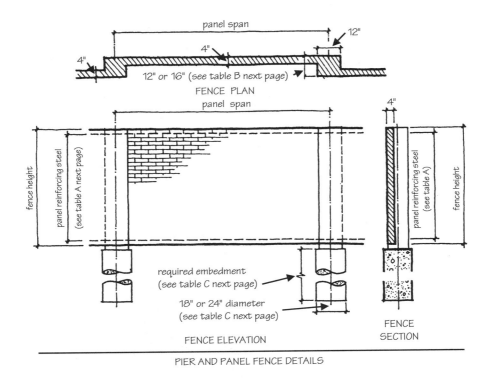

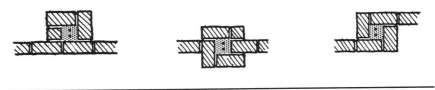

ALTERNATE METHODS OF CONSTRUCTING PIERS (see table B for reinforcing steel)

FIGURE 12-11 Single-wythe brick pier and panel fences. (*From BIA, Technical Note 29A.*)

of curvature from front to back no less than one-half of the height. A maximum height of 15 times the thickness is recommended for the CMU wall, and depth-to-curvature ratios are slightly different. Free ends of a serpentine wall should be supported by a pilaster or short-radius return for added stability. Thicker sections and taller walls may be built if proper design principles are applied to resist lateral wind loads.

Masonry screen walls and fences must be supported by an adequately designed concrete footing to prevent uneven settlement or rotation. *Figure 12-14* shows rule-of-thumb

TABLE A PANEL WALL REINFORCING STEEL

Wall span (ft)	Vertical Spacing* (in.)								
	Wind load 10 psf			Wind load 15 psf			Wind load 20 psf		
	A	B	C	A	B	C	A	B	C
8	45	30	19	30	20	12	23	15	9.5
10	29	19	12	19	13	8.0	14	10	6.0
12	20	13	8.5	13	9.0	5.5	10	7.0	4.0
14	15	10	6.5	10	6.5	4.0	7.5	5.0	3.0
16	11	7.5	5.0	7.5	5.0	3.0	6.0	4.0	2.5

*A, two - No. 2 bars; B, two - $\frac{3}{16}$-in. diam wires; C, two - 9 gauge wires.

TABLE B PIER REINFORCING STEEL*

Wall span (ft)	Wind load 10 psf			Wind load 15 psf			Wind load 20 psf		
	Wall height (ft)			Wall height (ft)			Wall height (ft)		
	4	6	8	4	6	8	4	6	8
8	2#3	2#4	2#5	2#3	2#5	2#6	2#4	2#5	2#5
10	2#3	2#4	2#5	2#4	2#5	2#7	2#4	2#6	2#6
12	2#3	2#5	2#6	2#4	2#6	2#6	2#4	2#6	2#7
14	2#3	2#5	2#6	2#4	2#6	2#6	2#5	2#5	2#7
16	2#4	2#5	2#7	2#4	2#6	2#7	2#5	2#6	2#7

*Within heavy lines 12 by 16-in. pier required. All other values obtained with 12 by 12-in. pier.

TABLE C REQUIRED EMBEDMENT FOR PIER FOUNDATION*

Wall span (ft)	Wind load 10 psf			Wind load 15 psf			Wind load 20 psf		
	Wall height (ft)			Wall height (ft)			Wall height (ft)		
	4	6	8	4	6	8	4	6	8
8	2'-0"	2'-3"	2'-9"	2'-3"	2'-6"	3'-0"	2'-3"	2'-9"	3'-0"
10	2'-0"	2'-6"	2'-9"	2'-3"	2'-9"	3'-3"	2'-6"	3'-0"	3'-3"
12	2'-3"	2'-6"	3'-0"	2'-3"	3'-0"	3'-3"	2'-6"	3'-3"	3'-6"
14	2'-3"	2'-9"	3'-0"	2'-6"	3'-0"	3'-3"	2'-9"	3'-3"	3'-9"
16	2'-3"	2'-9"	3'-0"	2'-6"	3'-3"	3'-6"	2'-9"	3'-3"	4'-0"

*Within heavy lines 24-in. diam. foundation required. All other values obtained with 18-in. diam. foundation.

FIGURE 12-12 Brick and pier panel fence design tables. (*From BIA, Technical Note 29A.*)

sizes and proportions for both panel and pilaster sections. Where the ground under a screen wall or fence slopes slightly, the footing should be placed deeper in the ground at one end so that its entire length is level and below the frost line (*see Fig. 12-15A*). Where the ground slopes more steeply, both the footing and the fence must be stepped in a series of level sections, always keeping the bottom of the footing below the frost line (*see Fig. 12-15B*).

Special Wall Types

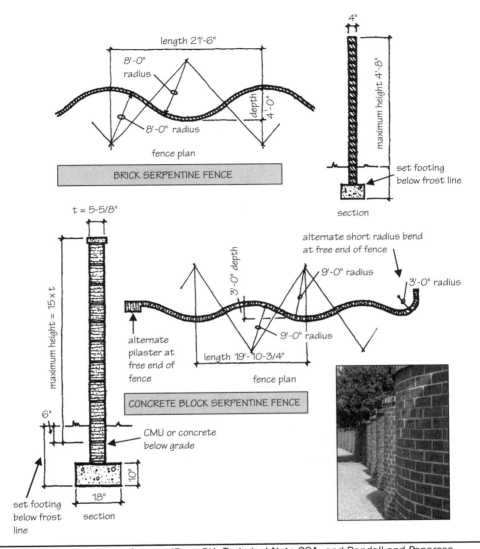

FIGURE 12-13 Serpentine fences. (*From BIA*, Technical Note 29A, *and Randall and Panarese*, Concrete Masonry Handbook, *Portland Cement Association.*)

All free-standing masonry walls and fences, regardless of thickness, must be properly capped to prevent excessive moisture infiltration from the top. The appearance and character of a wall are substantially affected by the type of coping selected, including natural stone, cast stone, metal, brick, or concrete masonry (*see Fig. 12-16*). The thermal and moisture expansion characteristics of the wall and coping materials should be similar. Control and expansion joint locations should

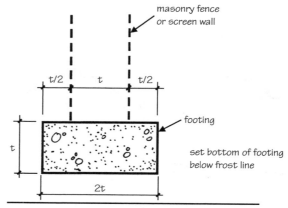

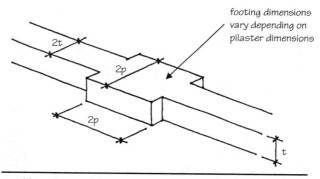

Figure 12-14 Footings for masonry fences and screen walls. (*From C. Beall,* Masonry and Concrete for Residential Construction, *McGraw-Hill Complete Construction Series, 2001.*)

be calculated (refer to Chapter 6), and joints should be tooled concave to compress the mortar against the face of the units and decrease porosity at the joint surface. Masonry and precast copings should slope to shed water and should project beyond the face of the wall a minimum of ½ in. on both sides to provide a positive drip and prevent water from flowing back under the coping. Through-wall flashing should be installed immediately below the coping to prevent excessive water penetration. Mortar joints in masonry copings should be raked back and sealed with a backer rod or bond breaker tape and sealant. Grouting of hollow units, cavities, and hollow sections also increases the durability and strength of the wall by eliminating voids where water can accumulate and cause freeze-thaw damage or efflorescence. The combined use of masonry piers and metal fence panels should allow for differential thermal expansion and contraction between the two materials (*see Fig. 12-17*).

Special Wall Types 229

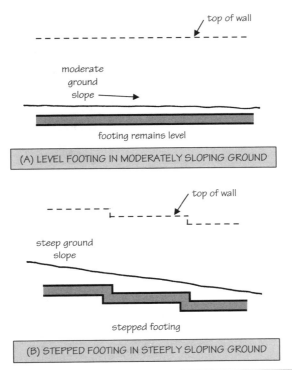

FIGURE 12-15 Footings for masonry fences and screen walls on sloping ground. (*From C. Beall, Masonry and Concrete for Residential Construction, McGraw-Hill Complete Construction Series, 2001.*)

Natural stone is used to build free-standing dry-stack and mortared walls. *Dry-stack walls* laid without mortar are generally 18 to 24 in. wide and depend only on gravity for their stability. Trenches are dug to below the frost line and if the ground slopes may take the form of a series of flat terraces. A concrete footing may be poured in the trench, but walls are often laid directly on undisturbed soil. Two rows of large stones laid with their top planes slightly canted toward the center will provide a firm base. All stones placed below grade should be well packed with soil in all the crevices. Stones should be well fitting, requiring a minimum number of shims. A bond stone equal to the full wall width should be placed every 3 or 4 ft in each course to tie the inner and outer wythes together. All of the stones should be slightly inclined toward the center of the wall so that the weight leans in on itself (*see Fig. 12-18*). Greater wall heights require more incline from base to cap. Wall ends and corners are subject to the highest stress and should be built with stones tightly interlocked for stability. Relatively flat slabs of roughly rectangular shape work best for cap stones. The top course should be as level as possible for the full length of the wall. Large stones make dramatic walls and may be combined with smaller stone shims for stability.

Chapter Twelve

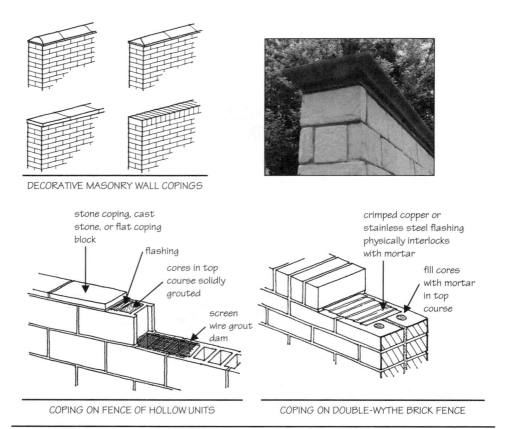

FIGURE 12-16 Copings and flashing for masonry screen walls and fences. (*From C. Beall, Masonry and Concrete for Residential Construction, McGraw-Hill Complete Construction Series, 2001.*)

FIGURE 12-17 Allow for expansion and contraction of metal between stone piers.

Special Wall Types

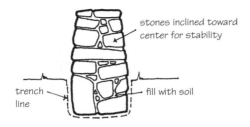

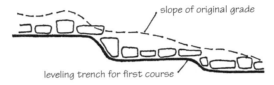

FIGURE 12-18 Dry-stack stone walls. (*From Duncan*, The Complete Book of Outdoor Masonry.)

Mortared stone walls are laid on concrete footings poured below the frost line. Rubble stone or fieldstone walls are laid up in much the same way as dry-stack walls except that the voids and cavities are filled with mortar. Type N mortar should be used, and each course should be laid in a full mortar bed for maximum bond and strength. Building codes generally require that bond stones be uniformly distributed and account for no less than 10% of the exposed face area. Mortared rubble stone walls less than 24 in. thick must have bond stones at a maximum of 3 ft on center vertically and horizontally. For thicknesses greater than 24 in., one bond stone for each 6 sq ft of wall surface is required. The minimum thickness of the wall must be sufficient to withstand all horizontal forces and the vertical dead load of the stone itself. For relatively low mortared walls, a thickness of as little as 8 in. may be adequate, but 12-in.-thick walls are more commonly used.

12.3 Glass Block Panels

Glass block is used in non-loadbearing interior and exterior applications and is most often installed as single-wythe, stack bond panel walls. The compressive strength of the units is sufficient to carry the dead load of the material weight for a moderate height. Intermediate supports at floor and roof slabs require care in detailing to allow expansion and contraction of dissimilar materials (*see Fig. 12-19*). Deflection of supporting members above or below glass block panels should be limited to L/600. Movement joints at the perimeter of the panels should be at least 3/8 to ½ in. Glass blocks are normally laid in Type N cement-lime mortar and often with a bonding agent. Bed joints are reinforced with ladder-type horizontal joint reinforcement spaced a maximum of 16 in. on center vertically. Because the bond between mortar and glass block is relatively weak, head and jamb recesses or channel-type supports are usually required

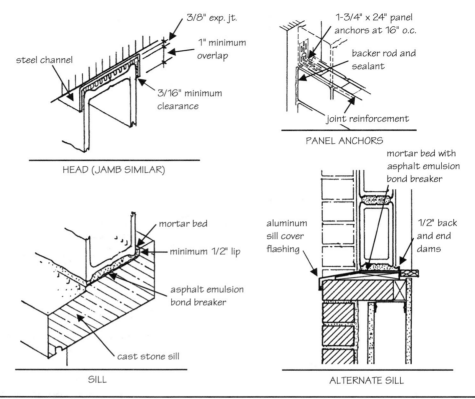

FIGURE 12-19 Typical glass block panel details.

to increase the lateral resistance of the panel section. If jamb recesses or channels are not provided in the adjacent wall, jamb anchors are required at a maximum spacing of 16 in. on center.

Size and area limitations for glass block wall panels prescribed by code are shown in *Fig. 12-20*. Whenever panels exceed code requirements for area limitations, they must be subdivided by metal stiffeners and/or supports. Vertical stiffeners should also be installed at the intersection of curved and straight sections and at every change in direction in multi-curved panels. All metal accessories, including joint reinforcement, jamb anchors, and stiffeners, should be hot-dip galvanized after fabrication in accordance with ASTM A153, *Standard Specification for Zinc Coating (Hot-Dip) on Iron or Steel Hardware*. Panels constructed of solar reflective block must be protected from runoff of rainwater from concrete, masonry, or metal materials located above the panel. Harmful substances may stain or etch the reflective block surface, so panels

Special Wall Types   233

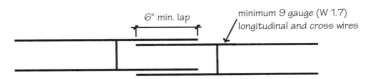

JOINT REINFORCEMENT SPLICE DETAIL

joint reinforcement is required in bed joints at 16" on center, and in the first bed joint above and below openings

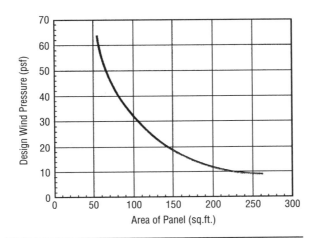

PANEL AREA ADJUSTMENTS FOR 3-7/8 IN. THICK STANDARD GLASS BLOCK UNITS FOR WIND PRESSURES OTHER THAN 20 PSF

	Maximum Glass Block Panel Sizes					
	Exterior Walls			Interior Walls		
Unit	Area (sq.ft.)	Height (ft.)	Length (ft.)	Area (sq.ft.)	Height (ft.)	Length (ft.)
Standard, 3-7/8" thick	144 §	20	25	250	20	25
Thin, 3-1/8" hollow	85	10	15	150	20	25
Thin, 3" solid	85	10	15	100	20	25

§ Maximum area limit for standard units is for 20 psf wind pressure. For other wind pressures, see graph below.

FIGURE 12-20 Code requirements for glass block panels. (*Based on* International Building Code, 2003.)

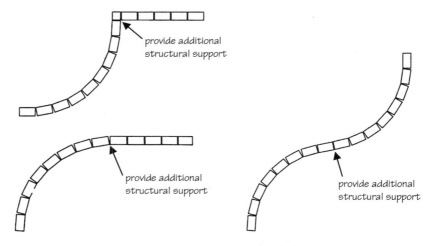

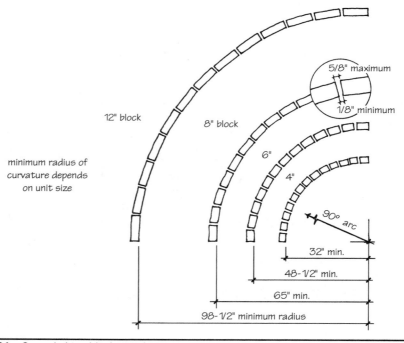

Figure 12-21 Curved glass block panels.

Special Wall Types 235

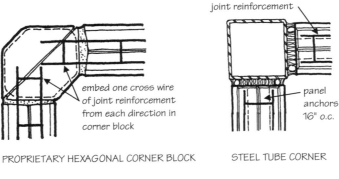

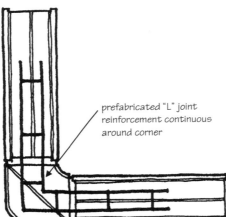

Figure 12-22 Glass block corner details.

should be recessed a minimum of 4 in. and a drip provided at the edge of the wall surface above.

Using wedge-shaped head joints, panels can be curved at various radii depending on the size of the units. *Figure 12-21* shows the smallest achievable radius for each of four different block lengths. Ninety-degree corners may be laid to a corner post of wood or steel or may incorporate special-shaped bullnose or hexagonal units (*see Fig. 12-22*).

CHAPTER 13
Lintels and Arches

There are two ways to span openings in masonry walls. *Beams and lintels* are horizontal elements that carry loads as flexural members. Masonry *arches* may be flat or curved but carry loads in compression because of the shape or orientation of the individual units.

This chapter describes the design of steel, concrete, and masonry lintels and masonry arches. Structural masonry beams for large openings or heavy loads are beyond the scope of this book. For detailed methods of analysis, design formulas, and sample calculations for structural beams, the reader should consult the engineering texts listed at the beginning of Chapter 14.

13.1 Lintels

Lintels of steel, reinforced masonry, stone, concrete, precast concrete, and cast stone can be used to span door and window openings in masonry walls. Lintels must resist compressive, bending, and shear stresses (*see Fig. 13-1*). Lintels must be analyzed to determine the actual loads that must be carried and the resulting stresses that will be created in the member.

13.1.1 Determining Loads

Regardless of the material used to form or fabricate a lintel, the first step in design is the determination of applied loads. When masonry is laid in a running bond pattern, it creates a natural corbeled arch, which transfers much of the vertical load to either side of the opening (*see Fig. 13-2*). The area inside a triangle with sides at 45° angles to the lintel represents the weight of the masonry that must be supported by the lintel itself (*see Fig. 13-3*). Outside this area, the weight of the masonry is assumed to be carried to the supporting abutments by natural arching. For this assumption to be true, however, the arching action must be stabilized by a minimum of 8 to 16 in. of masonry above the top of the load triangle. There must also be sufficient mass on both sides of the opening to resist the horizontal thrust, and there cannot be a movement joint at either side of the opening.

If arching action cannot be assumed to occur because of inadequate height above the load triangle, inadequate thrust resistance at the sides, movement joint locations, or because the masonry is not laid in running bond, the lintel must be sized to carry the full weight of the masonry above its entire length (*see Fig. 13-4*).

The table in *Fig. 13-5* lists allowable spans for steel, concrete, and masonry lintels supporting masonry veneer. For single-wythe concrete block walls and for loadbearing masonry, engineering analysis should be used to determine lintel or beam sizing and reinforcement.

Chapter Thirteen

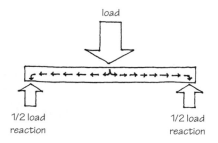

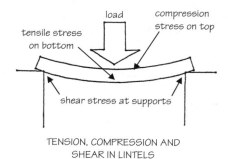

FIGURE 13-1 Lintels must resist compressive, bending, and shear stresses. (*From C. Beall and R. Jaffe,* Concrete and Masonry Databook, *McGraw-Hill, 2003.*)

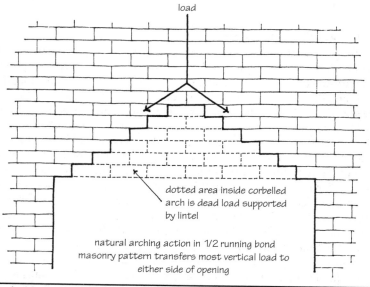

FIGURE 13-2 Arching action in running bond masonry. (*From C. Beall and R. Jaffe,* Concrete and Masonry Databook, *McGraw-Hill, 2003.*)

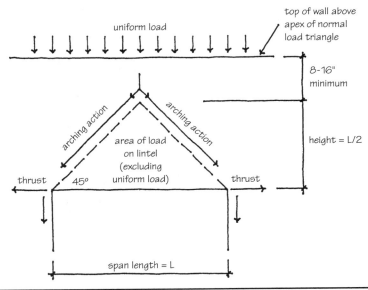

FIGURE 13-3 Area of lintel load with arching action. (*From C. Beall and R. Jaffe,* Concrete and Masonry Databook, *McGraw-Hill, 2003.*)

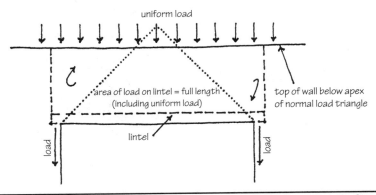

FIGURE 13-4 Area of lintel load without arching action. (*From C. Beall and R. Jaffe,* Concrete and Masonry Databook, *McGraw-Hill, 2003.*)

13.1.2 Steel Lintels

Structural steel shapes are commonly used to span masonry openings. Steel angles are the simplest shapes and are suitable for openings of moderate width where superimposed loads are not excessive (*see Fig. 13-6*). For wider openings, thicker walls, or heavy loads, multiple angles or steel beams with plates or angles may be required (*see Fig. 13-7*). The horizontal leg of a steel angle must support at least two-thirds of the thickness of the masonry, and the lintel must have a minimum of 4 in. of bearing on either side of the opening for brick and a minimum of 8 in. bearing for concrete block.

Chapter Thirteen

Allowable Spans for Steel, Concrete and Masonry Lintels Supporting Masonry Veneer				
Size of Steel Angle Lintel§, Vertical X Horizontal X Thickness (in.)	Number of 1/2" or Equivalent Reinforcing Bars in Masonry or Concrete Lintels†	Less Than One Story of Masonry Above Lintel	Lintel Supporting One Story of Masonry Above Opening	Lintel Supporting Two Stories of Masonry Above Opening
3 x 3 x 1/4	1	6'-0"	4'-6"	3'-0"
4 x 3 x 1/4	1	8'-0"	6'-0"	4'-6"
6 x 3-1/2 x 5/16	2	14'-0"	8'-0"	6'-0"
5 x 3-1/2 x 5/16	2	10'-0"	9'-6"	7'-0"
two 6 x 3-1/2 x 5/16	4	20'-0"	12'-0"	9'-6"

§ Steel lintels indicated are adequate typical examples. Other steel lintels meeting structural design requirements may be used.
† Depth of reinforced lintels shall not be less than 8 inches, and all cells or cores of hollow masonry lintels shall be grouted solid. Reinforcing bars shall extend not less than 8 inches into the support.

FIGURE 13-5 Allowable spans for lintels in masonry veneer. (*From* International Residential Code for One- and Two-Family Dwellings, *2003.*)

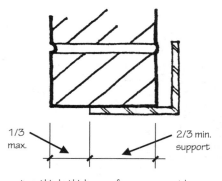

two-thirds thickness of masonry must be supported by the lintel

To avoid mortar joint spalling, hold loose lintels at window and door heads back for deeper mortar joint or use sealant and backer rod at toe of angle.

FIGURE 13-6 Steel lintel.

The steel used for lintels should meet the minimum requirements of ASTM A36, *Standard Specification for Carbon Structural Steel*. The minimum size of steel angle lintels should be ¼ in. thick with a horizontal leg of at least 3½ in. for use with nominal 4-in.-thick face brick. Deflection should not exceed $L/600$.

For single openings in masonry veneers, steel lintels are typically installed "loose." That is, they are laid across the top of the opening but not attached to the backing wall. For multiple openings, it may be practical to install a single larger lintel to span the openings. Where openings are continuous, support of the masonry above is usually provided by a continuous steel shelf angle rather than a lintel.

Lintels and Arches 241

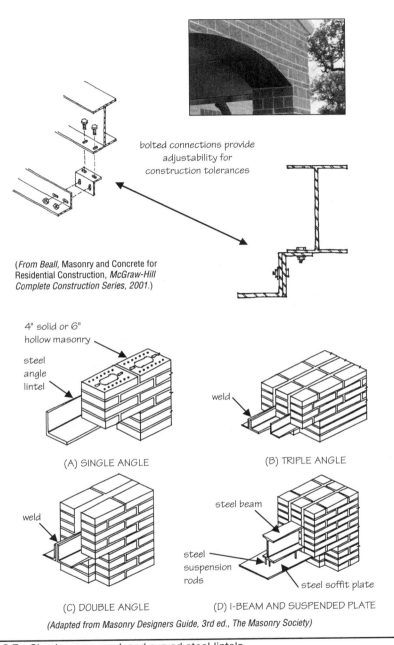

FIGURE 13-7 Simple, compound, and curved steel lintels.

Using steel lintels to span openings in masonry walls requires careful attention to flashing and drainage details and to provisions for differential movement of the steel and masonry. Code requirements for fireproofing of steel members should also be thoroughly investigated. If fireproofing is required, it may be simpler to design the lintel as reinforced concrete or masonry. Steel lintels should be galvanized to prevent corrosion when they will be exposed to the weather.

13.1.3 Concrete and Concrete Masonry Lintels

Openings in concrete masonry walls are more commonly spanned with U-shaped lintel blocks grouted and reinforced with steel reinforcing bars. Reinforced CMU lintels not only cost less than structural steel lintels but also eliminate the danger of steel corrosion and subsequent masonry cracking, as well as the painting and maintenance of exposed steel. Cast-in-place, precast concrete, or cast stone lintels are also often used.

Cast-in-place lintels are subject to drying shrinkage and have surface textures that are not always compatible with the adjoining masonry. *Precast concrete lintels* and *cast stone lintels* are better in some respects because they are delivered to the job site ready for use, do not require temporary shoring, and can carry superimposed loads as soon as they are in place. Precast and cast stone lintels can be produced with colors and surface textures to match or contrast with the surrounding masonry. Mortar for bedding precast lintels should be the same quality as that used in laying the wall, and at least equal to ASTM C270, *Mortar for Unit Masonry* Type N.

Reinforced concrete masonry lintels are constructed with special-shaped lintel units, bond beam units, or standard units with depressed, cut-out, or grooved webs to accommodate the steel bars (*see Fig. 13-8*). Individual units are laid end to end to form a channel in which continuous reinforcement and grout are placed. Among the major advantages of CMU lintels over steel lintels are low maintenance and the elimination of differential movement between dissimilar materials. Concrete masonry lintels are often designed as part of a continuous bond beam course, which helps distribute shrinkage and temperature stresses in the masonry above openings. This type of installation is more satisfactory than steel lintels in areas subject to seismic activity.

Units used for lintel construction should comply with the requirements of ASTM C90, *Standard Specification for Loadbearing Concrete Masonry Units*, and should have a minimum compressive strength adequate to provide the masonry compressive strength (f'_m) used in the design. Mortar should be equal to that used in constructing the wall and should meet the minimum requirements of ASTM C270, Type N. Grout for embedment of reinforcing steel should comply with ASTM C476, *Grout for Reinforced and Nonreinforced Masonry*, and maximum aggregate size is dependent on the size of the grout space (*see* Chapter 3). The first course of masonry above the lintel should be laid with full mortar bedding so that the cross-webs and the face shells of the units bear on the lintel and reduce the shear stress between the grout-filled core and the face shells.

A minimum end bearing of 8 in. is recommended for reinforced CMU lintels with relatively modest spans. For longer spans or heavy loads, bearing stresses should be calculated to ensure that the allowable compressive stress of the masonry is not exceeded. High stress concentrations may require the use of solid units or solidly grouted hollow units for one or more courses under the lintel bearing so that loads are distributed over a larger area.

The *National Concrete Masonry Association* (NCMA) design table in *Fig. 13-9* is based on typical equivalent uniform loads of 200 to 300 lb/lin ft for wall loads and 700 to 1000 lb/lin ft for combined floor and roof loads. The table can be used to determine required lintel size and reinforcing for various spans subject to this type of loading.

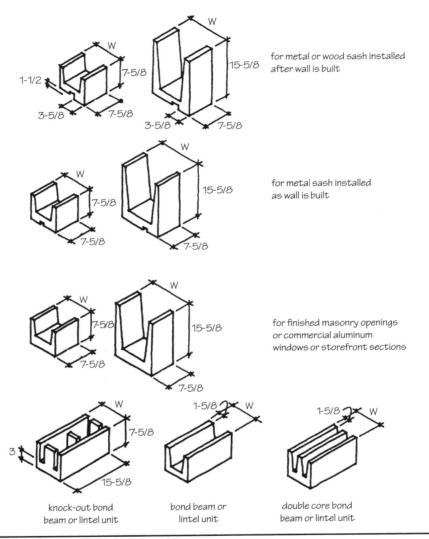

FIGURE 13-8 Concrete masonry lintel units.

Reinforced CMU lintels must be shored in place until the mortar and grout have cured enough to carry superimposed loads. Reinforced CMU lintels may also be prefabricated, which eliminates the need for shoring.

13.1.4 Reinforced Brick Lintels

Standard brick masonry units are also adaptable to reinforced lintel design even though they do not have continuous channels for horizontal steel. Reinforcing may be located in bed joints or in a widened collar joint created by using half-units (*see Fig. 13-10*).

Required Reinforcement for Simply Supported CMU lintels									
		Clear Span							
Type of Loading	Nominal Size of Lintel Section (in.)	3'-4"	4'-0"	4'-8"	5'-4"	6'-0"	6'-8"	7'-4"	8'-0"
Wall loads 200-300 lb/lin.ft.	6 x 8	1 #3	1 #4	1 #4	2 #4	2 #5	—	—	—
	6 x 16	—	—	—	—	1 #4	1 #4	1 #4	1 #4
Floor and roof loads 700-1000 lb/lin.ft.	6 x 16	1 #4	1 #4	2 #3	1 #5	2 #4	2 #4	2 #5	2 #5
Wall loads 200-300 lb/lin.ft.	8 x 8	1 #3	2 #3	2 #3	2 #4	2 #4	2 #5	2 #6	—
	8 x 16	—	—	—	—	—	—	2 #5	2 #5
Floor and roof loads 700-1000 lb/lin.ft.	8 x 8	2 #4	—	—	—	—	—	—	—
	8 x 16	2 #3	2 #3	2 #3	2 #4	2 #4	2 #4	2 #5	2 #5

FIGURE 13-9 Steel reinforcement for CMU lintels. (*From NCMA,* TEK Bulletin 25.)

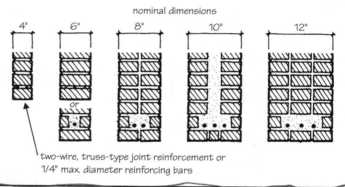

FIGURE 13-10 Reinforced brick lintels.

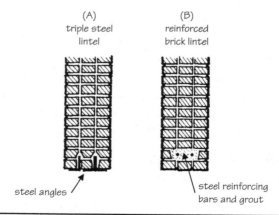

FIGURE 13-11 Steel angle lintel (A) is less efficient because it requires more steel than a reinforced brick lintel (B) with same load-carrying capacity. (*From BIA,* Technical Note 17H.)

Figure 13-11B shows a reinforced brick lintel capable of carrying the same loads as the three steel angles in *Fig. 13-11A*. The reinforced brick lintel is more economical because less steel is required, so it is a more efficient use of structural materials. The combined action of the masonry and the steel reinforcing bars is more efficient than support provided by steel alone. Reinforced brick lintels must be shored in place until the mortar and grout have cured enough to carry superimposed loads. Reinforced brick lintels may also be prefabricated to eliminate the need for temporary shoring.

13.2 Arches

Masonry arches may be constructed in a variety of historical forms, such as *segmental, elliptical, Tudor, Gothic, semicircular,* and *parabolic* to *flat* or *jack* arches (see *Fig. 13-12*). Masonry arches create an entirely different aesthetic than lintels, but the primary structural advantage is that under uniform loading conditions, the induced stress in an arch is principally compression rather than tension (*see Fig. 13-13*). For this reason, an arch will frequently provide the most efficient structural span. Because masonry's resistance to compression is greater than its resistance to tension, it is an ideal material for the construction of arches.

Arches are divided structurally into two categories. *Minor arches* are those whose spans do not exceed 6 ft with a maximum rise-to-span ratio of 0.15, with equivalent uniform loads of the order 1000 lb/ft. *Major arches* are those with spans greater than 6 ft or rise-to-span ratios of more than 0.15.

13.2.1 Minor Arch Design

In a fixed masonry arch, three conditions must be maintained to ensure the integrity of the arch action:

- The length of span must remain constant.
- The elevation of the ends must remain unchanged.
- The inclination of the skewback must be fixed.

If any of these conditions is altered by sliding, settlement, or rotation of the abutments, critical stresses can develop and may result in structural failure. Adequate foundations and high-quality mortar and workmanship are essential to proper arch construction. Mortar joints must be completely filled to ensure maximum bond and even distribution of stresses.

Arches are designed by assuming a shape and cross section and then analyzing the shape to determine its adequacy to carry the superimposed loads. The following discussion of arch design is taken from the Brick Industry Association's *Technical Notes,* Series 31.

13.2.2 Graphic Analysis

The primary theory of arch analysis is the *line-of-thrust method,* which considers the stability of the arch ring to be dependent on friction and the reactions between the several arch sections or voussoirs (*see Fig. 13-14* for arch terminology).

246 Chapter Thirteen

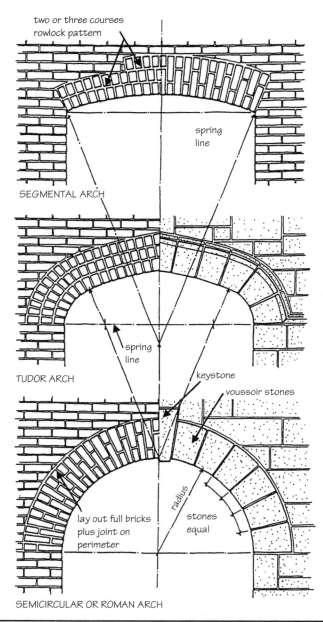

Figure 13-12 Masonry arch forms. (*From BIA*, Technical Note 31.)

Lintels and Arches 247

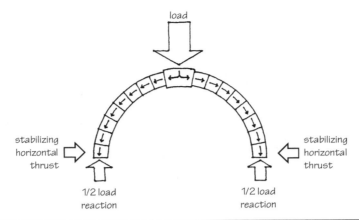

FIGURE 13-13 Load distribution in masonry arches. (*From C. Beall and R. Jaffe,* Concrete and Masonry Databook, *McGraw-Hill, 2003.*)

The simplest and most widely used line-of-thrust method is based on the hypothesis of "least crown thrust," which assumes that the true line of resistance of an arch is that for which the thrust at the crown is the least possible consistent with equilibrium. Normally, the direction of the crown thrust is assumed as horizontal and its point of application as the upper extremity of the middle one-third of the section.

The joint of rupture is the joint for which the tendency of the arch to open at the extrados is the greatest and which therefore requires the greatest crown thrust applied to prevent the joint from opening. At this joint, the line of resistance of the arch will fall on the lower extremity of the middle third of the section. For minor segmental arches, the joint of rupture is ordinarily assumed to be the skewback of the arch. Based on the joint of rupture at the skewback and the hypothesis of least crown thrust, the magnitude and direction of the reaction at the skewback may be determined graphically (*see Fig. 13-15*).

In this analysis, only one-half of the arch is considered, as it is symmetrical and uniformly loaded over the entire span. *Figure 13-15A* shows the external forces acting on the arch section. For equilibrium, the lines of action of these three forces ($W/2$, H, and R) must intersect at one point as shown in *Fig. 13-15B*. Because the crown thrust (H) is assumed to act horizontally, this determines the direction of the resisting force (R). The magnitude of the resistance may be determined by constructing a force diagram as indicated in *Fig. 13-15D*. The arch is divided into voussoirs and the uniform load transformed into equivalent concentrated loads acting on each section (*see Fig. 13-15C*). Starting between the reaction and the first load segment past the skewback, numbers are placed between each pair of forces, so that each force can subsequently be identified by a number (i.e., 1–2, 5–6, 7–1, and so on). The side of the force diagram that represents $W/2$ (*see Fig. 13-15D*) is divided into the same number of equivalent loads, and the same numbers previously used for identification are placed as shown

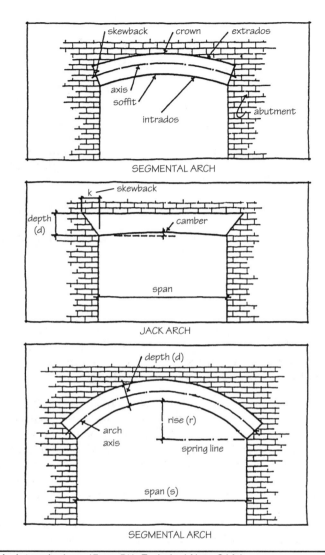

FIGURE 13-14 Arch terminology. (*From BIA, Technical Note 31A.*)

in *Fig. 13-15E* to identify the forces in the new force diagram. Thus, the line 7–1 is the skewback reaction, 6–7 the horizontal thrust, and so on. From the intersection of H and R (7–1 and 6–7), a line is drawn to each intermediate point on the leg representing $W/2$.

The equilibrium polygon may now be drawn. First extend the line of reaction until it intersects the line of action of 1–2 (see *Fig. 13-15F*). Through this point, draw a line parallel to the line 7–2 until it intersects the line of action of 2–3. Through this point,

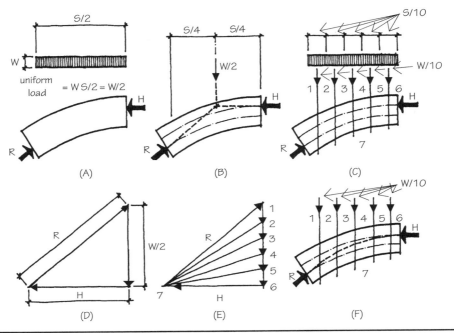

FIGURE 13-15 Graphic arch analysis. (*From BIA, Technical Note 31.*)

draw a line parallel to the line 7–3, and so on, and complete the polygon in this manner. If the polygon lies completely within the middle third of the arch section, the arch is stable. For a uniformly distributed load, the equilibrium polygon, which coincides with the line of resistance, will normally fall within this section, but for other loading conditions it may not.

CHAPTER 14
Structural Masonry

Extensive structural engineering design is beyond the intended scope of this text. This chapter describes only the general concepts of masonry bearing wall design. For detailed methods of analysis, design formulas, and sample calculations, the reader should consult Klingner's *Masonry Structural Design* (2010) or the *Masonry Designers' Guide*, which is based on the Masonry Standards Joint Committee code and specification.

14.1 Masonry Structures

Contemporary masonry is very different from the traditional, empirically designed loadbearing masonry construction of the past. Contemporary masonry buildings have thinner, lighter-weight, more efficient structural systems, and structures designed in compliance with modern code requirements are strong and durable, even in significant seismic activity.

Loadbearing masonry may be reinforced or unreinforced and either empirically or analytically designed. Empirical designs are based on arbitrary limits of wall height and thickness. Unreinforced masonry may still be designed by empirical methods but is applicable only to low-rise structures with modest loads. Unreinforced masonry is strong in compression but weak in tension and flexure (*see Fig. 14-1*). Small lateral loads and overturning moments are resisted by the weight of the walls. Shear and flexural stresses are resisted only by the bond between mortar and units.

Reinforced loadbearing masonry is based on allowable stresses or on engineering analysis of the loads and the strength of the masonry. Modern *strength design* is based on standard engineering analysis used to determine the actual compressive, tensile, and shear stresses on any given building, and the masonry is then designed specifically to resist these calculated forces. Where lateral loads are higher than allowed by unreinforced masonry, flexural strength can be increased by solidly grouting reinforcing steel into hollow unit cores or wall cavities wherever design analysis indicates that tensile stress is developed. The cured grout binds the masonry and the steel together to act as a single load-resisting element. Buildings are often designed as a combination of reinforced and unreinforced masonry elements. Multi-wythe walls may be designed with composite or noncomposite action between the wythes (*see Fig. 14-2*).

Design requirements for both the empirical and analytical methods are governed by the *Masonry Standards Joint Committee* (MSJC) *Building Code Requirements for Masonry Structures* (ACI 530/ASCE 5/TMS 402) and *Specification for Masonry Structures*

Chapter Fourteen

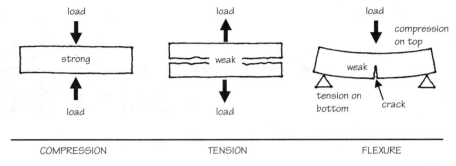

Masonry is strong in compression. It resists tensile and flexural stresses through the incorporation of reinforcing steel in the same way that concrete does.

FIGURE 14-1 Compressive, tensile, and flexural strength of masonry.

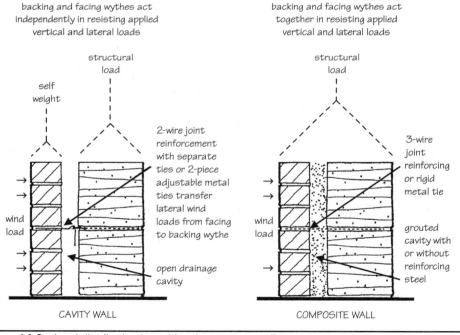

FIGURE 14-2 Load distribution in multi-wythe masonry walls.

(ACI 530.1/ASCE 6/TMS 602). Sometimes referred to simply as ACI 530, the MSJC Code also forms the basis of the *International Building Code* (IBC) and is referenced in the IBC throughout.

14.1.1 Differential Movement

Buildings are dynamic structures whose successful performance depends on allowing the differential movement of adjoining or connecting elements to occur without excessive stress or its resulting damage. All building materials experience volumetric changes from temperature fluctuations, and some experience moisture-related movements as well. Structural movements include column shortening, elastic deformation, creep, and wind sway. The differential rates and directions of movements must be accommodated by flexible connections and movement joints (refer to Chapter 6).

Structural connections may be required to permit movement in some directions and to transfer applied loads in others. For example, at the joint between interior bearing walls and exterior nonbearing wall shear flanges, the exterior wall will undergo more thermal movement and less elastic and creep-induced movement than the interior wall. The connection between these elements must accommodate such movement or must transfer shear stress. Each building must be analyzed for differential movement characteristics and provisions made to relieve the resulting stresses. The MSJC Code includes design coefficients for masonry thermal expansion, moisture expansion, shrinkage, and creep.

14.1.2 Load Distribution

Loadbearing masonry supports its own weight, the dead and live loads of the floor and roof structure, and all lateral wind and seismic forces. Normal axial load distribution in masonry is based on the units being laid in running bond pattern with a minimum overlap between units of one-fourth the unit length. A one-half unit overlap is more common. Units laid in stack bond without overlapping units must be reinforced with bond beams or joint reinforcement to achieve the same distribution of axial loads.

Bearing pads or plates are used to distribute concentrated load stresses and to permit any slight differential lateral movement that might occur. When a joist, beam, or girder bears directly on a masonry wall, the reaction will not generally occur in the center of the bearing area because deflection of the bearing member moves the reaction toward the inner face of the support. If significant eccentricity develops, the addition of reinforcing steel may be required to resist the tensile bending stresses that result.

Loadbearing masonry buildings are classified as rigid box frames in which lateral forces are resisted by shear walls. A box frame structural system must provide a continuous and complete path for all of the assumed loads to follow from the roof to the foundation. This is achieved through the interaction of floors and walls securely connected along their planes of intersection. Lateral forces are carried by the floor diaphragms to vertical shear walls parallel to the direction of the load.

A roof or floor diaphragm must have limited deflection in its own plane to transmit lateral forces without inducing excessive tensile stress or bending in the walls perpendicular to the direction of the force. The stiffness of the diaphragm also affects the distribution of lateral forces to the shear walls parallel to the direction of the force. Rigid concrete and semirigid steel joist diaphragms distribute lateral loads to the shear walls in proportion to their relative rigidity compared with that of the walls.

Shear walls resist horizontal forces acting parallel to the plane of the wall through resistance to overturning and shearing resistance. The location of shear walls in relation to the direction of the applied force is critical. Because ground motion may occur in perpendicular directions, the location of resisting elements must coincide with these forces. Shear walls may be loadbearing or non-loadbearing, but they perform best when they are loadbearing, because the added compressive load counteracts some of the bending stresses. It is best to combine functions whenever possible and design the building with both transverse and longitudinal loadbearing shear walls. Designing all the loadbearing walls to resist lateral forces improves design efficiency because the increased number of shear walls distributes the load and lowers unit stresses. Loadbearing shear walls also have greater resistance to seismic forces because of the stability provided by increased axial loads.

If loadbearing walls also function as shear walls, then general building layout becomes a very important aspect of seismic design. Several compartmented floor plans are shown in *Fig. 14-3*. The regular bay spacing lends itself to apartment, hotel, hospital, condominium, and nursing home occupancies, where large building areas are subdivided into repetitive smaller areas. By changing the span direction of the floor, shear walls in two or more directions can become loadbearing.

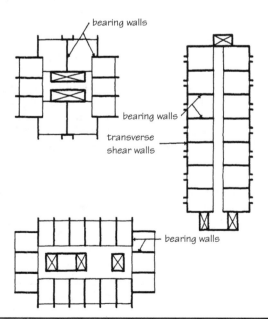

FIGURE 14-3 Examples of loadbearing walls oriented in two directions and functioning as shear walls to resist lateral loads in those directions.

Although the weight of a loadbearing structure is greater than that of a similar frame building, the required soil-bearing capacity is often less because the bearing walls distribute the weight more evenly. Bearing wall structures are compatible with all of the common types of foundations, including grade beams, spread footings, piles, caissons, and mats. Foundation walls below grade may be of either concrete or masonry but must be doweled to the footing to ensure combined action of the wall and the foundation.

14.1.3 Beams and Girders

The use of reinforcing steel in masonry construction permits the design of flexural masonry members such as lintels, beams, and girders to span horizontal openings without the need for steel plates and angles (*see Fig. 14-4*). Reinforced masonry flexural members are more efficient structurally because the amount of steel required for reinforcement is significantly less than that required for steel supporting plates or angles.

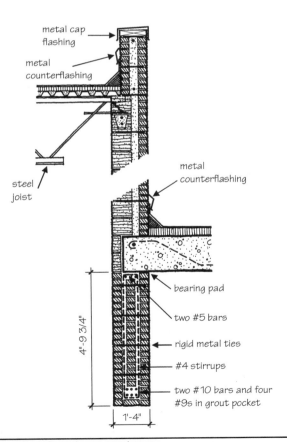

Figure 14-4 Example of reinforced masonry beam.

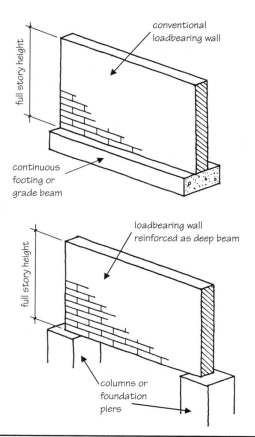

FIGURE 14-5 Deep wall beams. (*From R. Schneider and W. Dickey*, Reinforced Masonry Design, *2nd ed., Prentice-Hall, 1987.*)

The concept of deep masonry wall beams is based on a wall spanning between columns or footings instead of having continuous line support at the bottom as in conventional loadbearing construction (*see Fig. 14-5*). If soil-bearing capacities permit this type of concentrated load, the wall may be designed as a flexural member and must resist forces in bending rather than in direct compression.

Deep wall beams may also be used to open up the ground floor of a load-bearing structure. The bearing wall on the floor above can be supported on columns to act as a deep wall beam and transfer its load to the columns. Bearing walls, nonbearing walls, and shear walls may all use this principle to advantage in some circumstances.

14.1.4 Connections

The box frame system of lateral load transfer requires proper connection of shear walls and diaphragms. Connections can be made with anchor bolts, reinforcing dowels, mechanical devices, or welding and may be either fixed or hinged. Each individual condition will dictate the type of connection needed, and a variety of solutions

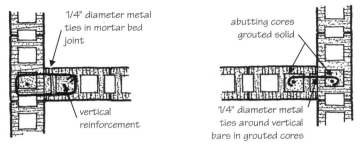

SHEAR CONNECTIONS AT CORES WITH VERTICAL REINFORCING

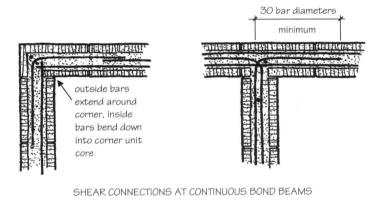

SHEAR CONNECTIONS AT CONTINUOUS BOND BEAMS

Figure 14-6 Steel reinforcing at intersecting single-wythe masonry walls.

can usually be designed for a given problem. The design and detailing of structural connections are covered in-depth in the engineering texts listed at the beginning of this chapter. *Figures 14-6 through 14-12* show some examples of floor and roof system connections.

14.2 Empirical Design

Masonry buildings built before the twentieth century, including all historic masonry buildings throughout the world, are unreinforced, empirically designed structures. These traditional loadbearing designs used massive walls and buttresses to resist lateral loads, including those induced by roof thrusts, arches, and large domes. Empirical design is based on historical precedent and rules of thumb rather than engineering analysis of loads and stresses. Empirically designed contemporary buildings do not incorporate reinforcing steel for load resistance but may include joint reinforcement for control of shrinkage cracking and thermal movement as well as reinforcing steel in lintels.

Today, empirical design methods may be used for low-rise buildings where wind loads are low and seismic loading is not a consideration. Empirical requirements are

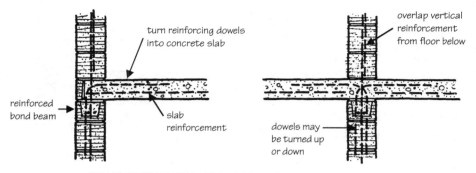

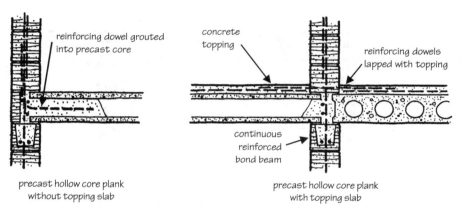

EXAMPLES OF CMU WALL CONNECTIONS TO PRECAST CONCRETE

FIGURE 14-7 Connecting single-wythe masonry walls to precast and cast-in-place concrete floors and roofs. (*From R. Schneider and W. Dickey,* Reinforced Masonry Design, *2nd ed., Prentice-Hall, 1987.*)

essentially only rules of thumb and are very simplistic in their application. Height- or length-to-thickness ratios are used in conjunction with minimum wall thicknesses to determine the required section of a given wall. The height of empirically designed buildings that rely on masonry walls for lateral load resistance is limited to 35 ft above the foundation or supporting element.

Prescriptive requirements are given for the ratio of the unsupported height or length to the nominal thickness of masonry bearing walls and nonbearing partitions. Lateral support must be provided in *either* the horizontal or the vertical direction within the limits shown in *Fig. 14-13*. Cross-walls, pilasters, buttresses, columns, beams, floors, and roofs may all be used to provide the required support. *Figures 14-14 and 14-15* show alternative methods of lateral support connections.

Multi-wythe brick walls may be bonded with masonry headers (*see Fig. 14-16*), metal ties, or prefabricated joint reinforcement (*see Fig. 14-17*). Masonry headers are

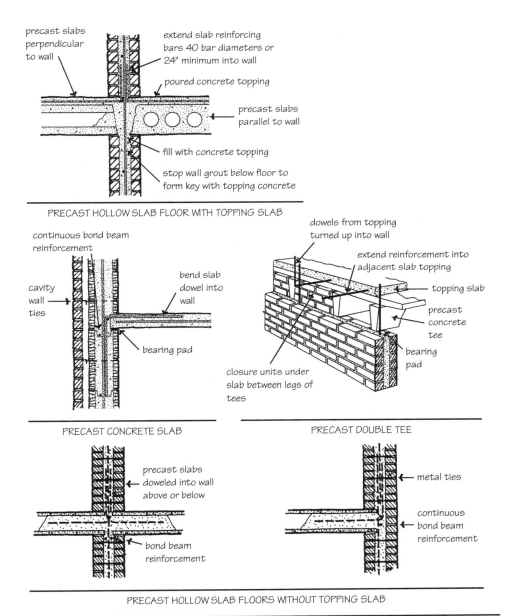

FIGURE 14-8 Connecting double-wythe masonry walls to precast concrete floors and roofs (flashing omitted for clarity). (*Adapted from Amrhein,* Reinforced Masonry Engineering Handbook, *5th ed., Masonry Industry Advancement Committee, 1992*)

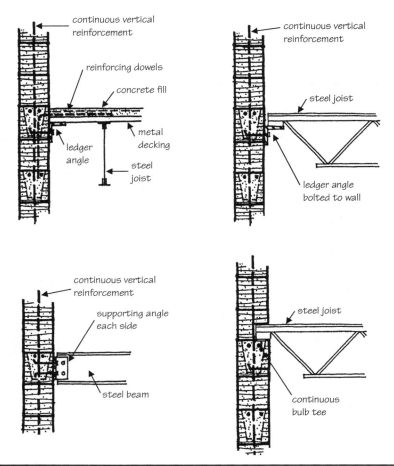

FIGURE 14-9 Examples of single-wythe masonry wall connections to steel beams and open web steel joists (flashing omitted for clarity).

seldom used except in garden walls and historic restorations. You will occasionally see some of the old bond patterns in veneer walls, but what looks like through-wall headers are actually just saw-cut brick set in the wall with the end of the brick facing out instead of to the side. Metal ties and joint reinforcement are used now almost exclusively. Spacing requirements are different for rigid and adjustable ties (*see Fig. 14-18*). Ties in alternating courses must be staggered horizontally. Additional ties must be provided at all openings, spaced not more than 3 ft apart around the perimeter, and within 12 in. of the opening itself. Prefabricated wire joint reinforcement used to provide bond between multiple wythes must have cross-wires of 9-gauge steel that are spaced a maximum of 16 in. on center. Spacings for joint reinforcement with three rigid wires are the same as for rigid ties, and spacings for joint reinforcement with eye-and-pintle or loop-and-tab type ties are the same as for adjustable ties.

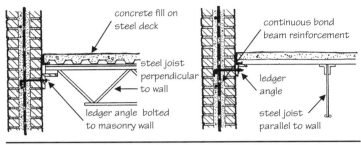

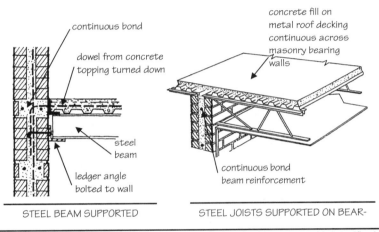

Figure 14-10 Connecting double-wythe masonry walls to steel beams and joists (flashing omitted for clarity). (*Adapted from Amrhein*, Reinforced Masonry Engineering Handbook, *5th ed., Masonry Industry Advancement Committee, 1992*)

Only solid masonry units may be used for corbeling. The maximum corbeled projection beyond the face of the wall is limited to one-half the wall thickness for solid walls or one-half the wythe thickness for cavity walls. The maximum projection of any individual unit may not exceed half the unit height or one-third its thickness (*see Fig. 14-19*).

14.3 Analytical Design

Current building codes provide for two methods of analytical design. For small, simple loadbearing structures, the *allowable stress method* is still appropriate. The allowable stress method is based on the calculation and design of actual design stresses that may not exceed code-dictated maximum allowable stresses. For greater efficiency in design and materials, the *strength design method* is more appropriate.

262 Chapter Fourteen

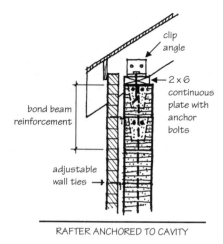

Figure 14-11 Connecting double-wythe masonry walls to wood frame floors and roofs (flashing omitted for clarity).

Strength design determines the actual wall thickness, beam depth, or column size required to resist compressive, flexural, lateral, and shear loads, eccentricity of vertical load, nonuniform foundation pressure, deflection, and thermal and moisture movements on buildings of any desired height. Flexural, shear, and axial stresses resulting from wind or earthquake forces must be added to the stress of dead and live loads, and connections must be designed to resist such forces acting either inward or outward. Strength design calculates loads and provides lateral support as needed to resist specific forces and provide stability.

Construction inspection is required by code to verify that the quality and acceptability of materials, equipment, and procedures meet code requirements. When using analytical design, the code does not allow uninspected work.

Masonry structural systems are designed based on a certain "specified compressive strength of masonry" (f'_m). Under IBC and MSJC Code requirements, the contractor

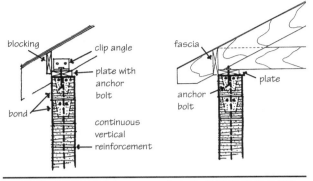
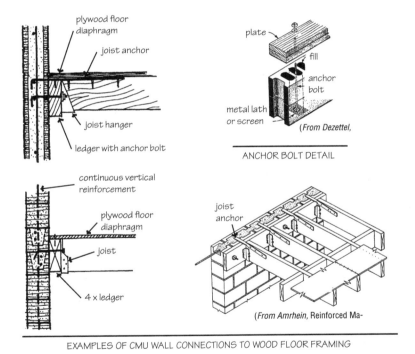

FIGURE 14-12 Connecting single-wythe masonry walls to wood floor and roof framing (flashing omitted for clarity).

must verify to the engineer by one of two methods that the proposed materials and construction will meet or exceed this strength. The contractor may elect to use the

- Unit strength method based on the combined strength of the specific masonry units and mortar type as determined by tables in the code
- Prism test method

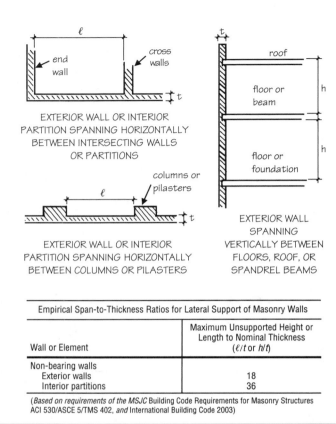

FIGURE 14-13 Lateral support requirements for empirically designed masonry.

Projects that are not large enough to justify the cost of prism testing generally use the unit strength method. Through submittals, the contractor certifies that the proprietary masonry units that will be used in the construction are of sufficient strength to produce the "specified compressive strength" (f'_m) when combined with the ASTM C476 grout mix and ASTM C270 mortar type (M, S, or N) that will be used (*see Fig. 14-20*). The proportion specification of ASTM C270 (the default) governs, as well as the proportion specification of ASTM C476. The proportion specifications of both ASTM C270 and ASTM C476 and the unit strength method of determining f'_m are very conservative and produce mortar and masonry of greater strength than the minimum required by the design. This is a built-in high safety factor, but it does not produce the most cost-efficient design.

For large projects where efficiency in design can produce a substantial cost savings by optimizing the use of materials in mortar and grout designs, the contractor may elect to use the prism test method to verify f'_m (*see Fig.14-21*). Using this method, the contractor hires a testing laboratory to produce mortar and grout mix designs in accordance with the minimum property specification of ASTM C270 and ASTM C476, respectively, which, when combined with the specified or selected masonry units, will produce the specified f'_m when prisms are laboratory tested in accordance with ASTM

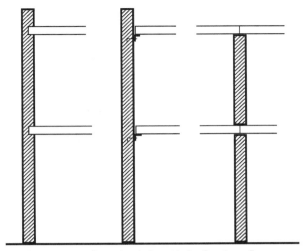
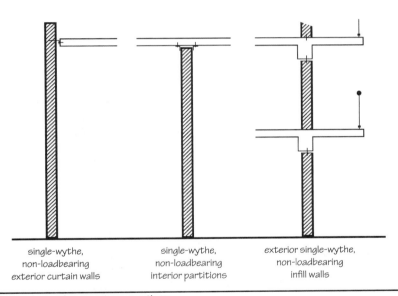

FIGURE 14-14   Lateral support connections.

C1314, *Constructing and Testing Masonry Prisms Used to Determine Compliance with Specified Compressive Strength of Masonry*.

The difference between the two methods of verifying the compressive strength of masonry is similar to the two methods of specifying concrete. Small projects are often specified by requiring a certain number of sacks of cement per cubic yard and a certain water-cement ratio (comparable with the unit strength method for masonry). For larger projects, it is more common to specify concrete by requiring a minimum compressive

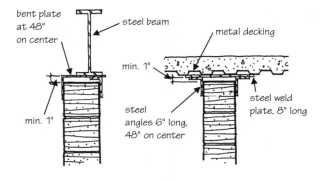

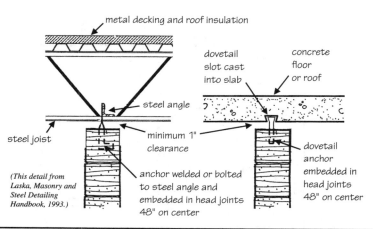

FIGURE 14-15 Examples of lateral support at the tops of interior, non-loadbearing partitions.

strength for which the contractor proposes a laboratory mix design (comparable with preconstruction masonry prism testing) that is verified by cylinder tests of field-sampled concrete (comparable with field-constructed masonry prism testing).

If the prism test method is used in masonry, ASTM C780, *Preconstruction and Construction Evaluation of Mortars for Plain and Reinforced Unit Masonry* is used for preconstruction and construction evaluation of mortar mixes, ASTM C1019, *Sampling and Testing Grout* is used for preconstruction and construction evaluation of grout mixes, and ASTM C1314, *Constructing and Testing Masonry Prisms Used to Determine Compliance with Specified Compressive Strength of Masonry* is used for prism tests.

14.3.1 Unreinforced Masonry

Although it may contain steel reinforcement at lintels, "unreinforced masonry" contains no reinforcing steel that is designed to resist applied loads. Unreinforced masonry is very strong in compression but weak in tension and shear. Resistance to shear and flexural stresses is limited by the bond between mortar and units and the precompression

Structural Masonry 267

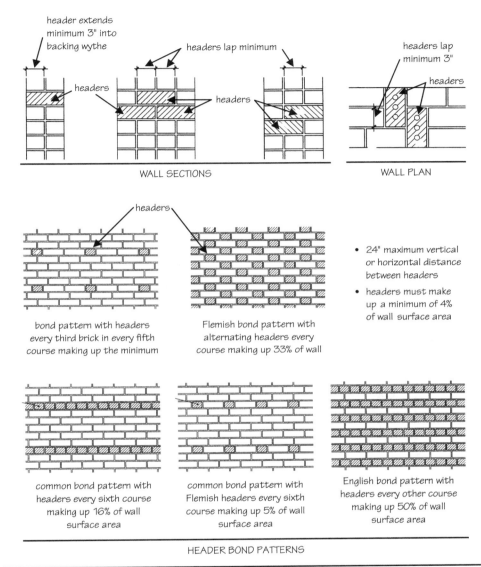

FIGURE 14-16 Header spacing for masonry bonded multi-wythe walls. (*Adapted from T. Patterson, Illustrated 2000 Building Code Handbook, McGraw-Hill, 2001.*)

effects of compressive loading. The lateral stability of loadbearing masonry walls and their resistance to shear and flexural stresses are greater than that of nonbearing or lightly loaded walls. Applied vertical loads produce compressive stresses that must be overcome by the tensile stress of the lateral load before failure can occur. In the lower stories of loadbearing buildings, compressive stresses generally counteract tension, but in the upper stories of taller buildings, where wind loads are higher and axial loads

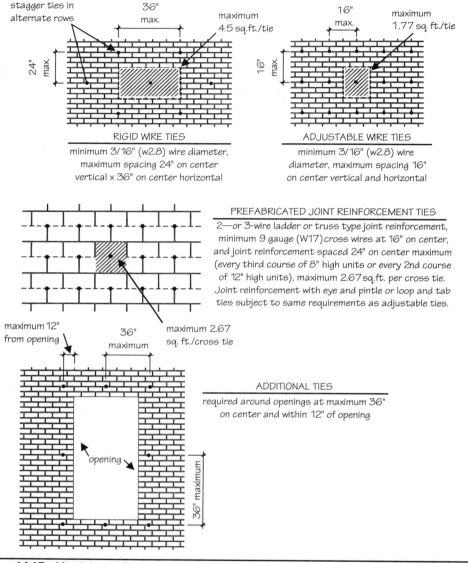

FIGURE 14-17 Metal tie spacing in multi-wythe masonry walls. (*Adapted from T. Patterson,* Illustrated 2000 Building Code Handbook, *McGraw-Hill, 2001.*)

smaller, the allowable flexural tensile stresses may be exceeded, thus requiring steel reinforcement to be added at those locations.

The flexural strength of unreinforced masonry depends on three things:

- Type and design of the masonry unit
- Type of mortar and its materials
- Quality of workmanship

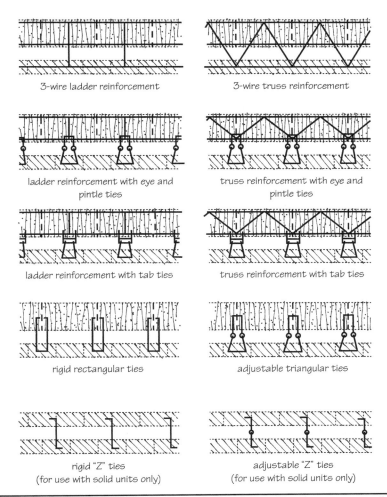

FIGURE 14-18 Metal ties and joint reinforcement for bonding multi-wythe masonry walls. (*From C. Beall and R. Jaffe,* Concrete and Masonry Databook, *McGraw Hill, 2003.*)

Higher flexural strengths are developed with solid masonry units because the full mortar bed provides more bonding surface than the face shell bedding of hollow units. Failure in lateral loading usually results from bond failure at a ruptured bed joint, so factors that affect mortar bond also affect the flexural strength of the wall. Full, unfurrowed mortar bed joints, good mortar flow and consistency, proper unit absorption, and moist curing all improve mortar-to-unit bond (*see* Chapter 15).

14.3.2 Reinforced Masonry

Reinforced masonry is used where compressive, flexural, and shearing stresses are higher than those allowed for unreinforced masonry. Designs that incorporate reinforcing steel neglect the flexural bond strength of the masonry altogether and rely on the steel to resist 100% of the tensile loads.

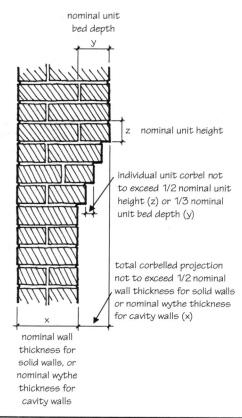

Figure 14-19 Empirical design limitations for corbeled masonry.

The flexural strength of masonry walls can be increased by placing steel reinforcement in mortar bed joints, bond beams, grouted cells, or cavities. The hardened mortar and grout bind the masonry units and steel together so that they act as a single element in resisting applied forces. Reinforcement may be added to resist stresses wherever design analysis indicates that excessive flexural tension is developed. The reinforcing steel is then designed to resist all of the tensile stresses, and the flexural resistance of the masonry-to-mortar bond is neglected entirely.

Reinforced masonry walls may be of double-wythe construction with a grouted cavity to accommodate the steel reinforcing or of single-wythe hollow units with grouted and reinforced cores. Steel wire reinforcement may be used in horizontal mortar joints as long as code requirements for wire size and minimum mortar cover are met. Horizontal reinforcing bars may be used in bond beams and lintels. Vertical reinforcing bars may be placed in the vertical cores of hollow units or in the cavity of double-wythe construction. Vertical reinforcement must be held in position at the top and bottom of the bar and at regular vertical intervals (*see* Chapter 4 for information on masonry accessories such as reinforcing bar positioners).

Required Net Area Compressive Strength of Clay Masonry Units (psi)		For Net Area Compressive Strength of Masonry (psi)
When Used With Type M or S Mortar	When Used With Type N Mortar	
1,700	2,100	1,000
3,350	4,150	1,500
4,950	6,200	2,000
6,600	8,250	2,500
8,250	10,300	3,000
9,900	—	3,500
13,200	—	4,000

Required Net Area Compressive Strength of Concrete Masonry Units (psi)		For Net Area Compressive Strength of Masonry (psi)
When Used With Type M or S Mortar	When Used With Type N Mortar	
1,250	1,300	1,000
1,900	2,150	1,500
2,800	3,050	2,000
3,750	4,050	2,500
4,800	5,250	3,000

The *unit strength method* of verifying specified compressive strength uses the tables below to show the net area compressive strength produced by combining units of a specific strength with either Type M, S or N mortar. The unit strength method of verifying f'_m may be used instead of laboratory prism testing when:
- units conform to ASTM requirements
- bed joint thickness does not exceed 5/8 in., and
- grout meets ASTM C476 requirements, or grout compressive strength is equal to f'_m but not less than 2000 psi

FIGURE 14-20 MSJC and IBC tables for unit strength method of verifying masonry compressive strength (f'_m). (*From* International Building Code, *2003, and Masonry Standards Joint Committee,* Specification for Masonry Structures, *ACI 530.1/ASCE 6/TMS 602.*)

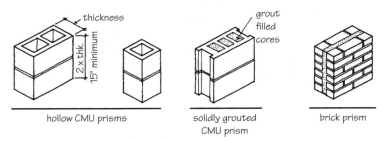

FIGURE 14-21 ASTM C1314 test prisms for masonry assemblages of units, mortar, and grout.

14.3.3 Wind and Seismic Loads

Masonry structures designed to withstand seismic and high wind loads are rigid but with high damping characteristics. The capacity of masonry structures to absorb seismic energy through damping is such that unit stresses remain extremely low, factors of safety very high, and damage negligible. Reinforced masonry structures designed in compliance with current building code requirements have successfully withstood substantial seismic forces, and the rigidity inherent in the masonry systems often reduces or eliminates secondary damage. Reinforced masonry buildings as tall as 10 stories survived near the epicenter of the 7.1 Loma Prieta, California, earthquake in 1989 and the 6.7 Northridge, California, earthquake in 1994 with little or no collateral damage to windows, partitions, piping, mechanical equipment, and other costly and potentially dangerous elements. Such secondary safety is critical in the construction of essential facilities such as hospitals, fire stations, communications centers, and other facilities required for emergency response. The only masonry buildings to sustain significant damage in earthquakes are older unreinforced masonry structures built before modern building code requirements and not yet retrofitted to meet newer, stricter performance criteria.

The MSJC Code includes prescriptive requirements for minimum reinforcing steel for structures in different Seismic Design Categories. The minimum amount of steel required is based on test results and empirical judgment rather than engineering analysis of stress or performance.

Seismic Design Category A or B

- No minimum reinforcing requirements. Walls must be anchored to walls, floors, or roofs providing lateral support. For Seismic Design Category B, shear walls may not be empirically designed and must meet minimum requirements for "ordinary plain" shear walls.

Seismic Design Category C

- Comply with requirements for Seismic Design Categories A and B plus the following requirements:
    - Masonry elements that are *not* part of the lateral-force-resisting system must be reinforced in *either* the vertical or horizontal direction, depending on location of the lateral supporting elements.

- Horizontal joint reinforcement with two longitudinal 9-gauge (W1.7) wires spaced 16 in. on center maximum, or two #3 or one #4 bar spaced 48 in. on center maximum. Must include horizontal reinforcement within 16 in. of the top and bottom of wall.
- Vertical reinforcement of two #3 or one #4 bar 48 in. on center maximum and within 16 in. of ends of walls.
- In addition to minimum reinforcing requirements, two #3 or one #4 bar on all sides of and adjacent to every opening larger than 16 in. in either direction and extending 40 bar diameters or 24 in. minimum beyond the corners of the opening.
  - For masonry elements that *are* part of the lateral-force-resisting system, shear walls must be reinforced in *both* the vertical and horizontal direction to comply with the minimum requirements for "ordinary reinforced" shear walls as follows: Horizontal joint reinforcement with two longitudinal 9-gauge (W1.7) wires, spaced 16 in. on center, or two #3 or one #4 bar spaced not more than 10 ft on center. Must include horizontal reinforcement at top and bottom of wall openings and extending 40 bar diameters or 24 in. minimum beyond the corners of the opening, within 16 in. of the tops of walls, and continuously at structurally connected roofs and floors.
  - Vertical reinforcement of two #3 or one #4 bar at corners, within 16 in. of each side of openings, within 8 in. of each side of movement joints, within 8 in. of ends of walls, and at 10 ft on center maximum.

Seismic Design Category D

- Comply with requirements for Seismic Design Category C plus the following requirements:
  - Masonry elements that *are* part of the lateral-force-resisting system must be reinforced in *both* the vertical and horizontal direction. The sum of the cross-sectional area of horizontal and vertical reinforcement must be at least 0.002 times the gross cross-sectional area of the wall, with a minimum of 0.0007 times the gross cross-sectional area in each direction. Reinforcement must be evenly distributed. Maximum spacing of reinforcement is 48 in. for other than stack bond masonry. For stack bond masonry, units must be solid, solidly grouted hollow open-end units, or solidly grouted hollow units with full head joints and reinforcement spaced a maximum of 24 in. on center.
  - Shear walls must comply with the minimum requirements for "special reinforced" shear walls. Reinforcement spacing must be the smaller of one-third the height or length of the shear wall or 48 in. on center. Minimum cross-sectional area of vertical reinforcement must be one-third of the required shear wall reinforcement. Shear reinforcement must be anchored around vertical reinforcing bars with standard hooks. Hooks for lateral tie anchorage shall be either 135° or 180° standard hooks. Columns must have lateral ties at 8 in. on center, minimum 3/8 in. diameter embedded in grout.

Seismic Design Category E or F

- Comply with requirements for Seismic Design Category D plus the following requirements:
  - For stack bond walls that are *not* part of the lateral-force-resisting system, solid units or solidly grouted hollow open-end units, horizontal reinforcement

with a cross-sectional area at least 0.0015 times the gross cross-sectional area of the masonry, maximum spacing 24 in. on center.
- For stack bond walls that *are* part of the lateral-force-resisting system, solid units or solidly grouted open-end units, horizontal reinforcement with a cross-sectional area at least 0.0025 times the gross cross-sectional area of the masonry, maximum spacing 16 in. on center.

Where analytical design indicates that more steel is required, the prescriptive minimum may be included as part of the total. If analysis indicates that less steel is required, the prescriptive minimum seismic reinforcing must still be provided. Joint reinforcement cannot be used for seismic resistance. All reinforcing steel designed to resist seismic loads must be fully embedded in grouted cores, cavities, or bond beams.

CHAPTER 15
Installation and Workmanship

Masons and bricklayers belong to one of the oldest crafts in history. The rich architectural heritage of many civilizations attests to the skill and workmanship of the trade. The advent of modern technological methods and sophisticated engineering practices has not diminished the importance of workmanship in masonry construction. The best intentions of the architect or engineer will not produce a masterpiece unless the workmanship is of the highest order and the field practices are as exacting and competent as the detailing.

15.1 Moisture Resistance

Workmanship has a greater effect on the moisture resistance of masonry than any other single factor. Key elements in the quality of workmanship include:

- Proper storage and protection of materials
- Consistent proportioning and mixing of mortar ingredients
- Full mortar joints
- Complete mortar-to-unit bond
- Continuity of flashing
- Unobstructed weep holes
- Tooled joint surfaces
- Protection of unfinished walls

Among these elements, mortar placement ranks high in limiting the amount of moisture that penetrates through the wall face. Such leakage can usually be traced to either capillary passages at the mortar-to-unit interface, partially filled mortar head joints, or cracks caused by unaccommodated building movements. All masonry walls suffer some moisture penetration because of joint defects and other design, construction, or workmanship errors. It is for this reason that the installation of flashing and weep holes is critical in collecting and draining any water that does enter the wall. This backup drainage system provides redundancy in moisture control and allows the construction to be somewhat forgiving of defects. Because it is the backup system, however, the flashing installation itself cannot tolerate defects or discontinuities without providing avenues for moisture penetration directly to the interior of the building.

15.2 Preparation of Materials

Field quality control begins with the proper storage and protection of materials. Preparations necessary prior to construction will vary according to the specific materials and conditions involved.

15.2.1 Material Storage and Protection

Proper storage and protection of masonry materials at the project site are critical to the performance and appearance of the finished construction. Materials properly stored and covered will remain in good condition, unaffected by weather. Improper procedures, however, can result in damage to units and contamination or degradation of mortar and grout ingredients.

Masonry units should be stored off the ground to prevent staining from contact with the soil and absorption of moisture, soluble salts, or other contaminants that might cause efflorescence in the finished work. Stored units should be covered for protection against the weather. Handling methods should avoid chipping or breakage of units.

Mortar and grout aggregates should also be covered to protect against contamination from rain, snow, ice, and blowing dust and debris. Different aggregates should be stockpiled separately. Packaged mortar and grout ingredients should be received in their original containers with labels intact and legible for easy identification. Broken packages, open containers, and materials with missing or illegible identification should be rejected. All packaged materials should be stored off the ground and covered to prevent moisture penetration, deterioration, and contamination (*see Fig. 15-1*).

Figure 15-1 All packaged materials should be stored off the ground and covered to prevent moisture penetration, deterioration, and contamination.

15.2.2 Mortar and Grout

The mortar mix required in the project specifications must be carefully controlled at the job site to maintain consistency in performance and appearance. Consistent measurement of mortar and grout ingredients should ensure uniformity of proportions, yields, strengths, workability, and mortar color from batch to batch. Volumetric rather than weight proportioning is most often called for, and most often miscalculated because of variations in the moisture content of the sand. Common field practice is to use a shovel as the standard measuring tool for aggregate proportioning. However, moisture in the sand causes a "bulking" effect, and for equal weights of wet sand and dry sand, the wet sand occupies more volume than that occupied by the dry sand. Such variables often cause over- or under-sanding of the mix, which affects both the strength and bonding characteristics of the mortar. Over-sanded mortar is harsh and unworkable, provides a weak bond with the masonry units, and performs poorly during freeze-thaw cycles.

For consistency, a 1 cu ft site-constructed batching box can be set to discharge directly into the mixer (*see Fig. 15-2*).

The other dry ingredients in masonry mortar are normally packaged and labeled only by weight. Regardless of weight, however, these cementitious materials are usually charged into the mixer in whole- or half-bag measures. Each bag of portland cement or masonry cement equals 1 cu ft regardless of its labeled weight, and each bag of hydrated mason's lime equals 1¼ cu ft regardless of its weight. In some regions, additional convenience is provided by preblended and bagged portland cement–lime mixes.

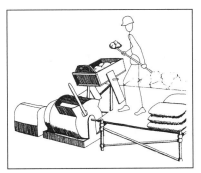

batching boxes for accurately and continuously measuring sand volume

Increases in moisture content cause a "bulking effect" in sand. Any given weight of wet sand occupies more volume than the same weight of dry sand, so sand volume may vary throughout the day and from day to day as its moisture content changes.

MEASURING SAND SIMPLY BY COUNTING SHOVELS IS NOT AN ACCURATE METHOD OF BATCHING MASONRY MORTAR AND IS NOT RECOMMENDED.

The drawing and photo above show a batching box in which the sand is shoveled into a 1 cubic foot measure and then discharged into the mixer from the box. This method is more accurate and accounts for continuous volume changes in the sand as it dries or bulks with moisture changes.

FIGURE 15-2 Measuring and batching sand for masonry mortar. (*From BIA, Technical Note 8B.*)

The amount of mixing water required is not stated as part of the project specifications. Unlike concrete, masonry mortar and grout require the maximum amount of water consistent with characteristics of good flow and workability. Excess water is rapidly absorbed by the masonry units, reducing the water-cement ratio to normal levels and providing a moist environment for curing. Optimum water content is best determined by the mason's feel of the mortar on the trowel. A mortar with good workability has the proper amount of water.

Mortar with good workability should spread easily, cling to vertical unit surfaces, extrude easily from joints without dropping or smearing, and permit easy positioning of the unit to line, level, and plumb. Dry mixes do not spread easily, they produce poor bond, and they may suffer incomplete cement hydration. Mixes that are too wet are difficult to trowel and allow units to settle after placement. Thus, mixing water additions are self-regulating. The water proportion will vary for different conditions of temperature, humidity, unit moisture content, unit weight, and so on.

The necessary water content for grout is significantly higher than that for mortar, because grout must flow readily into the unit cores and wall cavities and around reinforcement and accessories. Grout consistency should produce a slump of 8 to 11 in. (*see Fig. 15-3*).

Ready-mixed mortars and prebatching of dry mortar ingredients eliminate the field variables that often affect the quality and consistency of job-mixed mortar. This moves the mixing operation to a controlled batching plant where ingredients can be accurately weighed and mixed, then delivered to the job site. Ready-mixed mortars are delivered trowel-ready in trucks or sealed containers, without the need for additional materials or mixing. Extended-life set retarders, which keep the mix plastic and workable for up to 72 hours, must be absorbed by the masonry units before cement hydration can begin, so unit suction can affect set time and construction speed. Prebatched dry ingredients are delivered to the site in weather-tight silos ready for automatic mixing. Both methods improve uniformity and offer greater convenience and efficiency, but sand bulking can still be a problem with dry-batched mixes unless the sand is oven-dried. Ready-mixed mortars are governed by ASTM C1142, *Standard Specification for Ready-Mixed Mortar for Unit Masonry*.

There are two traditional methods of mixing mortar on the job site. For very small installations, *hand mixing* may be most economical. It is accomplished using a mason's hoe and a mortar box. Sand, cement, and lime are spread in the box in proper proportions and mixed together until an even color is obtained. Water is then added, and mixing continues until the consistency and workability are judged to be satisfactory.

More commonly, *machine mixing* is used to combine mortar ingredients. The mechanical drum or paddle-blade mixers used are similar to but of lighter duty than concrete mixers. Normal capacity ranges from 4 to 7 cu ft. Three-fourths of the mixing water, half the sand, and all of the cement and lime are first added and briefly mixed together. The balance of the sand is then added, together with the remaining water. After all the materials and water have been combined, grout should be mixed a minimum of 5 minutes and mortar a minimum of 3 and a maximum of 5 minutes. Less mixing time may result in nonuniformity, poor workability, low water retention, and less than optimum air content. Overmixing causes segregation of materials and entrapment of excessive air, which may reduce bond strength.

Specified admixtures and pigments should be added in the approved quantities only after all other ingredients are mixed. Pigments should always be prebatched for consistency in color.

Installation and Workmanship

Grout with a water-to-cement ratio and slump similar to concrete is too dry and stiff for use in masonry. It will not flow into unit cores and wall cavities to properly fill voids and encapsulate reinforcing steel and connectors.

Masonry grout should have a fluid consistency for pouring or pumping into small spaces that are often congested with reinforcing steel.

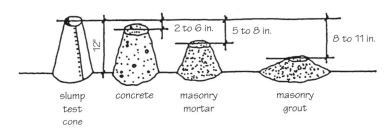

Concrete	Masonry Mortar	Masonry Grout
Concrete is generally mixed with the *minimum* amount of water required to produce workability appropriate to the method of placement. The amount of water is determined by laboratory mix design.	Masonry mortar is mixed with the *maximum* amount of water required to produce good workability with a given unit. The amount of water is determined by the mason based on masonry unit absorption and field conditions.	Masonry grout is mixed with the *maximum* amount of water required to produce good flow. The amount of water is determined by the mason based on field conditions, including the absorption of the masonry units.

FIGURE 15-3 Masonry grout should be a fluid consistency with a slump of 8 to 11 in.

To avoid excessive drying and stiffening, mortar batches should be sized according to the rate of use. Loss of water by absorption and evaporation can be minimized on hot days by wetting the mortar board and covering the mix in the mortar box. Within the first 1½ to 2½ hours of initial mixing, the mason may add water to replace evaporated moisture (refer to Chapter 3). *Retempering* is accomplished by adding water to the mortar batch and thoroughly remixing. Mortars containing added color pigment should not be retempered, as the increased water will lighten the color and cause variation from batch to batch.

15.2.3 Masonry Units

Concrete masonry units (CMU) are cured and dried at the manufacturing plant and should never be moistened before or during placement because they will shrink as they dry out. If this shrinkage is restrained, as it normally is in a finished wall, stresses can develop that will cause the wall to crack. Once the wall is *completed*, it is both acceptable and desirable (especially in hot and dry weather) to use moist curing techniques to ensure adequate cement hydration.

One exception to the caveat against prewetting of CMU is adhered manufactured stone masonry veneer. Even though these products are cement-based, installation in hot and dry or windy conditions often calls for wetting both the "stones" themselves and the substrate to ensure complete cement hydration and good mortar bond. Both the units and the substrate should be moist but surface dry to ensure that excessive suction does not draw too much water out of the mortar before cement hydration is complete.

When brick is manufactured, it is fired in a high-temperature kiln, which drives virtually all of the moisture out of the wet clay. Fired bricks remain extremely dry until they absorb enough moisture from the air to achieve a state of moisture equilibrium with their surroundings. Brick that is very dry when it is laid causes rapid and excessive loss of mixing water from the mortar, which results in poor adhesion, incomplete bond, and water-permeable joints of low strength. Brick that is very dry and absorptive is said to have a high initial rate of absorption (IRA) or high suction. Optimum mortar bond is produced with units having initial rates of absorption between 5 and 25 $g \cdot min^{-1} \cdot 30$ sq in.$^{-1}$ (refer to Chapter 2). If the IRA is higher than 30 $g \cdot min^{-1} \cdot 30$ sq in.$^{-1}$, the units should be wetted with a hose the day before they will be used so that moisture is fully absorbed into the units but the surfaces are dry to the touch before being laid. Visual inspection of a broken brick will indicate whether moisture is evenly distributed throughout the unit (*see Fig. 15-4*). Where prewetting of units is not possible, the time lapse between spreading the mortar and laying the unit should be kept to a minimum. Some experts recommend that brick not be wetted in winter because some high-suction units produce better bond strength in cold weather than low-suction units.

A simple field test can be performed to determine whether brick should be prewetted. Place 12 drops of water in a single spot on the bed surface of the brick and time how long it takes for them to be absorbed. If the water is completely absorbed in less than 1 minute, the brick is too dry and should be prewetted.

Brick and architectural CMU must also be properly blended for color to avoid uneven visual effects. Units from four different cubes or pallets should be used at the same time, and brick manufacturers often provide unstacking instructions for even color distribution. For single-color units, this takes advantage of the subtle shade variations produced in the manufacturing process, and for a blend of colors, this will prevent stripes or patchy areas in the finished wall (*see Fig. 15-5*). The wider the range of colors, the more difficult it is to get a uniform blend.

All masonry units should be clean and free of contaminants such as dirt, oil, or sand, which might inhibit bond.

15.2.4 Accessories

Steel reinforcement, anchors, ties, and other accessories should be cleaned to remove oil, dirt, ice, and other contaminants that could prevent good bond with the mortar or grout. Careful storage and protection will minimize cleaning requirements. Flashing materials should be protected from damage or deterioration prior to placement and insulation materials protected from wetting.

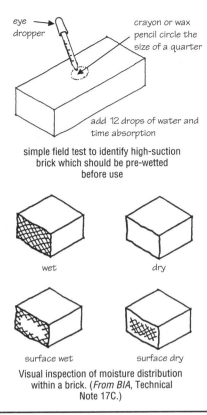

FIGURE 15-4 Field testing for brick with a high IRA.

FIGURE 15-5 Masonry units must be blended for uniform color distribution.

281

15.2.5 Layout and Coursing

The design of masonry buildings should take into consideration the size of the units involved. The length and height of walls and the location of openings and intersections will greatly affect both the speed of construction and the appearance of the finished work. The use of a common module in determining dimensions can reduce the amount of field cutting required to fit the building elements together.

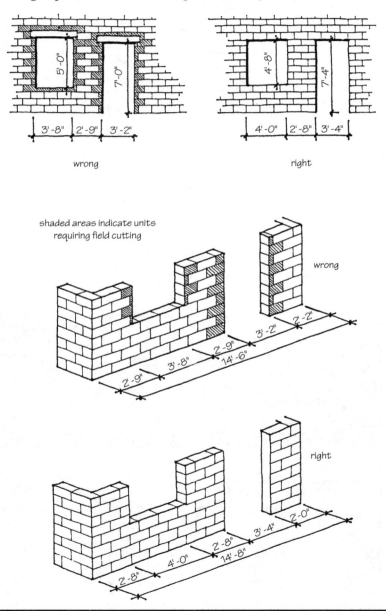

Figure 15-6 Wall openings based on 8 × 8 × 16 in. concrete block. (*From NCMA, TEK Bulletin 14.*)

A number of the common brick sizes available are adaptable to a 4-in. module, and dimensions based on these standards will result in the use of only full- or half-size units. Similarly, a standard 16-in. concrete block layout may be based on an 8-in. module with the same reduction in field cutting (*see Fig. 15-6*). In composite construction of brick and concrete block, unit selection should be coordinated to facilitate the anchorage of backing and facing wythes, as well as the joining and intersecting of the two systems. Three courses of standard modular brick equal the height of one concrete block course. As shown in *Fig. 15-7*, the brick and block units work together in both plan and section, but in a cavity wall the units are offset at corners and the head joints do not align anywhere in the wall.

Units must remain in both vertical and horizontal alignment throughout the height and length of the structure in order for the coursing to work out with opening locations, slab connections, anchorage to other structural elements, and so on. The first course must provide a solid base on which the remainder of the walls can rest. The base course at the foundation must always be laid in a full bed of mortar even if face shell bedding (for hollow brick or CMU) is to be used in the rest of the wall.

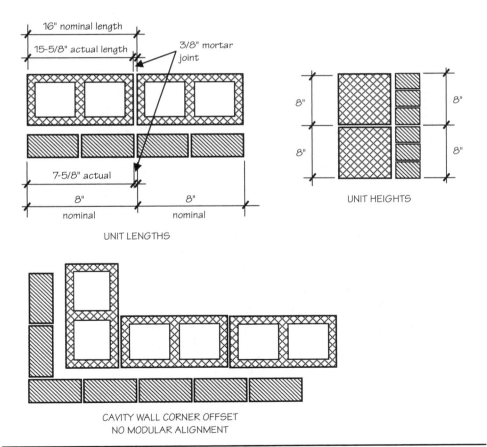

FIGURE 15-7 Unit coursing layout. (*From C. Beall*, Masonry and Concrete for Residential Construction, *McGraw-Hill Complete Construction Series, McGraw-Hill.*)

15.3 Installation

Masonry construction includes the placement of mortar, units, anchors, ties, reinforcement, grout, and accessories. Each element of the construction performs a specific function and should be installed in accordance with recommended practices.

15.3.1 Mortar and Unit Placement

Masonry walls with full head and bed joints are stronger and less likely to leak than walls with furrowed bed joints and lightly buttered head joints. Partially filled mortar joints reduce the flexural strength of masonry by as much as 50 to 60%, offer only minimal resistance to moisture penetration, and may contribute to spalling and cracking if freezing occurs when the units are saturated. Bed joints should be laid full and unfurrowed, only slightly beveled away from the cavity to minimize mortar extrusion and droppings (*see Fig. 15-8*). The ends of the units should be fully buttered with mortar so

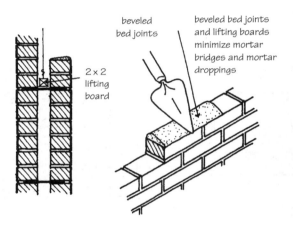

Figure 15-8 Beveled bed joints minimize mortar extrusions into the drainage cavity, and lifting boards can be used to remove mortar droppings. (*From BIA*, Technical Note 21C.)

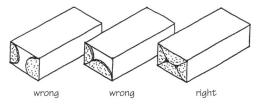

Fill head joints for better resistance to rain penetration.
(*From BIA*, Technical Note 17C.)

FIGURE 15-9   Fill head joints for better resistance to rain penetration. (*From BIA*, Technical Note 17C.)

that when they are shoved into place, mortar is extruded from the joint (*see Fig. 15-9*). Concrete block should always be laid with the thicker end of the face shell up to provide a larger mortar bedding area. For face shell bedding of hollow CMU, only the end flanges of the face shells are buttered with mortar (*see Fig. 15-10*). Because of their weight and difficulty in handling, masons often stand several units on end and apply

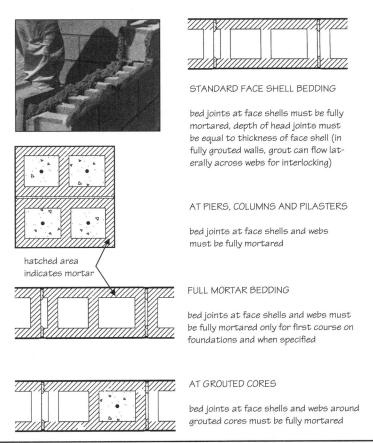

FIGURE 15-10   Mortar bedding of hollow masonry units. (*From T. Patterson*, Illustrated 2000 Building Code Handbook, *McGraw-Hill, 2001.*)

mortar to the flanges of three or four units at one time. Each block is then individually placed in its final position, tapped down into the mortar bed, and shoved against the previously laid block, thus producing well-filled vertical head joints at both faces of the masonry.

In cavity wall and veneer wall construction, it is important that the cavity between the outer wythe and the backing wall be kept clean to ensure proper moisture drainage. If mortar clogs the cavity, it can form bridges for moisture passage or it may block weep holes. Some masons use a removable wooden strip to temporarily block the cavity as the wall is laid up and prevent mortar droppings. However, beveling the mortar bed as shown previously in *Fig. 15-8* allows little mortar to extrude into the cavity. Any mortar fins that do protrude into the cavity should be cut off or flattened to prevent interference with the placement of reinforcing steel, grout, or insulation. A cavity with a minimum clear dimension of 2 in. is not as easily bridged by mortar extrusions and can be kept clean much more easily than a narrow cavity. Codes generally require only a 1-in. minimum cavity width, and corrugated anchors cannot be used when the cavity exceeds 1 in. A 2-in.-wide cavity, however, is preferable for better drainage.

Project specifications and many industry recommendations restrict mortar droppings in absolute terms. In reality, no masonry wall can be built under typical job-site conditions without some amount of mortar droppings. The *Brick Industry Associaton* (BIA) *Technical Note 7B* phrases its recommendations about mortar droppings in less than absolute terms:

- *"To the greatest extent possible*, mortar droppings should be prevented from falling into the air space or cavity."
- "[T]he amount of mortar droppings should be *limited as much as possible."*

Mortar droppings are to be expected in all masonry walls. Mortar droppings are "excessive" when they cause damage or obstruct water flow to the weeps. Drainage mats and mortar collection devices can help maintain an unobstructed flow of water to the weeps. It is always a good idea to randomly water-test wall drainage as described in Chapter 17.

To add visual interest to masonry walls, units may be laid in different positions as shown in *Fig. 15-11* and arranged in a variety of patterns originally conceived in connection with masonry wall bonding techniques that are not widely used today. In older work constructed without metal ties or reinforcement, rowlock and header courses were used structurally to bond the wythes of a wall together. Most contemporary buildings use one-third or one-half running bond or stack bond with very little decorative pattern work.

Lipped bricks used at shelf angles are challenging, especially at building corners and offsets. Manufactured lipped brick may or may not have corner units available, and it is impossible to field-cut a corner piece from a typical stretcher brick. One option that has been used at corners is two separate lipped bricks with mitered 45° angle cuts, but they may be susceptible to fracturing at the sharp corners unless there is a sealant joint between the two pieces. If the bricks are turned upside down and placed below the shelf angle, there is less likelihood of breakage because the shelf angle is no longer shoving against the lip as it expands and contracts with temperature changes.

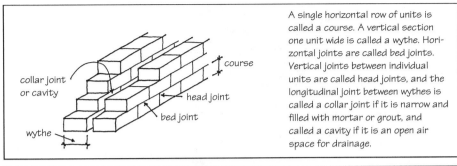

A single horizontal row of units is called a course. A vertical section one unit wide is called a wythe. Horizontal joints are called bed joints. Vertical joints between individual units are called head joints, and the longitudinal joint between wythes is called a collar joint if it is narrow and filled with mortar or grout, and called a cavity if it is an open air space for drainage.

A unit laid lengthwise in the wall is called a stretcher. Standing upright with the narrow side facing out, it is called a soldier—with the wide side facing out, a sailor. A stretcher unit that is rotated 90° in a wall so that the end is facing out is called a header. If the unit is then stood on its edge, it's called a rowlock.

A unit whose length is cut in half is called a bat. One that is halved in width is called a soap, and one that is cut to half height is called a split.

Whichever way you turn modular brick, they lay out to a 4-in. module. Turning a brick stretcher crosswise in a two-wythe wall, the header unit is exactly the same width as two wythes of stretcher brick with a 3/8" collar joint between. Two header units

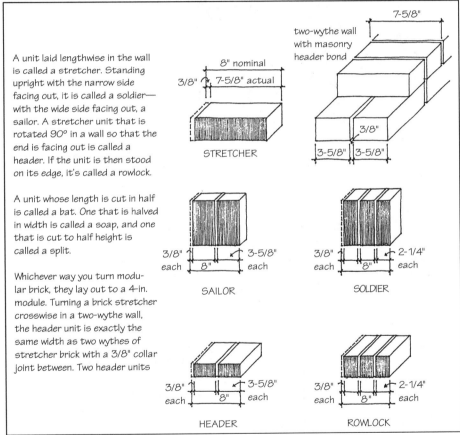

FIGURE 15-11 Masonry unit orientation and nomenclature. (*From C. Beall and R. Jaffe,* Concrete and Masonry Databook, *McGraw-Hill, 2003.*)

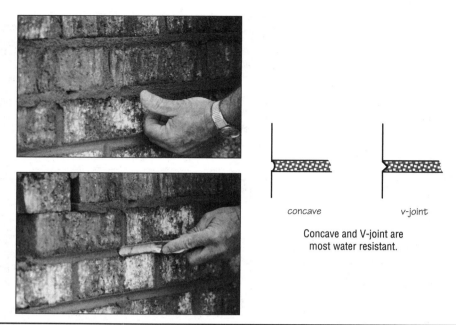

Figure 15-12 Concave tooling when the mortar is "thumbprint hard" should produce joints of uniform color and appearance that are well compacted at the bond line to maximize resistance to water penetration. (*Photos courtesy BIA.*)

The most moisture-resistant mortar joints are the concave and V-shaped tooled joints. Mortar squeezes out of the joints as the masonry units are set in place, and the excess is struck off with a trowel. After the mortar has become "thumbprint" hard (i.e., when a clear thumbprint can be impressed and the cement paste does not stick to the thumb), joints are finished with a jointing tool slightly wider than the joint itself (*see Fig. 15-12*). As the mortar hardens, it has a tendency to shrink slightly and separate from the edge of the masonry unit. Proper tooling compresses the mortar against the unit and compacts the surface, making it more dense and more resistant to moisture penetration (*see Fig. 15-13*). Mortar joints should always be tooled at a consistent moisture content or color variations may create a blotchy appearance in the wall. Drier mortar tools darker than mortar with a higher moisture content. Joint tooling is especially critical in single-wythe walls where there is little or no secondary defense against water penetration. Ribbed block joints are especially difficult to tool, and it is virtually impossible to get a weather-resistant joint (*see Fig. 15-14*).

15.3.2 Flashing and Weep Holes

Flashing must be installed in continuous runs, with all seams and joints lapped 4 to 6 in. and sealed. Metal flashing laps sealed with a nonhardening butyl tape or caulk can accommodate thermal expansion and contraction while preventing lateral moisture flow. Unsealed lap joints will allow water to flow underneath the flashing. Inside and outside corners can be fabricated of metal, or preformed rubber corner boots can

Installation and Workmanship

well compacted surface,
tight bond line

rough surface, bond
line separations

rough surface, bond
line separations, voids

FIGURE 15-13 Good joint tooling is the first line of defense against water penetration and is especially critical in single-wythe walls.

FIGURE 15-14 Ribbed block joints are very difficult to tool.

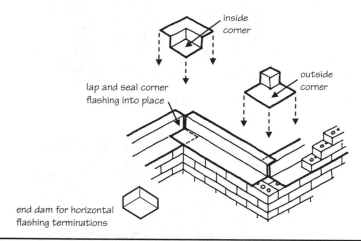

FIGURE 15-15 Corner flashing can be field or shop fabricated of metal, or prefabricated self-adhering rubberized asphalt flashing boots can be used.

be used, even with metal flashing systems (*see Fig. 15-15*). At lateral terminations where the flashing abuts other construction elements, and at terminations on each side of door and window lintels and window sills, flashing must be turned up to form an end dam. Metal flashing can be cut, folded, and soldered or sealed with mastic to form a watertight pan, and flexible flashing can be folded into place. Without end dams, water that collects on the flashing is free to run off the ends and into the drainage cavity of the wall or into adjacent door jambs, windows, curtain walls, or other cladding systems (*see Fig. 15-16*).

Installation and Workmanship 291

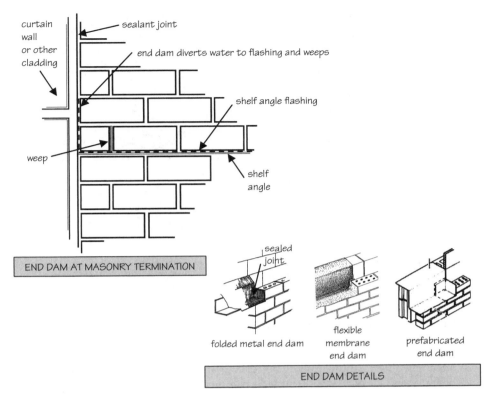

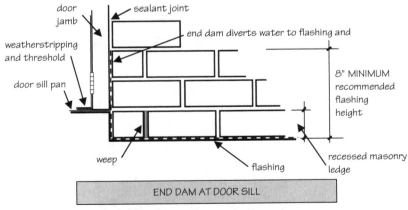

FIGURE 15-16 Form end dams wherever flashing terminates at windows, doors, and against adjacent construction.

FIGURE 15-17 Flashing that does not extend through the outside wall face can allow moisture to flow around the front edge of the membrane and back into the wall.

Flashing should never be stopped short of the face of the wall. If the flashing is not fully adhered to the substrate, water may flow around the front edge, under the flashing, and back into the wall (*see Fig. 15-17*). Metal flashing should be brought out beyond the wall face and turned down to form a drip. A hemmed edge will give the best appearance. Flexible flashing cannot be formed in the same way, but should be extended beyond the face of the wall and later trimmed flush with the joint (*see Fig. 15-18*). Some

FIGURE 15-18 Trim flexible flashing flush with outside face of wall. (*Photo courtesy of BIA.*)

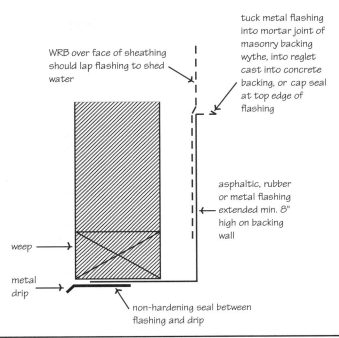

FIGURE 15-19 Lap flashing over a separate metal edge to form a drip, to avoid ultraviolet exposure, or to accommodate variations in cavity width.

designs may call for flexible membrane flashing such as rubberized asphalt to be lapped and sealed over the top of a separate metal drip edge (*see Fig. 15-19*). Two-piece flashing can also be used, even with all-metal flashing systems, to accommodate construction tolerances in the necessary length of the horizontal leg. The vertical leg of the flashing should be turned up a minimum of 8 in. to form a back dam and be placed in a mortar joint in the backing wythe, in a reglet on concrete walls, or behind the weather-resistive barrier on stud walls (*see Fig. 15-20*). Flexible flashing should follow the substrate profile so that it is fully supported. Voids at the 90° angle leave the flashing vulnerable to puncture. The back leg of the flashing must be taller than any drainage mat that might be used in the cavity so that mortar droppings do not accumulate above the top edge of the flashing.

Weep holes are required in masonry construction at the base course and at all other flashing levels (such as shelf angles, sills, and lintels) so that water that is collected on the flashing may be drained from the wall as quickly and effectively as possible. Weep holes should be spaced 16, 24, or 32 in. on center, depending on the type of weep and type of unit:

- Open-head joints, large rectangular weep tubes, cellular grids, woven filament, or vented weep covers at 24 in. on center in brick or 32 in. on center in block.
- Oiled rods, rope, or pins placed 16 in. on center in the head joints and removed before final set of the mortar.

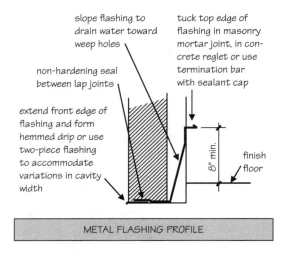

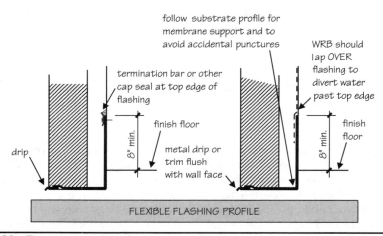

FIGURE 15-20 The top edge of the flashing should be terminated in such a way that water cannot flow behind it.

- Cotton sash cord or other suitable wicking material placed 16 in. on center in the head joint.
- Small plastic weep tubes are not recommended because they clog too easily both during and after construction (*see Fig. 15-21*).

To function properly, weep holes must be unobstructed by mortar droppings or other debris. Blocked or missing weep holes can cause saturation of the masonry just above the flashing as moisture is dammed in the wall for longer periods of slow evaporation. Efflorescence, staining, corrosion of steel lintels or studs, mold growth, and freeze-thaw damage can result. Weep hole tubes are most vulnerable to blockage. Chapter 4 illustrates several proprietary drainage accessory products, all of which are intended to maintain the free flow of moisture to the weep holes. Some are

FIGURE 15-21 Small plastic weep tubes are not recommended because they clog too easily both during and after construction.

more effective than others. Use of a drainage accessory, however, does not eliminate the need for proper construction procedures to minimize mortar droppings. It is not possible to eliminate mortar droppings entirely, but every effort should be made to minimize droppings and maintain a drainage path to the weeps.

15.3.3 Control and Expansion Joints

Control joints and expansion joints are used to relieve stresses caused by differential movement between materials and by thermal and moisture movement in the masonry itself (refer to Chapter 6).

Control joints are continuous vertical head joints constructed with or without mortar to accommodate the permanent moisture *shrinkage* that all CMU experience. When shrinkage stresses are sufficient to cause cracks, the cracking will occur at these weakened joints rather than at random locations. Shear keys are used to provide lateral stability against wind loads, and elastomeric sealants are used to provide a watertight seal. Mortared control joints must be raked out to a depth that will allow placement of a backer rod or bond-breaker tape and sealant. Shrinkage always exceeds expansion in concrete masonry because of the initial moisture loss after manufacture. So even though control joints contain hardened mortar or hard rubber shear keys, they can accommodate reversible thermal expansion and contraction because it occurs after the initial curing shrinkage. Joint reinforcement should be stopped on either side of control joints.

Expansion joints are used in *brick* construction to accommodate the permanent moisture *expansion* that all clay masonry products experience as they reabsorb atmospheric moisture after firing. Clay masonry moisture expansion always exceeds reversible thermal expansion and contraction, so expansion joints cannot contain mortar or other hard materials. Lateral support is provided by placing an anchor or tie on either side of expansion joints. During construction, mortar must be kept out or cleaned out of brick expansion joints. In cavity wall construction of brick with block backup, control joints and expansion joints, in the backing and facing wythes, respectively, should occur at approximately the same locations but need not align exactly. Joint reinforcement should not continue across movement joints.

15.3.4 Accessories and Reinforcement

Joint reinforcement, anchors, and ties are usually laid directly on the units. When the mortar is placed, it surrounds and encapsulates the wire. All metals should be protected by a minimum 5/8 in. mortar cover at exterior joint faces. With less mortar coverage, corrosion and joint spalling will occur (*see Fig. 15-22*).

Vertical reinforcement in masonry walls is held in place with bar positioners at periodic intervals to hold the reinforcing bars in vertical alignment. If horizontal steel is required, it is tied to the vertical bars or may rest on spacers at the proper intervals (*see Fig. 15-23*).

For reinforced CMU walls, special open-end units are made so that the block may be placed around the vertical steel rather than lifted and threaded over the top of the bar (*see Fig. 15-24*). Some specially designed blocks have been produced that can accommodate both vertical and horizontal reinforcing without the need for spacers. The proprietary block shown in *Fig. 15-24* not only has open ends but also incorporates notches in the webs for placement of horizontal bars.

Reinforcing steel in masonry construction is required by code to have certain minimum clearances between bars and cavities so that grout can easily flow around and encapsulate the steel. Reinforcing steel is also required to have minimum distances from the outside face of elements to protect the metal from moisture and from fire exposure (*see Fig. 15-25*). The *Masonry Standards Joint Committee* (MSJC) Code prescribes placement tolerances for reinforcing steel as shown in *Fig. 15-26*.

Installation and Workmanship 297

anchors too short for proper embedment

joint reinforcement too close to face of wall

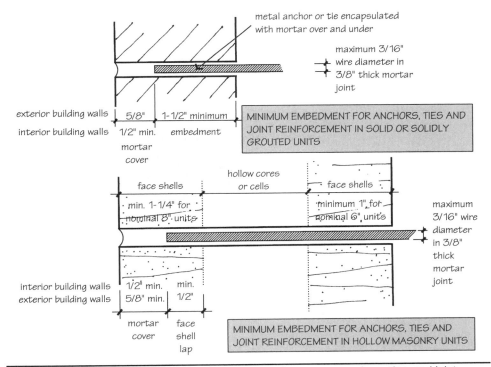

FIGURE 15-22 Minimum embedment and mortar cover for masonry anchors, ties, and joint reinforcement.

Masonry anchors and ties must transfer lateral wind and seismic loads to the backing wall and structure. Bending, stretching, or otherwise distorting them to accommodate variances in cavity width decreases their strength (*see Fig. 15-27*). It is prudent to anticipate construction tolerances that affect cavity width and have a variety of anchor lengths on the job site. Using the right sizes of anchors and ties will ensure that code requirements for pull-out strength, minimum embedment, and mortar cover are met.

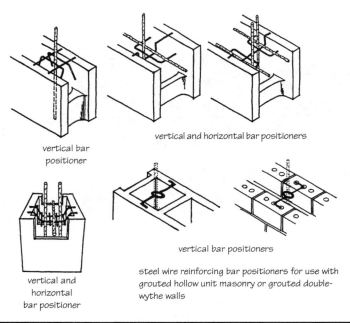

Figure 15-23 Bar positioners for vertical reinforcing steel.

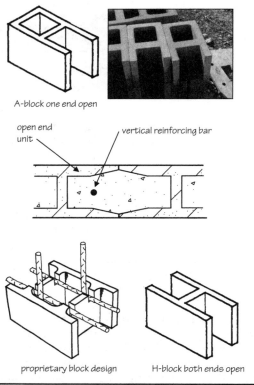

Figure 15-24 Special open-end block designs make it easier to place units around vertical reinforcing steel rather than threading over the top of the bar.

Reinforcement Cover for Masonry Structures	
Reinforced Masonry	Minimum Cover (in.)§

Masonry exposed to earth or weather
    No. 6 and larger
    No. 5 and smaller
	
	2
	1-1/2

Masonry not exposed to earth or weather	1-1/2

§ Minimum cover includes thickness of masonry unit.

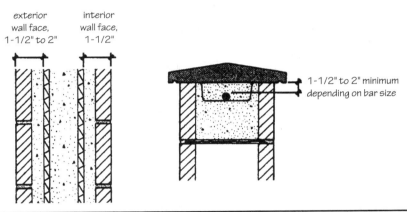

FIGURE 15-25 Code requirements for minimum masonry and grout cover for reinforcement. (Drawings from J. Amrhein, *Reinforced Masonry Engineering Handbook*, 5th ed. Los Angeles: Masonry Institute of America, 1992.)

Element	Distance From Centerline of Steel to the Opposite Face of Masonry		
	≤ 8 in.	> 8 in. but ≤ 24 in.	> 24 in.
Walls and Flexural Elements	± 1/2 in.	± 1 in.	± 1-1/4 in.
Walls	For vertical bars, within 2 in. of location along length of wall		

FIGURE 15-26 MSJC code placement tolerances for masonry reinforcement.

15.3.5 Grouting

In reinforced masonry construction, the open cavity of a double-wythe wall or the vertical cells of hollow masonry units must be pumped with grout to secure the reinforcing steel and bond it to the masonry.

The MSJC Code prescribes minimum grout space requirements based on grout type (fine or coarse) and pour height (1 to 24 ft). In both brick and CMU work, the importance of keeping mortar out of the wall cavity or unit cores has been stressed before but

Wires splayed because anchors are too long. Mortar coverage at face of wall still inadequate. Does not meet code.

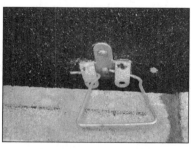

Tabs bent and pintles unhooked because anchors are too long. Does not meet code.

Wire stretched and improperly hooked because anchors are too short. Does not meet code.

Wire rotated because anchors are too long. Mortar coverage at face of wall still inadequate. Does not meet code.

FIGURE 15-27 Do not distort masonry ties or anchors. Use the right size.

should be re-emphasized here. Protrusions or fins of mortar that project into the grout space will interfere with proper flow and distribution of the grout and could prevent complete bonding. An unobstructed cavity and the spacers used to maintain alignment of vertical reinforcing will ensure that the rebar is completely encapsulated in the grout for proper structural performance (*see Fig. 15-28*). Each grout lift should be stopped

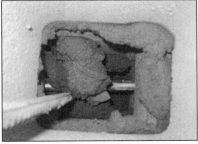

CMU cores obstructed by mortar droppings and protrusions prohibit good grout flow

unobstructed CMU cores allow good grout flow

Figure 15-28 Mortar protrusions into the wall cavity or unit cores will inhibit the flow of grout and create voids within the section.

about 2 in. below the top of the masonry to form a key with the next lift (*see Fig. 15-29*). Grouting of concrete masonry should be performed as soon as possible after the units are placed so that shrinkage cracking at the joints is minimized and so that the grout bonds properly with the mortar.

Grout should be well mixed and carefully poured or pumped to avoid segregation of materials. Grout in contact with the masonry hardens more rapidly because the mixing water is quickly absorbed by the masonry units (*see Fig. 15-30*), so it is important that consolidation take place immediately after the lift and before this hardening begins. Grout must be consolidated by vibration to minimize voids. Grout consolidation can be performed by puddling with a piece of reinforcing bar if the lifts do not exceed 12 in., but for higher lifts, a mechanical vibrator must be used. Five to ten minutes after the grout is placed, the vibrator should be inserted into the grout cavity or cores for a few seconds in each location. Within 30 minutes of consolidation, the grout must be reconsolidated to ensure proper bond to the masonry and reinforcement. Reconsolidation

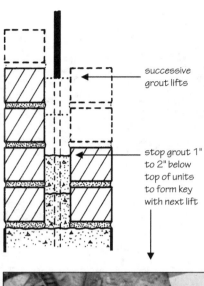

FIGURE 15-29 Grout key between lifts.

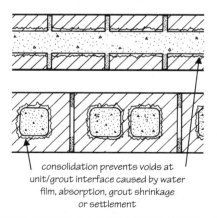

FIGURE 15-30 Grout that is in contact with the masonry hardens more rapidly because the mixing water is absorbed by the masonry units. (*From* Informational Guide to Grouting Masonry, *Masonry Institute of America, 1992.*)

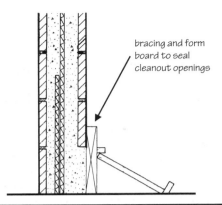

FIGURE 15-31 Grouting cleanouts and plugs. (*From* Informational Guide to Grouting Masonry, *Masonry Institute of America, 1992.*)

prevents separations from developing between the grout and the masonry after shrinkage, settlement, and absorption have occurred. For grout pours greater than 5 ft, cleanouts must be provided at the base of the wall to clear debris from the grout space and to allow for inspection (*see Fig. 15-31*).

15.4 Construction Tolerances

Historically, most construction was done on site and custom fitted. Within generous limits, brick and stone could be laid to fit existing conditions; roof timbers were cut to fit whatever the masons built; and hand-made doors and windows could be made to accommodate the peculiarities of any opening. Today, we have metals that are fabricated at the mill, stone that is cut and dressed at the quarry, and concrete that is cast before erection. These prefabricated components are not easily customized on the job, and they must be fitted to site-built frames. Suddenly, construction tolerances become very important in ensuring that the puzzle pieces fit together with reasonable accuracy—puzzle pieces that may come from a dozen different manufacturers in a half-dozen competing industries.

Little is exact in the manufacturing, fabrication, and construction of buildings and building components. Tolerances allow for the realities of fit and misfit of the various parts as they come together in the field and ensure proper technical function such as structural safety, joint performance, secure anchorage, moisture resistance, and acceptable appearance. *Webster's* defines tolerance as "the allowable deviation from a standard, especially the range of variation permitted in maintaining a specified dimension."

Each construction trade or industry develops its own standards for acceptable tolerances based on economic considerations of what is reasonable and cost effective. Few, if any, construction tolerances are based on hard data or engineering analysis. There has also never been any coordination among various industry groups even though steel and concrete are used together, masonry is attached to or supports both, windows must fit into openings in all three, and sealants are expected to fill all the gaps left between adjacent components.

Different materials and systems, because of the nature of their physical properties and manufacturing methods, have greater or lesser relative allowances for the manufacture or fabrication of components and the field assembly of parts. Masonry includes a variety of materials and unit types, each of which has its own set of tolerances.

15.4.1 Masonry Size Tolerances

Allowances for the variation in sizes of face brick are covered in ASTM C216, *Facing Brick*. Face brick tolerances are divided into Type FBX and Type FBS. The strength and quality of the units is the same, but Type FBX is required to have tighter size tolerances so that when a designer wishes to create a crisp, linear appearance with, for example, a stack bond, the units and mortar joint variations are kept to a minimum (*see Fig. 15-32*). Type FBS bricks have slightly larger size tolerances but are more popular for both commercial and residential masonry using running bond and other patterns. Type FBA bricks are not governed by size tolerances because they are supposed to vary significantly from one unit to the next so that they often look like rough, hand-molded brick. Type FBA bricks are very popular for residential masonry and for projects in historic areas.

Concrete block dimensional variations are covered in ASTM C90, *Loadbearing Concrete Masonry Units* and C129, *Nonloadbearing Concrete Masonry Units* for loadbearing and non-loadbearing units, respectively. For both types, the standards permit a maximum size variation of ±1/8 in. from the specified standard dimensions (defined as the "manufacturer's designated dimensions").

15.4.2 Mortar Joints

Unit masonry size tolerances are accommodated by varying the thickness of the mortar joints. Modular units are designed to be laid with standard 3/8-in. joints. The MSJC *Building Code Requirements for Masonry Structures* (ACI 530/ASCE 6/TMS 402) and *Specification for Masonry Structures* (ACI 530.1/ASCE 6/TMS 602) set allowable variation in joint thickness based on the structural performance of the masonry, not on aesthetics (*see Fig. 15-33*).

Type FBS

for traditional
or contemporary
architectural styles

Type FBX

for crisp, linear,
contemporary styles
of architecture

Type FBA

for rustic styles
of architecture

FIGURE 15-32 Types FBS, FBX, and FBA have the same quality but differ in appearance for variations in architectural style.

Installation and Workmanship

Joint	Allowable Tolerance (in.)
Bed joint	± 1/8
Head joint	−1/4 to + 3/8
Collar joint	−1/4 to + 3/8

FIGURE 15-33 Structural tolerances permitted by code for masonry mortar joints.

The type of brick or block and its allowable tolerances will dictate how much variation there must be in the mortar joints. Type FBX brick can be laid with the most uniform joint thickness because FBX size tolerances are very tight. This characteristic lends itself to stack bond patterns where alignment of the head joints is critical to appearance. Usually, all of the units in a shipment are either oversized or undersized, but not both. Type FBA brick will require considerable variation in joint thickness because of the greater FBA size variations, but this is part of the reason for the popularity of this brick type. Even FBA brick's considerable joint size variation can be acceptable aesthetically if the joints are more or less uniformly variable. A wall that has a wide range of joint sizes that differ from one another significantly, however, is aesthetically unacceptable (*see Fig. 15-34*).

15.4.3 Wall Cavity Width Variations

Conflicts between structural frame and masonry veneer tolerances affect anchor embedment, support at shelf angles, and flashing details. To accommodate minor field adjustments, specify the following:

- Bolted rather than welded connections for steel shelf angles, with slotted holes for field adjustments and wedge inserts at points of attachment to concrete frames.
- That the contractor provide a variety of anchor lengths as necessary to accommodate construction tolerances and provide minimum 5/8-in. mortar cover on outside wall face and minimum 1½-in. embedment in solid masonry units, or minimum ½-in. embedment into face shell of hollow units.

FBA brick mortar joints are more or less uniformly variable and aesthetically more pleasing

inconsistent joint width variations within the same wall are aesthetically unacceptable

FIGURE 15-34 Aesthetically, mortar joint width variations are more acceptable if the variations themselves are somewhat uniform.

- Two-piece flashing to accommodate varying cavity widths.
- Horseshoe shims that are the full height of the vertical leg of the shelf angle and of plastic or a compatible metal for shimming the angle up to a maximum of 1 in.

15.4.4 Grout and Reinforcement

For reinforced masonry, tolerances are allowed for the placement of the steel bars and the size of the grout spaces. The most important thing is to ensure complete embedment of the steel within the grout so that full strength is developed. To ensure that the reinforcement is not displaced during the grouting operation, specify reinforcing bar spacers or special units that hold the steel in place.

15.5 Cold Weather Construction

Cold weather causes special problems in masonry construction. Even though sufficient water may be present, cement hydration and strength development in mortar and grout will stop at temperatures below 40°F. Construction may continue during cold weather, however, if the mortar and grout ingredients are heated and the masonry units and structure are protected during the initial hours after placement. As temperatures drop, additional protective measures are required. Protective enclosures may range from a simple windbreak to an elaborate heated enclosure (*see Fig. 15-35*).

Cement hydration will resume only when the temperature of the mortar or grout is raised above 40°F and its liquid moisture content exceeds 75%. When these conditions are maintained, ultimate strength development and bond will be the same as those attained under moderate temperature conditions. During cold weather construction, it may be desirable to use a Type III, high-early-strength portland cement because of the greater protection it will provide the mortar. The MSJC Code requires cold weather

FIGURE 15-35 Cold weather enclosures keep the masonry above the required protection temperatures.

Installation and Workmanship

MSJC Cold Weather Construction Requirements		
	Temperature	Action Required
	During Construction	
Ambient Temperature	32 to 40°F or When temperature of masonry units is less than 40°F	Do not lay glass unit masonry. Do not lay units that have a temperature below 20°F. Remove visible ice on units before laying. Heat mortar sand or mixing water to produce mortar temperature between 40°F and 120°F at time of mixing. Maintain mortar above freezing until used in masonry.
	20 to 25°F	Perform actions required when ambient temperature is 32 to 40°F. Provide heat sources on both sides of the masonry. When wind velocity exceeds 15 mph, install wind breaks.
	Less than 20°F	Perform actions required when ambient temperature is 32 to 40°F. Enclose masonry under construction. Provide supplementary heat to maintain temperature within enclosure above 32°F.
	For 24 Hours After Construction	
Mean Daily Temperature	32 to 40°F	Cover completed masonry with weather-resistive membrane to protect from rain and snow.
	25 to 32°F	Completely cover completed masonry with weather-resistive membrane.
	20 to 25°F	Completely cover completed masonry with insulating blankets or equal protection.
	Less than 20°F	Enclose masonry. Provide supplementary heat to maintain temperature of masonry within enclosure above 32°F.
	For 48 Hours After Construction	
	All	Maintain temperature of glass unit masonry above 40°F.

FIGURE 15-36 Cold weather masonry construction. (*From Masonry Standards Joint Committee, Specification for Masonry Structures, ACI 530.1/ASCE 6/TMS 602.*)

protection measures during construction when the ambient temperature is below 40°F. The table in *Fig. 15-36* summarizes heating and protection requirements for various job site temperatures.

15.6 Hot Weather Construction

Hot weather conditions also pose special concerns for masonry construction (*see Fig. 15-37*). High temperatures, low humidity, and wind can adversely affect performance of the masonry. Rapid evaporation and the high suction of hot, dry units can quickly reduce the water content of mortar and grout mixes so that cement hydration actually stops.

When ambient temperatures are above 100°F or above 90°F with wind velocities greater than 8 mph, the MSJC Code requires that protective measures be taken to

MSJC Hot Weather Construction Requirements	
Temperature	Action Required
Preparation	
Ambient temperature above 100°F, or 90°F with wind velocity greater than 8 mph	Maintain sand piles in a damp, loose condition. Provide necessary conditions and equipment to produce mortar having a temperature below 120°F.
Ambient temperature above 115°F, or 105°F with wind velocity greater than 8 mph	Maintain sand piles in a damp, loose condition. Provide necessary conditions and equipment to produce mortar having a temperature below 120°F. Shade materials and mixing equipment from direct sunlight.
During Construction	
Ambient temperature above 100°F, or 90°F with wind velocity greater than 8 mph	Maintain temperature of mortar and grout below 120°F. Flush mixer, mortar transport container, and mortar boards with cool water before they come into contact with mortar ingredients or mortar. Maintain mortar consistency by retempering with cool water. Use mortar within 2 hrs. of initial mixing.
Ambient temperature above 115°F, or 105°F with wind velocity greater than 8 mph	Maintain temperature of mortar and grout below 120°F. Flush mixer, mortar transport container, and mortar boards with cool water before they come into contact with mortar ingredients or mortar. Maintain mortar consistency by retempering with cool water. Use mortar within 2 hrs. of initial mixing. Use cool mixing water for mortar and grout. Ice is permitted in the mixing water prior to use. Do not permit ice in the mixing water when added to the other mortar ingredients or grout materials.
Protection	
Mean daily temperature above 100°F or 90°F with wind velocity greater than 8 mph	Fog spray all newly constructed masonry until damp, at least three times a day until the masonry is 3 days old.

FIGURE 15-37 Hot weather masonry construction. (*From Masonry Standards Joint Committee, Specification for Masonry Structures, ACI 530.1/ASCE 6/TMS 602.*)

ensure continued hydration, strength development, and maximum bond. Whenever possible, materials should be stored in a shaded location and aggregate stockpiles covered with polyethylene sheets to retard moisture evaporation. High-suction brick can be wetted to reduce initial absorption, and metal accessories such as reinforcing steel, anchors and ties, mixers, mortar boards, and wheelbarrows can be kept cool by spraying with water.

Additional mixing water may be needed in mortar and grout, and additional lime will increase water retentivity (refer to Chapter 3). Increasing the cement content in the mix accelerates early strength gain and maximizes hydration before evaporative water loss. Water that is too hot can cause the cement to flash set. Approved set-retarding or water-reducing admixtures may also be used. Retempering should be limited to the first 2 hours after mixing. Mortar beds should not be spread more than 4 ft ahead of the masonry, and units should be set within 1 minute of spreading the mortar.

Sun shades and wind screens can also modify the effects of hot, dry weather.

15.7 Moist Curing

Cement hydration cannot occur if the temperature of the mortar or grout is below 40°F or if the internal moisture content of the mix is less than 75%. Both hot and cold weather can produce conditions that cause hydration to stop before curing is complete. These *dryouts* occur most frequently in concrete masonry construction and under winter conditions but may also occur in brick construction and in hot, dry weather. Dryouts are reactivated by higher temperatures and rain, but in the meantime, the construction is temporarily limited in compressive strength, bond, and weather resistance.

Moist curing methods similar to those used in concrete and stucco construction can help prevent masonry dryouts. Periodically wetting the finished masonry with a fine water spray for several days will usually ensure that adequate moisture is available for curing, strength development, and good bond. Covering the walls with polyethylene sheets will also retard evaporation and create a greenhouse effect that aids in moist curing.

CHAPTER 16
Specifications

Project specifications establish standards of quality to ensure structural integrity, weather resistance, and durability. To achieve quality workmanship and proper performance, the architect or engineer must carefully outline the products and standards of construction required.

16.1 Specification Guidelines

Reference standards should be used to govern the quality of specified products. *American Society for Testing and Materials* (ASTM) standards cover all of the mortars, unit types, and varieties of stone (*see* Appendix B) and are the accepted industry standards.

Lump-sum or unit-price allowances may be used for specifying masonry units, but the specifications should also include sufficient information (including unit size, grade, and type) so that the contractor can accurately bid the labor required for installation. Most ASTM standards for masonry products cover two or more grades and types of units, so the project specifications must identify what is required. Omitting this information makes it impossible for bidders to estimate cost accurately.

The size of unit required should always be included in the specifications, preferably giving actual rather than nominal dimensions to avoid ambiguity. In some industries, "nominal" means approximate, but in modular masonry, it means the manufactured dimension of the unit plus the thickness of one mortar joint. A nominal 8-in. modular brick can be manufactured at 7½ in. for use with ½-in. joints, or 7-5/8 in. for use with 3/8-in. joints. Dimensions should be listed with thickness first, followed by the face dimensions of height and then length. Color and texture are not included in ASTM standards, so requirements must be established by the specifications. If an allowance method is used, the final selection may be made from samples submitted by the contractor or supplier. If trade names are used to identify a color range and finish or if descriptions are not given in the project specifications, samples of acceptable materials should be available to the contractors for inspection prior to bidding.

The specification guidelines that follow may be used as a checklist for the primary items requiring attention in the specifications. If more than one masonry system is used on a project, sections should be combined to include the mortar, units, and accessories for each system under a separate heading (e.g., "Veneer Masonry System" or "Reinforced Unit Masonry System"). This makes it clear to the contractor which anchors or ties go where, what mortar type, flashing, and so on, for each system.

16.1.1 Mortar and Grout

- Portland cement: ASTM C150 Type I, or Type III for cold weather, low alkali content, nonstaining
- Masonry cement: ASTM C91 (list acceptable manufacturers)
- Mortar cement: ASTM C1329 (list acceptable manufacturers)
- Hydrated lime: ASTM C207, Type S
- Sand: ASTM C144, clean and washed
- Grout aggregates: ASTM C404
- Water: clean and potable
- Admixtures: no calcium chloride permitted (list others permitted or prohibited)
- Mortar type: ASTM C270, Type (M, S, N, O, or K), proportion specification (default) or property specifications (minimum compressive strength for structural masonry)
- Grout type: ASTM C476 (fine or coarse)

16.1.2 Masonry Accessories

- Metals
  - Cold-drawn steel wire: ASTM A82
  - Welded steel wire fabric: ASTM A185 or ASTM A497
  - Sheet metal: ASTM A366
  - Plate, headed, and bent bar ties: ASTM A36
- Reinforcing steel
  - Billet steel deformed bars: ASTM A615
  - Rail steel deformed bars: ASTM A616
  - Axle steel deformed bars: ASTM A617
- Corrosion protection
  - Stainless steel: ASTM A167, Type 304
  - Hot-dip galvanized steel: ASTM A153, Class B
- Masonry ties: manufacturer, model number, type of metal
- Veneer anchors: manufacturer, model number, type of metal
- Fasteners: list appropriate types
- Joint reinforcement: ASTM A951, wire gauge, type (ladder or truss), corrosion protection (see above)
- Accessories: through-wall flashing, weep-hole inserts, drainage accessories, control joint shear keys, compressible expansion joint filler, cleaning agents

16.1.3 Masonry Units

- Facing brick: ASTM C216, Grade, Type (FBX, FBS, or FBA), unit size, color and texture, manufacturer, minimum compressive strength
- Glazed brick: ASTM C1405, Class (exterior or interior), Grade (S or SS), Type (I or II), unit size, color and texture, manufacturer, minimum compressive strength

- Building brick: ASTM C62, Grade SW, unit size, minimum compressive strength
- Hollow brick: ASTM C652, Grade SW, Type (HBX, HBS, HBA, or HBB), unit size, color and texture, manufacturer, minimum compressive strength
- Hollow or solid loadbearing *concrete masonry units* (CMU): ASTM C90, weight (normal, medium, or light), unit size, color and texture (architectural block only), minimum compressive strength
- Non-loadbearing CMU: ASTM C129, weight (normal, medium, or light), unit size, color and texture (architectural block only), minimum compressive strength
- Concrete brick: ASTM C55, Grade (N or S), weight (normal or light), unit size, color and texture, manufacturer, minimum compressive strength
- Water-repellent CMU shall comply with the performance criteria of *National Concrete Masonry Association* (NCMA) TEK 19-7

16.1.4 Construction

- Preconstruction conference
- Submittals, sample panels, mock-ups, testing
- Storage and protection of materials, hot and cold weather protection requirements
- Tolerances for placement and alignment of masonry
- Mortar mixing, retempering, placement, joint tooling, and pointing
- Wetting of brick with high initial rate of absorption (IRA), unit blending, unit placement
- Installation of flashing and weep holes, connectors, joint reinforcement, control joints, and/or expansion joints
- Placing reinforcement, grouting methods
- Temporary bracing and shoring, protection during construction, protection of finished work, moist curing

16.1.5 Quality Control Tests

Laboratory testing of materials and assemblages is usually limited to structural masonry rather than veneer systems. Mortar, grout, and masonry prisms may all be tested before construction to establish quality standards and tested during construction to verify compliance (refer to Chapter 17). Tests may also be used as part of the material selection process.

When mortar is specified to have a certain minimum compressive strength for structural masonry, it is required to meet the property specification of ASTM C270, *Mortar for Unit Masonry* rather than the default proportion specification. To verify that the contractor's proposed mortar mix meets the strength requirements, a sample can be tested in accordance with ASTM C270, but the results will not be comparable for testing later field samples because the methods of preparing the laboratory sample are not the same as those used in the field. If subsequent testing of field samples will also be required, both preconstruction and construction testing should be done in accordance with ASTM C780, *Standard Test Method for Preconstruction and Construction Evaluation of Mortars for Plain and Reinforced Unit Masonry*. The preconstruction test sets a quality standard against which field samples may be compared. ASTM C780 actually includes several different types of tests, including compressive strength, board life, mortar-aggregate ratio, water content, air content, and tensile strength. Specify only those tests that are needed.

Grout testing before and during construction can be done by a single test, ASTM C1019, *Standard Method of Sampling and Testing Grout*, which applies to both laboratory-prepared and field-prepared samples.

The compressive strength (f'_m) of structural masonry may be verified by the unit strength method or by the prism test method (refer to Chapter 14).

If f'_m must be verified by the prism test method rather than the unit strength method, an assemblage of the selected units and mortar must be constructed and tested in accordance with ASTM C1314, *Standard Test Method for Constructing and Testing Masonry Prisms Used to Determine Compliance with Specified Compressive Strength of Masonry*. This test may be used both for preconstruction and construction evaluation of the masonry.

16.1.6 Sample Panels and Mock-Ups

A sample panel is defined as a site-constructed panel of masonry to be used as a basis of judgment for *aesthetic* approval of the appearance of the materials and workmanship. Judging the appearance of masonry can be very subjective, but there are several basic things that should be considered:

- Compliance with allowable unit chippage and warpage
- Compliance with allowable size tolerances
- Unit placement
- Mortar joints and tooling
- Overall workmanship

Typical sample panels range in size from 4 × 4 ft to 4 × 6 ft or larger. The *Masonry Standards Joint Committee* (MSJC) *Specification for Masonry Structures* (ACI 530.1/ASCE 6/TMS 602) requires a minimum sample panel size of 4 × 4 ft. Larger panels that incorporate technical as well as aesthetic criteria can be more effective in establishing project standards.

A mock-up panel goes a step beyond the sample panel because it includes other elements of the work not related to aesthetics (refer to Chapter 17). Mock-ups may be required instead of or in addition to sample panels. They may serve the dual purpose of setting criteria for both aesthetic and technical consideration, and they may also be built for testing purposes. Mock-ups should include all of the basic components of the masonry system and backing wall, including reinforcement, connectors, shelf angles, flashing, weep holes, and expansion and control joints. If more elaborate mock-ups are required to show specific areas or details of the work such as window detailing, the panels should be delineated on the drawings or described adequately in the specifications to clearly identify the work required. Mock-up panels are often larger than sample panels. The size will vary with complexity, but a basic panel without a window element or other special components should be at least 6 ft wide × 8 ft high.

Sample panels and mock-ups can be built freestanding or as part of the permanent construction. If freestanding, they should be located where they will not interfere with subsequent construction or other job-site activities because they must remain in place until the masonry work has been completed and accepted. Sample panels and mock-ups should be constructed early enough in the construction schedule to allow for rejection and reconstruction without delay to the work.

Because many of the items required in a mock-up will be concealed, and because acceptance is based on procedure as well as appearance, the architect or engineer should try to be present during construction of the panel to observe the work. Documentation of concealed elements and procedural items may best be accomplished by photographing the work in progress. A cursory examination of a completed mock-up panel will tell the observer nothing about what's inside the wall (or isn't inside the wall). Acceptance on such a basis does not give adequate criteria on which to accept or reject the project masonry.

Specifications typically say too little about sample panels and mock-ups. The construction documents should allow bidders to accurately estimate the cost of constructing the mock-up. Size and number of panels required and all of the components to be included should be specified. Complex mock-ups that include various design elements should be illustrated on the drawings in plan, elevation, and section and referenced to specific project details. The specifications should designate the accepted mock-up as the project standard. They should also clearly establish the aesthetic and technical criteria on which acceptance or rejection of the panel will be based, as well as the person who will be responsible for evaluation (i.e., architect, engineer, construction manager, independent inspector, owner, etc.). Only specified products and materials or accepted substitutes should be used to construct the mock-up. Units should represent the full range of color variation to be expected in the project. Mortar ingredients, including sand and water, should also be those that will be used for project construction, as they have a significant effect on mortar color. The specification should also stipulate that the panel be built by a mason whose work is typical of that to be expected in the finished wall. A mason contractor would not be wise to assign the best available bricklayer to build the sample, because if the rest of the crew cannot match that workmanship, there may be a basis for rejection of the finished work. Before construction of a sample panel or mock-up begins, all project submittals should be reviewed for conformance to contract document requirements, and any required preconstruction testing should be complete.

16.2 Specifying with the MSJC Code

The *International Building Code* (IBC) requirements for masonry construction are based primarily on the MSJC *Building Code Requirements for Masonry Structures* (ACI 530/ASCE 5/TMS 402), which is jointly written by the *American Concrete Institute* (ACI), the *American Society of Civil Engineers* (ASCE), and *The Masonry Society* (TMS). IBC 2000 is based on the 1999 MSJC code, IBC 2003 is based on the 2002 MSJC code, and so on. The MSJC Code includes ACI 530.1/ASCE 6/TMS 602, *Specification for Masonry Structures*, as part of the Code.

The MSJC Specification establishes a minimum quality standard for materials and construction and attempts to ensure a level of testing and inspection commensurate with that required for concrete and steel structures. The document, however, must be coordinated with individual project specifications to avoid overlaps, duplications, conflicts, and omissions.

The MSJC Specification is intended to be "modified and referenced" in the project specifications. Individual sections, articles, or paragraphs should not be copied into the project specifications, as taking them out of context may change their meaning.

The project specifications may stipulate more stringent requirements. They must supplement the MSJC Specification in order to customize its application to each particular project and design.

The MSJC Specification is not written as a guide specification with instructions or recommendations to the specifier. There is a commentary published with the MSJC Code and the Specification, which gives some background information and suggestions on using the standards. A much more comprehensive handbook has been written by The Masonry Society entitled *Masonry Designers' Guide*. A *Masonry Designers' Guide* has been published for each edition of the Code and Specification published. The MSJC Specification also includes a checklist of mandatory items to which the specifier must respond and a checklist of optional items where methods and materials other than the standard requirements may be specified. Items required in addition to the MSJC Specification must also be addressed in the project specifications. The MSJC Specification must be well coordinated with the project specifications (including Division 1 requirements) to avoid overlaps, duplications, conflicts, and omissions.

The Code mandates use of the MSJC Specification and at the same time states that the MSJC Specification does not govern where different provisions are specified. This allows the specifier to alter requirements through the project specifications. Although the intent is to permit the project specification to impose more stringent requirements, it is equally possible that less stringent requirements could be specified, and these would take precedence over the MSJC specification.

There are many items not mentioned in the MSJC Specification that still must be covered in the project specifications. Among these are delivery, storage, handling, and protection of materials; placement requirements for flashing and weep holes; and protection of walls during construction. Coordinate your office master specifications with the requirements of the MSJC Specification to make sure that all Specification material and workmanship requirements are covered.

The Masonry Society's *Annotated Guide to Masonry Specifications* is the best resource available for detailed description and discussion of typical masonry specification requirements. It is based on the MSJC Code and Specification and is intended to guide the specifier through the many decisions required in compiling masonry project specifications. The following discussion is intended to provide general guidance on preparing project specifications that must be coordinated with the MSJC Specification.

16.2.1 General

References

The correct title, document number, issuing body, and date of the MSJC Specification should be given in the list of references. The MSJC Specification includes a list of ASTM references, which should be checked for conflicts and omissions. The mandatory checklist then requires that sections, parts, and articles of the MSJC Specification excluded from the project specifications be indicated, and that articles at variance with the project specifications be listed. This list will vary for each project.

Quality Assurance

The checklists use the term "quality assurance" ambiguously to indicate both construction submittals, inspection, and preconstruction testing. The mandatory checklist asks that the specifier define the submittal reporting and review procedure, which should be the same as requirements outlined in Division 1 of the project specifications. The mandatory checklist also requires that the specifier designate the quality assurance level appropriate to the project (refer to Chapter 17 for a discussion of MSJC level A, level B, and level C quality assurance). The level of quality assurance designated includes minimum requirements for testing, submittals, and inspection. Check the articles on inspection agency and testing agency services and duties for conflicts with Division 1 requirements.

Loadbearing Masonry

The mandatory checklist requires that the specifier designate when grout strength must be verified by test.

16.2.2 Products

Materials

The mandatory checklist contains a number of product-related items. The MSJC Specification lists all of the ASTM clay, concrete, and stone masonry unit and material standards that are applicable to structural masonry systems. The specifier must indicate which units will be used and specify the required grade, type, size, and color as applicable. Mortar and grout ingredients must be specified, including any acceptable admixtures. The type and grade of reinforcement are required by the MSJC Code to be shown on the drawings and by the MSJC Specification to be given in the project specifications. Wire fabric, if used, must be designated as either smooth or deformed. Whereas the MSJC Specification does list ASTM requirements for the materials used for anchors and ties, it does not specify the anchors and ties themselves. The exact types and sizes required for the project, including any proprietary products, must be given in the project specifications.

Although the MSJC Code includes design requirements for masonry veneers, only passing reference is made to flashing and weep holes. The MSJC Specification does not include material or installation requirements for these items, so flashing and weep holes must still be covered in the project specifications. All required accessories, including flashing and weep-hole materials, must be specified, as well as the size and shape of joint fillers and the size and spacing of pipes and conduits to be furnished and installed by the mason. If prefabricated masonry elements are used, specify any requirements supplemental to ASTM C901, *Standard Specification for Prefabricated Masonry Panels*.

Mixes

Specify grout requirements at variance with the MSJC Specification.

16.2.3 Execution

Preparation

The optional checklist asks the specifier to note when wetting of the masonry units is required to ensure good bond between units and mortar. However, the wording in the

MSJC Specification itself prescribes these limits correctly as high-suction clay masonry units with initial rates of absorption in excess of $1 \text{ g} \cdot \text{min}^{-1} \cdot \text{sq in.}^{-1}$ when tested in accordance with ASTM C67, *Standard Method of Sampling and Testing Brick and Structural Clay Tile*. Units should not be wetted when the IRA is acceptable, nor during winter construction.

Installation

There are several items on the optional checklist that apply to installation of the masonry. The specifier must indicate, first of all, if the pattern of units in the project is anything other than one-half running bond and if the joints are other than 3/8 in. Collar joints ¾ in. wide or less are to be solidly filled with mortar unless otherwise required by the project specifications. Face shell bedding of hollow units also governs except in piers, columns, pilasters, starting courses at the foundation, and at grouted cells or cavities, where cross-webs must also be mortared. If there are other locations that require full mortar bedding, these should be identified in the project specification. Variations from the standard full bedding requirements for solid units should also be noted, such as beveling to minimize mortar droppings in the cavity. The location of embedded sleeves for pipes and conduits should be shown on the drawings and only the requirements for their installation covered in the specifications. Requirements for the size and spacing of both veneer anchors and rigid and adjustable wall ties, if different from those in the MSJC Specification, should be given. The location and types of expansion and control joints are required to be indicated on the drawings.

The construction tolerances listed in the MSJC Specification are structural tolerances intended to limit the eccentricity of applied loads. For veneer and other exposed masonry applications, tighter tolerances for aesthetic considerations may be included in the project specifications.

Cleaning

If acid or other caustic cleaning materials are permitted, the optional checklist requires that the project specification cover methods of neutralization after cleaning.

CHAPTER 17
Quality Assurance and Quality Control

Building owners want to be sure that they are getting what they pay for in terms of the quality of a building's construction. Quality assurance programs and quality control procedures are used toward that end.

17.1 Standard of Quality

In construction projects, the particular standard of quality that will apply in a given case is established by and measured in terms of the contract document requirements. The owner initially establishes a general standard of quality, which is then developed by the architect/engineer into specific terms and incorporated into the contract documents. The standard of quality required on a given project will vary depending on the needs of the owner, the project type, and the established schedule and budget.

A standard of quality may be established in different ways, depending on the method of specifying. *Descriptive specifications* identify exact properties of materials and methods of installation without using proprietary names. *Proprietary specifications* list specific products, materials, or manufacturers by brand name, model number, and other proprietary information. *Reference standard specifications* stipulate minimum quality standards for products, materials, and processes based on established industry standards. *Performance specifications* establish a standard of quality by describing required results, the criteria by which performance will be judged, and the method(s) by which it can be verified.

17.2 Quality Assurance/Quality Control in Masonry

The *International Building Code* (IBC) and the *Masonry Standards Joint Committee* (MSJC) *Building Code Requirements for Masonry Structures* both contain specific mandated requirements for quality assurance. Both are based on type of facility and defined risk categories. The MSJC requirements are tabulated in *Figs. 17-1 and 17-2*. Both the IBC and the MSJC code require that the specifier designate Level A, Level B, or Level C quality assurance as appropriate to the project type and function.

17.2.1 Industry Standards for Masonry

Industry standards such as those developed and published by ASTM International are an important part of quality assurance and quality control in masonry construction.

Description	Minimum Required Quality Assurance Level		
	Level A	Level B	Level C
Non-essential facilities§ designed in accordance with empirical requirements, masonry veneers and glass unit masonry	X		
Essential facilities§ designed in accordance with empirical requirements, masonry veneers and glass unit masonry		X	
Non-essential facilities§ designed in accordance with allowable stress, strength design, or prestressed masonry requirements		X	
Essential facilities§ designed in accordance with allowable stress, strength design, or prestressed masonry requirements			X

FIGURE 17-1 Minimum level of quality assurance tests, submittals, and inspections required by the *MSJC*, Building Code Requirements for Masonry Structures, *ACI 530/ASCE 5/TMS 402*, and Specification for Masonry Structures, *ACI 530.1/ASCE 6/TMS 602, 2005 edition*.

Some standards establish minimum requirements for products or systems, and others outline standardized testing procedures for verifying compliance with the requirements stated in the contract documents.

At last count, there were more than 80 ASTM standards on masonry and masonry-related products, with more in development. Most project specifications, however, require reference only to a core group of standards that apply to the most frequently used products and systems. Because there are so many different products and materials that fall under the umbrella of the term *masonry*, there are perhaps more standards than for other construction systems. Some standards, however, are embedded references within other standards and ordinarily do not require specific citation in project specifications. Others apply to specialty products such as sewer brick, chemical-resistant units and mortar, high-temperature refractory brick, and clay flue liners that are outside the scope of the typical design project. Still other standards are used primarily for research and product development rather than building construction.

Many ASTM standards cover more than one grade, type, or class of material or product from which the specifier must choose. Some also contain language designating which requirements govern by default if the project specifications fail to stipulate a preference. The following summary of standards should serve as a checklist in preparing project specifications and developing a quality assurance and quality control testing program.

17.2.2 Standards for Clay Masonry Units

ASTM C216, *Standard Specification for Facing Brick (Solid Masonry Units Made of Clay or Shale)*

Face bricks are solid clay units for exposed applications where the appearance of the brick is an important consideration in the design. "Solid units" are defined as those with a maximum cored area of 25%. ASTM C216 covers two grades and three types of

Quality Assurance	Minimum Tests and Submittals	Minimum Inspection
Level A	Certificates for materials used in masonry construction indicating compliance with the contract documents	Verify compliance with the approved submittals
Level B	Certificates for materials used in masonry construction indicating compliance with the contract documents Verification of f'_m prior to construction, except where specifically exempted by code	As masonry construction begins, verify the following are in compliance • proportions of site-prepared mortar • construction of mortar joints • location of reinforcement, connectors, and prestressing tendons and anchorages Prior to grouting, verify the following are in compliance • grout space • grade and size of reinforcement, prestressing tendons and anchorages • placement of reinforcement, connectors, and prestressing tendons and anchorages • proportions of site-prepared grout and prestressing grout for bonded tendons • construction of mortar joints Verify that the placement of grout and prestressing grout for bonded tendons is in compliance. Observe preparation of grout specimens, mortar specimens, and/or prisms Verify compliance with the required in spection provisions of the contract documents and the approved submittals.
Level C	Certificates for materials used in masonry construction indicating compliance with the contract documents Verification of f'_m • prior to construction • every 5000 sq.ft. during construction Verification of proportions of materials in mortar and grout as delivered to the site	From the beginning of masonry construction and continuously during construction of masonry, verify the following are in compliance • proportions of site-mixed mortar, grout, and prestressing grout for bonded tendons • grade and size of reinforcement, prestressing tendons and anchorages • placement of masonry units and construction of mortar joints • placement of reinforcement, connectors, and prestressing tendons and anchorages • grout space prior to grouting • placement of grout and prestressing grout for bonded tendons Observe preparation of grout specimens, mortar specimens, and/or prisms Verify compliance with the required in spection provisions of the contract documents and the approved submittals.

FIGURE 17-2 Requirements for quality assurance levels A, B, and C. (*From MSJC,* Building Code Requirements for Masonry Structures, *ACI 530/ASCE 5/TMS 402, and* Specification for Masonry Structures, *ACI 530.1/ASCE 6/TMS 602, 2005 edition.*)

face brick. Brick *type* designates size tolerance and allowable chippage and distortion based on desired appearance. Type FBS (standard) is the industry standard and the type of face brick used in most commercial construction. Type FBX (select) has tighter size tolerances and less allowable chippage for use in applications where a crisp, linear appearance is desired such as stack bond masonry. Type FBA (architectural) is nonuniform in size and texture, producing characteristic "architectural" effects such as those typical of, or required to simulate, hand-made brick. Type FBA is popular for residential masonry styles because of its softer profile and less commercial look. All three brick types must meet the same strength and physical property criteria, but brick type is not related to color or color range. If the project specifications do not identify a specific proprietary product or designate brick type, ASTM C216 states that type FBS standards shall govern.

Brick *grade* classifies units according to their resistance to damage from freezing when they are wet. The property requirements for Grades SW and MW are given in a table that covers minimum compressive strength, maximum water absorption, and minimum saturation coefficient. These properties are tested in accordance with ASTM C67, *Methods of Sampling and Testing Brick and Structural Clay Tile.* Because ASTM C67 is referenced in ASTM C216, it is not necessary for the specifier to list ASTM C67 as a separate reference standard. If the brick is specified to meet the requirements of ASTM C216 that automatically requires that the units be tested for compliance in accordance with ASTM C67 methods and procedures.

In general, Grade SW (severe weathering) should be specified when a "high and uniform" resistance to damage from cyclic freezing is required and when the brick is likely to be frozen when it is saturated with water. Grade MW (moderate weathering) should be specified where only moderate resistance to damage from cyclic freezing is required and when the brick may be damp but not saturated when freezing occurs. ASTM C216 includes a table of grade recommendations for various types of exposure and a related map of geographic weathering regions. If the project specifications do not designate the required grade, Grade SW is the default standard, and Grade SW may be substituted by the supplier if Grade MW is specified.

Grade SW brick is required by ASTM C216 to have a minimum average *gross area* compressive strength of 3000 psi and Grade MW brick a minimum average *gross area* compressive strength of 2500 psi. These strengths are more than adequate for most loadbearing and non-loadbearing applications, and the majority of brick produced in the United States and Canada is much stronger. If a specific unit strength requirement greater than the standard minimum is required, that compressive strength should be required by the project specifications.

ASTM C216 also requires that brick be tested for efflorescence in accordance with ASTM C67 and be rated "not effloresced." Color is not covered in this standard, so the specifier must designate the desired color by specifying a proprietary product, with color and color range verified with a sample panel or mock-up panel.

ASTM C62, *Standard Specification for Building Brick (Solid Masonry Units Made from Clay or Shale)*

Building brick (sometimes called common brick) is used primarily for utilitarian applications or as a backing for other finishes, where strength and durability are more important than appearance. ASTM C62 covers Grades SW and MW on the basis of the same physical requirements for durability and resistance to freeze-thaw weathering as those for face brick. Building brick is also available in Grade NW (no weathering), which is permitted only for interior work where there will be no weather exposure.

This standard lists permissible variations in size but does not classify units by various types. The size tolerances listed apply to all ASTM C62 brick. Because these units are generally used in unexposed applications, there is no requirement for efflorescence testing. The discussion of compressive strength requirements under ASTM C216 above also applies to building brick.

ASTM C652, *Standard Specifications for Hollow Brick (Hollow Masonry Units Made from Clay or Shale)*

ASTM C652 covers hollow brick with core areas between 25 and 40% (Class H40V) and between 40 and 60% (Class H60V). The two grades listed correspond to the same requirements for durability as for face brick—Grade SW (severe weathering) and Grade MW (moderate weathering). Types HBX (select), HBS (standard), and HBA (architectural) are comparable with face brick types FBX, FBS, and FBA, respectively. Another type, HBB, is for general use where appearance is not a consideration and greater variation in size is permissible. Type HBB is the hollow brick equivalent of ASTM C62 building brick. When the project specification does not designate brick type, requirements for Type HBS govern. The default standard is Grade SW. This standard does include requirements for efflorescence testing the same as for ASTM C216 face brick. The discussion of compressive strength requirements under ASTM C216 also applies to hollow brick.

Color is not covered in this standard, so the specifier must designate the desired color by specifying a proprietary product, with color and color range verified with a sample panel or mock-up panel.

ASTM C1405, *Standard Specification for Glazed Brick (Single-Fired, Solid Brick Units)*

Most glazed brick is single-fired with a glaze that is applied during the normal firing process rather than after the unit itself is fired. ASTM C1405 covers physical requirements for the brick body and includes Grade S (select) and Grade SS (select sized or ground edge), where a high degree of mechanical perfection and minimum size variation is required. Units may be either Type I, single-faced, or Type II, double-faced. Weathering properties are specified as Exterior Class or Interior Class. Properties of the glaze and tolerances on dimension and distortion are covered as well as strength and durability requirements.

17.2.3 Standards for Concrete Masonry Units

ASTM C90, *Standard Specification for Loadbearing Concrete Masonry Units*

Hollow and solid loadbearing concrete blocks are covered in this standard. Weight classifications are divided into lightweight (less than 105 lb/cu ft oven dry weight of concrete), medium weight (105 to less than 125 lb/cu ft), and normal weight (125 lb/cu ft or more). Unit weight affects water absorption, sound absorption, sound transmission, and thermal and fire resistance. There are no default requirements in ASTM C90, so the specifier must designate unit type and weight classification if these properties are important to the design.

The minimum *net area* compressive strength required for all three weight classifications for ASTM C90 loadbearing units is 1900 psi. Compressive strength is largely a function of the characteristics of the aggregate used in the units and may vary regionally according to the types of aggregates available. Aggregates in some areas may routinely produce units with much higher compressive strengths without a cost premium. If a specific unit strength requirement greater than the standard minimum is required, that compressive strength should be required by the project specifications.

Compliance with the requirements of ASTM C90 is verified by testing in accordance with ASTM C140, *Sampling and Testing Concrete Masonry Units.* ASTM C140 is referenced in the ASTM C90 standard and need not be listed separately in the project specification. ASTM C90 also references ASTM C33, *Aggregates for Concrete,* and ASTM C331, *Lightweight Aggregates for Concrete Masonry Units,* as well as standards for the cementitious materials that are permitted in these units. It is not necessary for the specifier to list these referenced standards separately.

Size tolerances and limits on chippage and cracking are covered in the text of the standard. These requirements are more liberal than those for clay brick because of the nature of the material and the method of manufacture. For exposed architectural units such as split-face, ribbed, or ground-face units, these requirements may not be appropriate. Rough units may require greater tolerances and ground-face units tighter tolerances. For such products, it may be more appropriate to consult local manufacturers for tolerance requirements. Color is not covered in this standard, so the specifier must designate the desired color by specifying a proprietary product, with color and color range verified with a sample panel or mock-up panel.

ASTM C129, *Standard Specification for Non-Loadbearing Concrete Masonry Units*

The requirements of this standard are similar to those of ASTM C90 except that the units are designed for non-loadbearing applications. Unit weight classifications are the same, as are referenced standards for aggregates, cements, sampling, and testing. Because the units are designated as non-loadbearing, the minimum requirements for net area compressive strength are lower than for ASTM C90 units at an average of only 600 psi. For typical non-loadbearing applications, this strength is more than adequate, but stronger units may be commonly available without a cost premium in some areas. Color requirements are not covered in the specification and should be specified in the same way as that recommended for ASTM C90 units.

ASTM C55, *Standard Specification for Concrete Building Brick*

Concrete brick can be loadbearing or non-loadbearing. Grading is based on strength and resistance to weathering. Grade N provides "high strength and resistance" to moisture penetration and severe frost action. Grade S has only "moderate strength and resistance" to frost action and moisture penetration. Minimum gross area compressive strength for Grade N units is 3500 psi and for Grade S units is 2500 psi. ASTM C55 does not include requirements for color, texture, weight classification, or other special features. These properties must be covered separately in the project specifications. Sampling and testing are referenced to ASTM C140, and standards for aggregates and cements are also referenced, so the specifier need not list these separately.

17.2.4 Standards for Masonry Mortar and Grout

ASTM C270, *Standard Specification for Mortar for Unit Masonry*

This standard covers four types of masonry mortar made from a variety of cementitious materials, including portland cement (ASTM C150), mortar cement (ASTM C1329), and masonry cement (ASTM C91), as well as blended hydraulic cement and slag cement (ASTM C595), quicklime (ASTM C5), and hydrated masonry lime (ASTM C207). These material standards are referenced in ASTM C270, so the specifier need not list them separately. If any materials are to be excluded for any reason, this should be noted in the project specifications. Requirements for mortar aggregates are referenced to ASTM C144.

Types M, S, N, and O mortar may be specified to meet either the proportion requirements or the property requirements of ASTM C270. If the project specifications do not designate which method the contractor must use, then the proportion method governs by default. The proportion method is the most conservative and will usually produce mortars with higher compressive strengths than those required by the property method. It is generally not desirable to use mortar that is stronger in compression than the application requires. To optimize mix design, property-specified mortar requires preconstruction laboratory testing in accordance with the test methods included in ASTM C270. These test methods are not suitable for testing of field-sampled mortar during construction and cannot be compared with the results of such tests. If field testing of mortar will be required, then *both* preconstruction and construction phase testing should be performed in accordance with ASTM C780 rather than ASTM C270. There is no test method for accurately measuring the compressive strength of hardened mortar removed from a completed masonry wall or structure.

Recommendations for appropriate use of the four basic mortar types are included in a nonmandatory Appendix X1 to ASTM C270 (*Selection and Use of Mortar for Unit Masonry*) and are summarized in Chapter 3.

ASTM C476, *Standard Specification for Grout for Masonry*

This standard covers two types of masonry grout—fine and coarse. The same standards for cementitious materials are referenced as those in ASTM C270, but aggregates must conform to ASTM C404. Fine grout is used for small grout spaces and coarse grout for economy in larger grout spaces (*see* Chapter 3). Masonry grout may be specified to meet the proportion requirements included in the standard or it may be required to have a minimum compressive strength of 2000 psi when sampled and tested in accordance with ASTM 1019. If higher compressive strength is required for structural masonry, the required strength should be indicated in the contract documents.

ASTM C476 permits the use of "pumping aid" admixtures in cases where the brand, quality, and quantity are approved in writing. Such admixtures are commonly used in high-lift grouting projects, as are certain other types of admixtures.

17.2.5 Standards for Masonry Accessories

ASTM A82, *Standard Specification for Cold Drawn Steel Wire for Concrete Reinforcement*

This standard covers steel wire that is used in prefabricated joint reinforcement and some types of masonry anchors and ties. It includes strength requirements and permissible variations in wire size but does not include any options, which the specifier must designate in the project documents.

ASTM A951, *Standard Specification for Joint Reinforcement for Masonry*

This standard covers material properties, fabrication, test methods, and tolerances for prefabricated wire joint reinforcement for masonry. The specifier must designate corrosion protection as basic, mill galvanized, Cass I mill galvanized (minimum 0.40 oz zinc per square foot of surface area), Class III mill galvanized (minimum 0.80 oz zinc per square foot of surface area), or hot-dipped galvanized (minimum 1.50 oz zinc per square foot of surface area). The hot-dip galvanizing is the same as that required for under ASTM A153 as listed below and is required by code for joint reinforcement used in exterior walls.

ASTM A153, *Standard Specification for Zinc Coating (Hot-Dip) on Iron or Steel Hardware*

This standard covers hot-dip galvanized coatings that are required to provide corrosion resistance in exterior wall applications for masonry accessories such as steel joint reinforcement, anchors, and ties. Minimum coating weight is given in four classes based on the size and type of item being coated. Masonry accessories of various sizes are covered under Class B.

ASTM A167, *Standard Specification for Stainless and Heat Resisting Chromium-Nickel Steel Plate, Sheet, and Strip*

This standard covers stainless steel of the type that is used for masonry anchors, ties, and flashing. There are more than two dozen types of stainless steel included in the standard, varying according to chemical and mineral composition. Type 304 is the type most commonly used in masonry construction. Type 316 is also sometimes used in masonry.

17.2.6 Standards for Laboratory and Field Testing

ASTM C780, *Standard Test Method for Preconstruction and Construction Evaluation of Mortars for Plain and Reinforced Unit Masonry*

ASTM C780 covers methods for sampling and testing mortar for its plastic and hardened properties either before or during construction. If construction-phase testing of mortar will be required, there must be some basis for comparison of the results of such tests. The compressive strengths and other requirements listed under the property specification of ASTM C270 or ASTM C1142 cannot be used for comparison because the test methods are different. The laboratory test methods used in ASTM C270 mix mortar samples with a relatively low water-cement ratio. Field-mixed mortars, however, use much higher water-cement ratios to overcome the initial absorption of the masonry units. When compared with one another, the field-mixed mortars would appear to be much weaker than the ASTM C270 test results. To provide an "apples-to-apples" comparison, the preconstruction design mix must also be tested with a high water-cement ratio to simulate that which will actually be prepared during construction. Using ASTM C780 to obtain a preconstruction benchmark for the mortar provides a basis for acceptance or rejection of field-sampled mortars during construction.

ASTM C1019, *Standard Test Method for Sampling and Testing Grout*

This standard covers both field and laboratory sampling for compressive strength testing of masonry grout. This standard should be referenced in the project specifications for loadbearing masonry construction if the compressive strength of the masonry construction is to be verified by either the unit strength method or prism test method.

ASTM C1314, *Standard Test Method for Constructing and Testing Masonry Prisms Used to Determine Compliance with Specified Compressive Strength of Masonry*

In structural masonry projects, the engineer must indicate on the drawings the required compressive strength of masonry ($f'm$) on which the design is based. The contractor must verify to the engineer that the construction will achieve this minimum compressive strength. This verification may be provided in one of two ways—the unit strength method or the prism test method. The unit strength method is very conservative and is based on the empirical assumption that the combination of certain mortar types with units of a certain compressive strength will produce masonry of a given strength. If the manufacturer submits certification of the unit compressive strength and the mortar is specified by the ASTM C270 proportion method, compressive strength verification can be provided by a table in the code without any preconstruction or construction testing of any kind. If the mortar was specified by the ASTM C270 property method, the mortar test discussed earlier, along with the manufacturer's certification of unit strength, is sufficient to verify compressive strength compliance. If $f'm$ must be verified by the prism test method, an assemblage of the selected units and mortar must be constructed and tested in accordance with ASTM C1314. This test may be used both for preconstruction and construction evaluation of the masonry. Although the ASTM C1314 test method is similar to other compressive strength test methods, ASTM C1314 does not require any extraneous information other than that required for verification of the specified compressive strength.

17.3 Masonry Submittals

Submittals are a time-consuming but important part of construction projects. Submittals are used to help ensure that the work meets the requirements of the contract documents, and that the contractor achieves the standard of quality established by the specifications. For each project, the architect or engineer must decide what submittals are needed for each portion of the work. Submittals require time and money to prepare and process (for both the architect/engineer and the contractor), so it is important that only those submittals that are appropriate and necessary to the work be required.

The types of submittals that are appropriate or necessary will vary from project to project according to the nature of the construction, both aesthetic and structural. For masonry projects designed under the MSJC *Building Code Requirements for Masonry Structures* (ACI 530/ASCE 5/TMS 402), some submittals are mandatory. Projects that are nonstructural but aesthetically important may lean more toward submittal of unit and mortar samples than test reports. Each project is unique in its requirements.

17.3.1 Specifying Submittals

According to the *Construction Specifications Institute* (CSI) *Manual of Practice*, administrative and procedural requirements for submittals should be specified in "Division 1—General Requirements" because they apply to all project submittals. Administrative requirements for all submittals would include information such as the number of copies required, how much time should be allowed for review, and to whom reviewed submittals should be distributed.

Specific submittals required for a masonry project should be specified in the appropriate technical section in "Division 4." Each of the technical sections should include in

SHOP DRAWINGS	QUALITY ASSURANCE / QUALITY CONTROL SUBMITTALS

• Fabrication dimensions and placement locations for reinforcing steel and accessories
• Sheet metal flashing details
• Stone fabrication and setting drawings
 	
PRODUCT DATA
• Proprietary mortar ingredients
   • portland cement
   • masonry cement
   • mortar cement
   • lime
   • admixtures (including pigments)
• Connectors
• Joint reinforcement
• Flashing materials
• Weephole and drainage accessories
• Cleaning agents
 	
SAMPLES
• Masonry units
• Stone
• Mortar colors
• Connectors
• Flashing materials
• Accessories
	• Design data
	   • mortar mix designs (property specification only)
	   • grout mix designs (for required compressive strength only)
	• Test reports
	   • prism test (alternate method of verifying f'_m)
	   • preconstruction testing
	   • field testing
	• Certifications
	   • compliance with specified ASTM standards
	   • brick IRA
	• Inspection reports
	   • materials
	   • construction procedures
	   • reinforcement
	   • grouting
	   • protection measures
	• Manufacturer's instructions
	   • mortar admixtures
	   • mortar pigments
	   • cleaning agents
	• Manufacturer's field reports
	   • cleaning operations
	• Proposed hot and/or cold weather construction procedures

FIGURE 17-3 Masonry submittals checklist.

"Part 1" a complete list of the submittals required for that portion of the work. Submittals may include shop drawings, product data, samples, and quality assurance/quality control submittals. Each type of submittal has a different function and is applicable to different types of materials, products, or systems. *Figure 17-3* lists all of the types of submittals and submittal information that might be included in a masonry specification. The list will vary as appropriate to the project, the type of construction, and the wishes of the architect or engineer (A/E).

17.3.2 Submittal Procedures

Submittals must be reviewed and approved before construction of the related work can begin. Material and equipment cannot be ordered or fabricated until specified submittals are approved by both the contractor and the A/E. The general contractor (GC) or construction manager (CM) is responsible for submitting required information to the A/E for review and approval. Many of the required submittals may actually be prepared by subcontractors, suppliers, fabricators, or manufacturers and passed through the GC or CM.

The general contractor must check all submittals, stamp and sign them, assemble them with transmittal forms, and submit them to the A/E for review. Submittals that are not approved must be resubmitted with the required changes, reviewed, and approved before construction can begin. Both the A/E and the contractor should maintain a submittal log to track the progress of all project submittals. A copy of all approved submittals should be kept with the record documents at the job site until the project is complete. Both the A/E and the contractor usually retain copies of approved submittals as part of their permanent project records.

In masonry construction, it is the responsibility of the masonry subcontractor to prepare or assemble the required masonry submittals and turn them over to the general contractor. Manufacturers' literature on masonry accessories, product certifications on masonry units, or metal flashing details may sometimes be prepared by the supplier, manufacturer, or fabricator, respectively, for submittal by the masonry subcontractor to the general contractor.

17.3.3 Shop Drawings

Shop drawings are prepared to illustrate some details of the construction. They are typically prepared by a manufacturer or fabricator for use in producing items and as an aid to the contractor in coordinating the work with adjacent construction.

For example, structural engineering drawings typically show reinforcing steel only diagrammatically in plans and sections. The shop drawings show each size, dimension, and type of rebar and its configuration and splice details, as well as a key to its plan location and the quantity required. These drawings are then used in the steel fabricator's shop to prepare the individual elements needed at the project site. The engineer reviews the shop drawings for conformance to design and contract document requirements but does not generally check the quantities. Projects under the jurisdiction of the MSJC code are required to have shop drawings for structural reinforcing steel.

The A/E may also wish to have shop drawings submitted to illustrate metal flashing details such as end dams, corners, lap seals, and abutments with other construction. These drawings can then be used to fabricate the required flashing sections in the sheet metal shop for installation at the project site by the masons. Requiring shop drawings for flashing can help ensure that the contractor has anticipated and planned for all field installation conditions and has properly interpreted the drawing and specification requirements.

Loose steel angle lintels and prefabricated concrete lintels should require the submittal of shop drawings for verification of dimensions and coordination with masonry coursing. Projects with cut stone may have extensive shop drawings that identify each size and shape of stone, its anchorage conditions, and placement location. In grouted construction, the engineer may also require shop drawings showing the type of temporary construction that will be used to brace uncompleted walls.

17.3.4 Product Data

Fabricated products such as the accessories used in masonry construction typically require the submittal of manufacturers' product data rather than shop drawings. Many specifications list the products of several different manufacturers that are acceptable for use in the construction. Others specify products only by description or by reference standard without mentioning proprietary names. These methods of specifying make it necessary to require the submittal of proprietary product data to verify that the products that the contractor proposes to use meet the specified requirements. Masonry product data might include catalog sheets or brochures for anchors, ties, rebar positioners, joint reinforcement, weep-hole inserts, and shear keys. The masonry contractor or supplier who prepares the submittal should clearly mark data sheets that include more than one item to show which item or items are proposed for use. If there are various model numbers, materials, sizes, and so forth, these too should be marked to show the appropriate selection.

Manufactured products such as cement, admixtures, mortar coloring pigments, and cleaning agents may also be included in the A/E's list of required submittals. If more than one brand of proprietary masonry cement or mortar cement is approved for use on the project, the manufacturer's product literature should be submitted to indicate which particular products the contractor is proposing to use and to verify their conformance to contract document requirements. Product data on approved types of admixtures should clearly indicate the chemical ingredients included to ensure that they contain no calcium chloride or other harmful substances. Product data on mortar coloring pigments and proprietary cleaning agents should also be submitted for review and approval.

17.3.5 Samples

Samples may be required for masonry units, colored mortars, and some selected accessory items. Unit samples are most often reviewed for color selection purposes during earlier design phases, but if the masonry has been specified on a unit price basis or only by ASTM reference standard, the A/E must approve samples submitted by the contractor. Cut stone, brick, and architectural CMU samples should indicate the full range of color, texture, shape, and size. Any project requirements for sample panels or mock-ups should be specified under the quality assurance article of "Part 1" rather than under this article, which is reserved for individual unit or material samples.

17.3.6 Quality Assurance/Quality Control Submittals

Quality assurance and quality control submittals include test reports, manufacturers "or contractors" certifications, and other documentary data. These submittals are usually for information only. They are processed in the same manner as shop drawings and product data but do not always require review and approval.

If mortar is specified by ASTM C270 property requirements for compressive strength rather than the default proportion specification, mix *design data* should be submitted for review, along with the results of *preconstruction tests* verifying compliance with the required compressive strength. Grout mixes that are required to attain a specified compressive strength should also require mix design and test result submittals. The results of preconstruction tests must be available for comparison with the results of any *field tests* that may be required because they are the only valid criteria against which field test results can be compared. For structural masonry projects on which the contractor chooses to verify the strength of masonry by *prism tests*, these results should also be submitted.

The A/E may sometimes require that a manufacturer or fabricator perform testing of a specific product lot, run, shipment, and so forth. For example, masonry unit manufacturers might be required to submit test results verifying compliance with specified properties such as compressive strength or absorption. For structural masonry projects on which the contractor chooses to verify the strength of masonry by the unit strength method, these unit strength test results should be compared with minimum requirements listed in the code tables. This type of submittal is called a *source quality control submittal*.

Instead of laboratory test results, some products may be submitted with written *certification* from the manufacturer that the item complies with specified requirements. Certifications are usually in the form of a letter and require the signature of an authorized company representative. The MSJC *Specification for Masonry Structures* requires that in addition to reinforcing steel shop drawings, certifications of compliance be submitted for each type and size of reinforcement, anchor, tie, and metal accessory to be used in the construction. Certification of unit, mortar, and grout materials may also be required instead of test results for some projects that do not involve structural masonry elements.

On structural masonry projects where field inspection is provided by someone other than the project engineer, the specifications should require submittal of *inspection reports* on materials, protection measures, construction procedures, reinforcing steel placement, and grouting operations. If the project engineer is doing field inspections, the same type of information may be kept on file as field notes.

For some products such as cleaning agents or mortar coloring pigments, the A/E may require submittal of *manufacturer's instructions* for application, mixing, or handling of materials. Hazardous materials should require submittal of material safety data sheets. *Manufacturer's field reports* are also sometimes required if the A/E wants to verify that a representative of the cleaning agent manufacturer, for instance, has visited the project site to inspect substrate conditions or to instruct the contractor in the application of certain materials or cleaning methods.

Finally, the A/E may require the submittal of proposed *hot and/or cold weather construction procedures* to meet the requirements stipulated by the MSJC ode and MSJC Specification. The contractor's submittal should describe the specific methods and procedures proposed to be used to meet these requirements.

17.3.7 Closeout Submittals

Closeout submittals include such things as record documents, extended warranty information, maintenance instructions, operating manuals, and spare parts. Masonry construction does not usually involve this type of documentation. If coatings or clear water repellents are used to reduce the surface absorption of the masonry, and if those materials carry a manufacturer's extended warranty, the information would be submitted by the applicator of the material rather than by the masonry contractor.

17.4 Sample Panels and Mock-Ups

Sample panels and mock-ups are an important part of quality assurance programs. They can also be an effective tool of communication between the design office and the job site in setting both technical and more qualitative aesthetic standards. For aesthetic

Figure 17-4 Traditional masonry sample panels for evaluating unit color, color range, mortar color, joint tooling, and general workmanship.

criteria, sample and mock-up panels are the *only* practical and effective method of establishing a fair and equitable procedure for evaluating the completed work. For technical criteria, mock-ups provide a well-defined yardstick for measuring performance without dispute.

17.4.1 Sample Panel

A sample panel is defined as a site-constructed panel of masonry to be used as a basis of judgment for *aesthetic* approval of the appearance of the materials and workmanship (*see Fig. 17-4*). Sample panels should not routinely be used to make design decisions on color, bond pattern, or joint type unless the work of constructing multiple sample panels has been contracted separately from the project construction based on a unit price per panel. Color matching masonry on renovation or rehabilitation projects may require numerous panels to make such decisions. Design panels should be constructed very early to allow time for procurement of the selected materials.

17.4.2 Mock-Up

A mock-up is more than just the units and mortar of the traditional masonry sample panel. Mock-ups also incorporate other elements of the project masonry including, as appropriate, backing wall, reinforcing steel, shelf angles or supports, ties or anchors, joint reinforcement, flashing, weep holes, and control or expansion joints (*see Fig. 17-5*). Design elements such as windows or parapets that may be considered critical aesthetically or from a performance standpoint can also be incorporated at the discretion of the architect or engineer.

Mock-ups should be used instead of sample panels whenever the acceptability of the masonry will be judged on more than just finished appearance, and construction observation or inspection will be provided to verify conformance. Mock-ups can be used not only to verify size, chippage, and warpage tolerances of units but also to establish aesthetic criteria such as unit placement, joint tooling, joint color uniformity, and the even distribution or blending of different color units or units with noticeable color variations. Because they incorporate other elements, however, mock-ups are perhaps most valuable for establishing acceptable workmanship and procedural requirements

FIGURE 17-5 Masonry mock-ups should include all elements important to the performance and appearance of the masonry such as backing wall, anchors, drainage mat, flashing, weeps, movement joints, or windows.

for such items as placement of reinforcement, embedment of connectors, installation of flashing, and prevention of mortar droppings in wall cavities.

Because many of the items required in the mock-up will be concealed, and because acceptance may be based on procedure as well as appearance, the architect or engineer should try to be present during construction of the panel to observe the work and to answer questions about specified requirements. Documentation of concealed elements and procedural items may best be accomplished by photographing the work in progress. A cursory examination of a completed mock-up panel will tell the observer nothing about what's inside the wall (or isn't inside the wall). Acceptance on such a basis does not give adequate criteria on which to accept or reject the project masonry. The proper evaluation and comparison of the project masonry with the standards of the mock-up require on-site observation or inspection by the architect, engineer, or independent inspector. The person who will evaluate and accept or reject the work, if different from the design architect or engineer, should also be present for construction of the mock-up. Depending on its size, construction of the mock-up could be incorporated into a preconstruction conference. Both the meeting and the mock-up can be instrumental in clarifying project requirements, understanding design intent, and resolving potential problems or conflicts.

The mock-up should be constructed by a mason or masons whose work is typical of that which will be provided in the project because it establishes the standard of workmanship by which the balance of the masonry will be judged.

17.4.3 Grout Demonstration Panel

The MSJC Code and Specification stipulate certain requirements for grout space geometry, grouting procedures, and construction techniques. Projects that use alternative methods or exceed the code limitations should require construction of a grout demonstration panel to determine the effectiveness of those methods.

17.5 Field Observation and Inspection

Field observation and inspection have become increasingly important with the explosion of construction litigation. The intent of these site visits is to ensure that the finished work complies with the contract documents and that the workmanship meets the required standards. It is the inspector's job to see that the work is in general conformance with the instructions and requirements of the drawings, specifications, and the approved sample panel or mockup. Safeguarding the quality of the work without impeding its progress is best achieved through cooperation with the contractor and workers. Good design and good intentions are not sufficient in themselves to ensure quality of the finished product. The inspector can facilitate the proper execution of the work, ensuring masonry structures that are as durable and lasting as the materials of which they are made.

Good workmanship affects masonry performance and is essential to high-quality construction. Masonry construction requires skilled craftsmen working cooperatively with the architect and engineer to execute the design. The goal of quality workmanship is common to all concerned parties for various reasons of aesthetics, performance, and liability.

Responsibility for construction in design-bid-build projects rests with the contractor. The A/E is not a party to the construction contract but acts solely as the owner's representative in the field. As part of the team, the architect can assist the contractor and offer expertise in solving or avoiding potential problems. The architect must also act as interpreter of design intent and safeguard the project quality by ensuring proper execution of the work according to the requirements of the contract documents.

Independent inspection agencies or testing laboratories serve a different function. If required by the specifications, it is their responsibility to test various materials and assemblies to verify compliance with reference standards, design strengths, and performance criteria. Field observation and inspection procedures are necessary to ensure the successful translation of the design, drawings, and specifications into a completed structure that functions as intended. An independent inspector's authority does not extend to supervision of the work or to revision of details or methods without the written approval of the architect, owner, and contractor.

The following is intended as a comprehensive guide to field observation of masonry construction. It is not intended for structural inspection of load-bearing masonry.

17.5.1 Materials

An inspector must be familiar with the project specifications and must verify compliance of materials at the job site with the written requirements. Manufacturers must supply test certificates showing that the material properties meet or exceed the referenced standards as to ingredients, strength, dimensional tolerances, durability, and so on.

Unit masonry may be visually inspected for color, texture, and size and compared with approved samples. Units delivered to the job site should be inspected for physical damage and storage/protection provisions checked. Stone, brick, or concrete masonry that has become soiled, cracked, chipped, or broken in transit should be rejected. If the manufacturer does not supply test certificates, random samples should be selected and sent to the testing agency for laboratory verification of minimum standards. The inspector should also check the moisture condition of clay masonry at the time of laying because initial rates of absorption affect the bond between unit and mortar and the strength of the mortar itself. Visual inspection of a broken unit can indicate whether field tests of absorption rates should be performed (refer to Chapter 15).

Mortar and grout ingredients should be checked on delivery for damage or contamination and to ensure compliance with the specified requirements. Packaged materials should be sealed with the manufacturer's identifying labels legible and intact. Cementitious ingredients that show signs of water absorption should be rejected. If material test certificates are required, check compliance with the specifications.

Acceptable mortar and grout mixing and batching procedures should be established at the preconstruction conference to ensure quality and consistency throughout the job. If field testing of mortar or grout prisms is required, preconstruction laboratory samples should be prepared and tested sufficiently in advance of construction to serve as a benchmark. Retempering time should be monitored to preclude the use of mortar or grout that has dried out or begun to set.

Accessories must also be checked for design compliance. The inspector must ensure use of proper anchoring devices, ties, inserts, flashing, weep inserts, drainage accessories, and reinforcement. Steel shelf angles and lintels should carry certification of yield strength and be properly bundled and identified for location within the structure.

17.5.2 Construction

Foundations, beams, floors, and other structural elements that will support the masonry should be checked for completion to proper line and grade before the work begins. Adequate structural support must be ensured and areas cleaned of dirt, grease, oil, laitance, or other materials that might impair bond of the mortar or grout. Overall dimensions and layout must be verified against the drawings and field adjustments made to correct discrepancies. Steel reinforcing dowels must be checked for proper location in relation to cores, joints, or cavities. The inspector should also keep a log of weather conditions affecting the progress or performance of the work. Inspectors should not interfere with the workers or attempt to direct their activities. If methods or procedures are observed that appear to conflict with the specifications or that might jeopardize the quality or performance of the work, they should be called to the attention of the contractor or foreman and adjustments made as necessary.

17.5.3 Workmanship

Perhaps the single most important element in obtaining strong, water-resistant masonry walls is full mortar joints and proper joint tooling. Partially filled head joints or furrowed mortar beds will produce voids that offer only minimal resistance to moisture infiltration. Poorly tooled joints allow excessive water infiltration (refer to Chapter 7). The first course of masonry must be carefully aligned vertically and horizontally and fully bedded to ensure that the remainder of the wall above will be plumb and level. Even if hollow CMU construction requires only face-shell bedding, this critical base course must have full mortar under face shells and webs. Head joints must be fully buttered with mortar and shoved tight against the adjacent unit to minimize water infiltration. Units must not be moved, tapped, or realigned after initial placement or the mortar bond will be destroyed. If a unit is displaced, all head and bed mortar must be removed and replaced with fresh material. Spot checks for proper bond can be made by lifting a fresh unit out of place to see if both faces are fully covered with adhered mortar.

The inspector should check for proper embedment and coverage of anchors, ties, and joint reinforcement and should monitor vertical coursing and joint uniformity. Differential widths or thicknesses of mortar joints can misalign the modular coursing and interfere with proper location of openings, lintels, and embedded items. Story poles, string lines, and tapes or templates should be used to construct coursing between corner leads. Nail and line pinholes must be filled with mortar when string lines are removed to avoid water penetration through these voids. Work of other trades that penetrates the masonry should be incorporated during construction of the wall and not cut in later. Drainage cavities must be kept as free as possible of mortar fins and droppings to avoid blocking weep holes, damaging flashing, or interfering with grout pours. When they become "thumbprint hard," joints should be concave tooled to compress the mortar surface.

The mason should place all vertical and horizontal reinforcement as the work progresses, holding the bars in correct alignment with spacers or wire. Minimum clearances should be maintained and bar splices lapped and securely tied. The inspector should check to see that reinforcement is free of rust, loose scale, or other materials that could impair bond to the mortar. Care should be taken to avoid moving or jarring vertical steel that is already embedded in lower grouted courses.

Inspection should also include proper installation of flashing, control joints, expansion joints, lintels, sills, caps, copings, and frames. Door frames must be adequately braced until the mortar has set and the masonry work surrounding them is self-supporting.

Grouting is important to the structural integrity of reinforced masonry walls. Cavities and cores should be inspected before the grout is placed and any remaining dirt, debris, mortar droppings, or protrusions removed before the work proceeds. Cleanout plugs left for high-lift pours allow visual inspection from below by use of a mirror inserted through the opening. Cleanout units should be fully mortared and shoved into place after inspection, then braced against blowout from the fluid pressure of the grout against the uncured mortar. The consistency of the grout should allow for easy pouring or pumping and complete filling of the space. Vibrating or consolidation to remove air bubbles and pockets also ensures that the grout covers fully around and between ties and reinforcement. Grout consolidation should occur 5 to 10 minutes after placement and reconsolidation after initial water loss and settlement. Reconsolidation should occur within 30 minutes of initial consolidation. A low-velocity electric vibrator placed into each grouted core or at 12- to 16-in. intervals in a grouted cavity for a few seconds is considered sufficient. Timing of grout pours should be monitored to avoid excessive lateral pressure on uncured joints.

17.5.4 Protection and Cleaning

Throughout the construction period, both the masonry materials and the work must be protected from the weather. Materials must be stored off the ground to prevent contamination or staining. Exposed tops of unfinished walls must be covered each night to keep moisture out of the cores and cavities by draping waterproof plastic or canvas 2 ft down each side. Cold weather may require heating of materials and possibly the application of heat during the curing period. Hot, dry climates cause rapid evaporation, and mortar mixes may have to be adjusted to compensate for premature drying. Both hot and cold weather may necessitate moist curing. The inspector must ensure that the required precautions are taken to avoid harmful effects and must also see that completed work is protected from damage during other construction operations.

Suitable cleaning methods must be selected on the basis of the type of stain involved and the type of material to be cleaned. Improper use of cleaning agents can create more problems than are solved by their application. Mortar smears on the face of the masonry should be removed daily before they are fully hardened and dry-brushed when powdery to prevent stains. Paints, textured coatings, or clear water repellents, if specified, should be applied carefully over clean, dry walls, and adjacent work protected against splatters and drips.

17.5.5 Moisture Drainage

Early in the construction, the drainage of cavity walls and veneers should be checked to ensure the unobstructed flow of water to the weeps and rapid drainage of moisture from the wall. A quick field check involves briefly placing a water hose in the drainage cavity after the mortar has had a chance to set (*see Fig. 17-6*). Water should

FIGURE 17-6 Checking cavity drainage with a water hose.

FIGURE 17-7 Water should drain immediately and freely from the weep hole directly below the hose location and then flow to adjacent weeps on either side.

drain immediately and freely from the weep just below the test location (*see Fig. 17-7*). As water travels laterally along the flashing, it should begin to drain from adjacent weeps as well. The test should be brief to avoid saturating the cavity but is a quick and effective means of detecting blocked weeps.

CHAPTER 18
Forensic Investigations

Forensic investigations of various scope may be performed on masonry walls for a variety of reasons including:

- Due diligence inspections in purchasing an existing building
- In preparation for repairs, remodeling, or renovation of an existing building
- Troubleshooting to find the source of problems on an existing building
- Litigation

In masonry construction, forensic investigations are most often related to water leakage and cracking.

18.1 Water Leakage

Water leak investigations must be systematic if they are to yield useful information in identifying the precise source of water penetration into a wall and its path to the building interior. A standardized protocol is outlined in ASTM E2128, *Standard Guide for Evaluating Water Leakage in Buildings*. The protocol itself can apply to almost any type of investigation because it is good practice. The seven recommended steps are as follows:

1. Review project documents (drawings, specs, construction photographs, RFIs, change orders, etc.)
2. Evaluate design concept (industry standards, accepted practices)
3. Determine service/maintenance/repair history
4. Investigate building
   - Known leak areas
   - Areas of similar design or construction
     Same orientation
     Different orientation
   - Corresponding indoor and outdoor areas
   - Areas above and below leak area
5. Test leak areas
   - Windows
   - Sealant joints
   - Roofing elements
   - Wall systems

6. Analyze inspection and test results
7. Prepare report with documentation

18.1.1 Diagnostic Water Testing

There are several methods of water testing building envelope components ranging from use of a simple garden hose to use of elaborate pressure chambers. The goal of investigative (or diagnostic) water testing is to determine the source and cause of *existing* leaks. ASTM E2128 states that the primary purpose "is to re-create leaks that are known to occur," that "investigative testing is not intended to demonstrate code compliance or compliance with project documents," and that "investigative testing is a diagnostic procedure, not a quality assurance procedure." In other words, investigative testing is not intended to determine *whether* there are leaks but to find the cause of leaks that have already occurred under normal in-service weather conditions.

The first step in diagnostic water testing is to open the wall or ceiling area inside the building where staining or other water damage has occurred. This will allow observation during the testing of any water penetration that might occur. It is possible to create leaks that did not exist prior to the water testing, but these should be discounted because they did not occur under normal service conditions, and the point of water entry associated with these new leaks is not necessarily the source of the existing leaks.

Diagnostic water testing must isolate each element of the building envelope that might be the source of the leak and water test them one at a time, working upward from the bottom of the wall. Starting at the bottom helps eliminate ambiguity about the origin of a leak that might result from water running down the surface of the test area. Plastic sheets and duct tape are the primary materials used to mask off one part of a test area from another (*see Fig. 18-1*).

Once testing reproduces an in-service leak, the water migration path must be traced from the identified entry point to the site of the existing water damage. A single entry point may lead to more than one water path or several entry points may merge together into a single migration path. Every water entry source that leads to each water migration path associated with the existing leak must be identified if an accurate diagnosis and appropriate repair is to be developed.

18.1.2 Types of Water Tests

Of the following commonly used tests, the first six are virtually identical and simulate wind-driven rain by using high air pressure on the outside of the test unit and low air pressure on the inside. They are rigorous tests but of very short duration. The seventh test is much simpler and less expensive, using only a water hose to simulate the flow of a sheet of water over a surface.

1. ASTM E331, *Test for Water Penetration of Exterior Windows, Curtain Walls, and Doors by Uniform Static Air Pressure Difference*: Laboratory test using pressurized air chamber with a rack of uniformly spaced water nozzles calibrated to deliver a minimum of 5 gal of water per sq ft/h, 15 minutes duration, at an air pressure difference equal to 15% of the design wind pressure.

2. AAMA 501.1, *Test Method for Metal Curtain Walls for Water Penetrations Using Dynamic Pressure*: Laboratory test using pressurized air chamber with a rack of uniformly spaced water nozzles calibrated to deliver a minimum of 5 gal of

FIGURE 18-1 Use of plastic sheeting and duct tape to isolate various elements of the building envelope for water testing.

water per sq ft/h, 15 minutes duration, at an air pressure difference equal to 15% of the design wind pressure.

3. ASTM E1105, *Field Determination of Water Penetration of Installed Exterior Windows, Skylights, Doors and Curtain Walls by Uniform or Cyclic Static Air Pressure Difference*: Pressurized air chamber test with a rack of uniformly spaced water nozzles calibrated to deliver a minimum of 5 gal of water per sq ft/h, 15 minutes duration, at an air pressure difference equal to 15% of the design wind pressure.

4. AAMA 501.3, *Field Check of Water Penetration Through Installed Exterior Windows, Curtain Walls and Doors by Uniform Static Air Pressure Difference*: Pressurized air chamber test with a rack of uniformly spaced water nozzles calibrated to deliver a minimum of 5 gal of water per sq ft/h, 15 minutes duration, at an air pressure difference equal to 15% of the design wind pressure.

5. AAMA 502-90, *Voluntary Specification for Field Testing of Windows and Sliding Glass Doors*: Pressurized air chamber test with a rack of uniformly spaced water

nozzles calibrated to deliver a minimum of 5 gal of water per sq ft/h, 15 minutes duration, at a static air pressure difference equal to 15% of the design wind pressure.

6. AAMA 503-92, *Voluntary Specification for Field Testing of Metal Storefronts, Curtain Walls, and Sloped Glazing Systems*: Pressurized air chamber test with a rack of uniformly spaced water nozzles calibrated to deliver a minimum of 5 gal of water per sq ft/h, 15 minutes duration, at a static air pressure difference equal to 15% of the design wind pressure.

7. AAMA 501.2, *Field Check of Metal Storefronts, Curtain Walls and Sloped Glazing Systems for Water Leakage*: Garden hose test at 30 to 35 psi, 12 in. from face of wall, 5 minutes duration for each 5-ft length, moving nozzle slowly back and forth.

Of the tests listed above, AAMA 501.2 is the only test whose stated purpose is diagnostic. It is also the least expensive test and is described by AAMA as "a simple, economical method of finding leaking joints." Although these tests were developed for the window industry, they are commonly used to test the wall areas surrounding windows, including masonry. There are other tests developed specifically to test masonry walls, including ASTM E514, *Water Permeance of Masonry*, and ASTM C1601, *Test Method for Field Determination of Water Penetration of Masonry Wall Surfaces*. These, however, are not diagnostic tests.

Simple hose tests can deliver a concentrated spray to isolated materials, elements, joints, cracks, and so on (*see Fig. 18-2*). The appropriate duration of the test is whatever it takes to find the point of water entry associated with the existing leak. Testing durations specified for the tests listed above may not be sufficient for a leak diagnosis if

FIGURE 18-2 Use of a hand-held garden hose for diagnostic water testing.

in-service leaks cannot be re-created within that time. The 5-minute or 15-minute wetting periods in ASTM and AAMA tests are typically too short for diagnostic purposes.

Other simple tests include the Rilem Tube test, typically used to test the water tightness of mortar joints under hydrostatic pressure, and spray rack tests, which can deliver a continuous film of water to a wall surface. *Figure 18-3* shows the Rilem Tube and spray

rilem tube test

spray rack test

pressure chamber test

FIGURE 18-3 More diagnostic water tests.

rack tests as well as a pressure chamber. Pressure tests simulate the negative air pressure that can result from improper HVAC operation or simply a wind-driven rain. Negative pressure pulls rainwater in through openings in the building envelope that might not otherwise leak. This type of test chamber is used in the first six test methods listed earlier.

18.2 Structural Performance

One of the most common structural issues with masonry is whether or not grout, reinforcing steel, and joint reinforcement were installed as required by the design. Two test methods are used. The first is thermographic imaging and the second is surface-penetrating radar (*see Figs. 18-4 and 18-5*). Both are capable of showing grouted and ungrouted unit

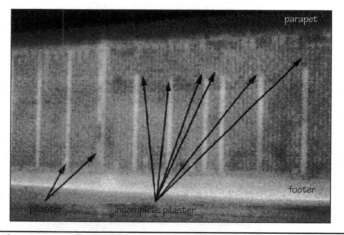

Figure 18-4 Infrared thermography identifies grouted and ungrouted CMU cores. (*From Stockton, "Using Infrared Thermography for Determining the Presence and Correct Placement of Grouted Cells in Single-Wythe Concrete Masonry Unit (CMU) Walls," in* RCI Interface, *February 2001.*)

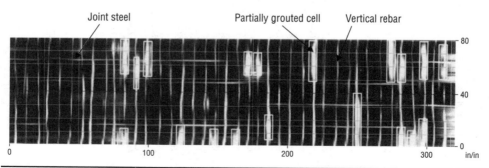

Figure 18-5 Surface-penetrating radar identifies grout, reinforcing bars, and joint reinforcement. (*From Radarview, LLC, 1036 First Street, Suite A3, Humble, TX 77338; www.radarviewllc.com.*)

FIGURE 18-6 Sounding hammer makes a hollow sound at ungrouted cores.

cores. The surface-penetrating radar also helps identify the location of steel reinforcing bars and metal joint reinforcement.

A sounding hammer can also be used for investigating small areas (*see Fig. 18-6*). Cores that are not grouted have a hollow sound.

18.3 Cracking

The first step in analyzing masonry cracks is visual evaluation. The location and shape of cracks provide clues to their cause (*see Figs. 18-7 through 18-10*). It is easy to visualize

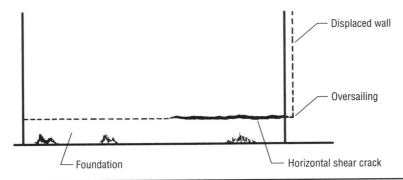

FIGURE 18-7 Horizontal shear cracking caused by longitudinal wall expansion. (*Illustrations excerpted from "Masonry Cracks: A Review of the Literature,"* Masonry: Materials, Design, Construction, and Maintenance, *ASTM STP 992, H. A. Harris, Ed., American Society for Testing and Materials, West Conshohocken, PA, pp. 257–280.*)

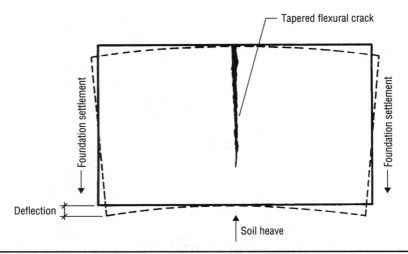

FIGURE 18-8 Tapered flexural cracking caused by foundation settlement at wall ends or soil heave in middle. (*Illustrations excerpted from "Masonry Cracks: A Review of the Literature,"* Masonry: Materials, Design, Construction, and Maintenance, *ASTM STP 992, H. A. Harris, Ed., American Society for Testing and Materials, West Conshohocken, PA, pp. 257–280.*)

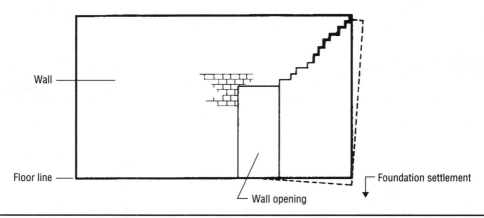

FIGURE 18-9 Tapered flexural cracking caused by foundation settlement at end of wall. (*Illustrations excerpted from "Masonry Cracks: A Review of the Literature,"* Masonry: Materials, Design, Construction, and Maintenance, *ASTM STP 992, H. A. Harris, Ed., American Society for Testing and Materials, West Conshohocken, PA, pp. 257–280.*)

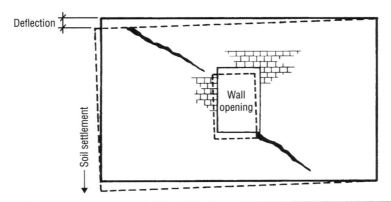

Figure 18-10 Tapered flexural cracking caused by foundation settlement at end of wall. (*Illustrations excerpted from "Masonry Cracks: A Review of the Literature,"* Masonry: Materials, Design, Construction, and Maintenance, *ASTM STP 992, H. A. Harris, Ed., American Society for Testing and Materials, West Conshohocken, PA, pp. 257–280.*)

what type of movement causes cracks that are narrow on one end and wider on the other. Once a hypothesis has been formed as to the cause of the cracks being investigated, the wall will have to be opened to verify the existing conditions and determine the appropriate remedial action.

APPENDIX A
Glossary

Boldface type within entries denotes terms for which there are glossary entries.

A

Absorption The amount of water that a masonry unit absorbs when immersed in water under specified conditions for a specified length of time.

Absorption Rate The weight of water absorbed when a brick is partially immersed for 1 minute, usually expressed in either grams or ounces per minute. Also called suction or initial rate of absorption.

Accelerator Ingredient added to mortar or grout to speed hydration of cementitious components to hasten set time.

Adhered Attached by adhesion rather than mechanical anchorage, as in adhered veneer.

Adhered Manufactured Stone Masonry Veneer (AMSMV) See **Concrete Masonry Unit.**

Adhered Veneer Masonry veneer secured to and supported by the backing through adhesion.

Admixture A material other than water, aggregates, hydraulic cement, or fiber reinforcement used as an ingredient of grout or mortar and added to the batch immediately before or during its mixing.

Adobe Soil of diatomaceous content mixed with sufficient water so that plasticity can be developed for molding into masonry units.

Aggregate Granular mineral material such as natural sand, manufactured sand, gravel, crushed stone, or air-cooled blast furnace slag.

Air Drying The process of drying block or brick without any special equipment, simply by exposure to ambient air.

Air Entraining The capability of a material or process to develop a system of minute bubbles of air in cement paste, mortar, or grout during mixing.

Anchor See **Connector, Anchor.**

Arch A vertically curved compressive structural member spanning openings or recesses; may also be built flat by using special masonry shapes or placed units.

Arching Action The distribution of loads in masonry over an opening. The load is usually assumed to occur in a triangular pattern above the opening extending from a maximum at the center of the span to zero at the supports.

Architectural Terra Cotta Hard-burned, glazed or unglazed clay building units, plain or orna-mental, machine extruded or hand molded, and generally larger in size than brick or facing tile.

Arris A sharp edge of brick, stone, or other building element.

Autoclave A type of curing system in the production of concrete masonry units that uses superheated steam under pressure to promote strength of units.

Autoclaved Aerated Concrete (AAC) A cementitious product based on calcium silicate hydrates in which low density is attained by the inclusion of an agent resulting in macroscopic voids and which is subjected to high-pressure steam curing.

B

Backing Surface or assembly to which veneer is attached.

Backup That part of a masonry wall behind the exterior facing.

Band Course A continuous, horizontal band of masonry marking a division in the wall elevation. Sometimes called belt course, string course, or sill course.

Bed Joint See **Joint, Bed.**

Block, Concrete See **Concrete Masonry Unit.**

Bond

Adhesion Bond Adhesion between masonry units and mortar or grout.

Masonry Bond Connection of masonry wythes with overlapping header units.

Metal Tie Bond Connection of masonry wythes with metal ties or joint reinforcement.

Mortar Bond or Grout Bond Adhesion between mortar or grout and masonry units, reinforcement, or connectors.

Pattern Bond Patterns formed by the exposed faces of the masonry units; for example, running bond or Flemish bond.

American Bond Bond pattern in which every sixth course is a header course and the intervening courses are stretcher courses.

Basketweave Bond Modular groups of units laid at right angles to those adjacent to form a pattern.

Blind Bond Bond pattern to tie the front course to the wall where it is not desirable that any headers should be seen in the face work.

Common Bond Bond pattern in which five to seven stretcher courses are laid between headers.

Cross Bond Bond pattern in which the joints of the stretcher in the second course come in the middle of the stretcher in the first course composed of headers and stretchers intervening.

Dutch Cross Bond A bond having the courses made up alternately of headers and stretchers. Same as an **English cross bond.**

English Bond Bond pattern with alternating courses of headers and stretchers. The headers and stretchers are situated plumb over each other. The headers are divided evenly over the vertical joints between the stretchers.

English Cross Bond A variation of English bond, but with the stretchers in alternate courses centered on the stretchers above and below. Also called **Dutch cross bond.**

Flemish Bond Bond pattern in which each course consists of alternate stretchers and headers, with the headers in alternate courses centered over the stretchers in intervening courses.

Flemish Garden Bond Bond pattern in which units are laid so that each course has a header to every three to five stretchers.

Header Bond Bond pattern showing only headers on the face, each header divided evenly on the header under it.

Herringbone Bond Arrangement of units in a course in a zigzag fashion, with the end of one unit laid at right angles against the side of a second unit.

Random Bond Masonry constructed without a regular pattern.

Running Bond Placement of masonry units such that head joints in successive courses are horizontally offset at least one-quarter of the unit length.

Stack Bond (1) Placement of units such that the head joints in successive courses are vertically aligned. (2) Units laid so that no overlap occurs; head joints form a continuous vertical line. Also called plumb joint bond, straight stack, jack bond, jack on jack, and checkerboard bond.

Bond Beam A course or courses of a masonry wall grouted and usually reinforced in the horizontal direction serving as an integral beam in the wall. May serve as a horizontal tie, as a bearing course for structural members, or as a flexural member itself.

Bond Beam Unit A hollow masonry unit with depressed sections forming a continuous channel in which reinforcing steel can be placed for embedment in grout.

Bond Breaker A material used to prevent adhesion between two surfaces.

Bond Strength The resistance of mortar or grout to separation from masonry units or reinforcement with which it is in contact.

Brick A manufactured masonry unit made from fired clay or shale, concrete, or sand-lime materials, which is usually formed in the shape of a rectangular prism and typically placed with one hand.

Adobe Brick An unfired, air-dried, roughly molded brick of earth or clay. When made with an emulsifier or fibers, called stabilized adobe.

Building Brick Brick for building purposes not especially produced for texture or color (formerly called common brick).

Calcium Silicate Brick Brick made from sand and lime, with or without the inclusion of other materials.

Clay Brick A solid or hollow masonry unit of clay or shale, usually formed into a rectangular prism while plastic and burned or fired in a kiln.

Common Brick See **Building Brick.**

Concrete Brick Brick made from portland cement, water, and suitable aggregates, with or without the inclusion of other materials.

Cored Brick A brick in which the holes consist of less than 25% of the section.

Facing Brick Brick made especially for facing purposes.

Fire Brick (1) Any type of refractory brick, specifically fire-clay brick. (2) Brick made of refractory ceramic material that will resist high temperatures.

Glazed Brick A brick prepared by fusing on the surface a ceramic glazing material; brick having a glassy surface.

Hollow Brick Brick whose net cross-sectional area (solid area) in any plane parallel to the surface containing the cores, cells, or deep frogs is less than 75% of its gross cross-sectional area measured in the same plane.

Jumbo Brick A generic term indicating a brick larger in size than the standard. Some producers use this term to describe oversize brick of specific dimensions manufactured by them.

Paving Brick, Heavy Vehicular Brick intended for use in areas with a high volume of heavy vehicular traffic.

Paving Brick, Light Traffic Brick intended for use as paving material to support pedestrian and light vehicular traffic.

Sand-Lime Brick See **Calcium Silicate Brick.**

Brick Grade Designation denoting durability of clay brick, expressed as SW (severe weathering), MW (moderate weathering), and NW (negligible weathering).

Brick Ledge A ledge on a footing or wall that supports masonry.

Brick Type Designation for clay brick that indicates qualities of appearance including tolerance, chippage, and distortion. Expressed as face brick standard (FBS), face brick extra (FBX), and face brick architectural (FBA) for solid brick; and hollow brick standard (HBS), hollow brick extra (HBX), hollow brick architectural (HBA), and hollow building brick (HBB) for hollow brick.

C

Cast Stone See **Stone, Cast.**

Cavity An unfilled space.

Cell See **Core.**

Cellular Concrete See **Autoclaved Aerated Concrete.**

Cement A general term for an adhesive or binding material. See specific terms such as **Portland Cement** or **Masonry Cement.**

Cement Mortar A mixture of cement, sand, or other aggregates and water used for plastering over masonry or to lay masonry units.

Cementitious Material When proportioning masonry mortars, the following are considered cementitious materials: portland cement, blended hydraulic cement, masonry cement, quicklime, and hydrated lime.

Cement-Lime Mortar A mixture of cement, lime, sand, or other aggregates and water used for plastering over masonry or to lay masonry units.

Cinder Block See **Concrete Masonry Unit.**

Cleanout An opening at the bottom of a grout space of sufficient size and spacing to allow the removal of debris.

CMU   Concrete masonry unit.

Cold Weather Construction   Procedures used in constructing masonry when ambient air temperature or the temperature of the masonry units is below 40°F.

Collar Joint   see **Joint, Collar**.

Composite Action   Transfer of stress between components of a member designed so that in resisting loads, the combined components act together as a single member.

Concrete Block   See **Concrete Masonry Unit**.

Concrete Brick   See **Brick, Concrete**.

Concrete Masonry Unit (CMU)   Manufactured masonry unit made from portland cement, mineral aggregates, and water, with or without the inclusion of other materials.

"A" Block   Cored masonry unit with one end closed by a cross-web and the opposite end open or lacking an end cross-web, typically forming two cells when laid in running bond. Also called open end block.

Adhered Manufactured Stone Masonry Veneer (AMSMV)   An assembly of thin masonry units adhered to a backing with a cementitious mortar.

Concrete Block   Hollow or solid unit consisting of portland cement and suitable aggregates combined with water. Other materials such as lime, fly ash, air-entraining agent, or other admixtures may be permitted.

Concrete Masonry Unit, Lightweight   Concrete masonry unit whose oven-dry density is less than 105 lb/cu ft.

Concrete Masonry Unit, Medium Weight   Concrete masonry unit whose oven-dry density is at least 105 lb/cu ft and less than 125 lb/cu ft.

Concrete Masonry Unit, Normal Weight   Concrete masonry unit whose oven-dry density is at least 125 lb/cu ft or more.

Ground Faced Block   Concrete masonry unit in which the exposed surface is ground to a smooth finish.

Jamb Block   Block specially formed for the jamb of windows or doors, generally with a vertical slot to receive window frames, and so forth.

Screen Block   Open-faced masonry units used for decorative purposes or to partially screen areas from the sun or outside viewers.

Slump Block   Concrete masonry unit produced so it will slump or sag in irregular fashion before it hardens.

Split-Faced Block   Concrete masonry unit with one or more faces purposely fractured to expose the rough aggregate texture to provide architectural effects in masonry wall construction.

Connector   Mechanical device, including anchors, wall ties, and fasteners, for securing two or more pieces, parts, or members together.

Connector, Anchor   Metal rod, wire, or strap that secures masonry to its structural support.

Connector, Fastener   Device used to attach nonmasonry materials to masonry.

Connector, Wall Tie   Metal connector that connects wythes of masonry walls together.

Control Joint   See **Joint, Control**.

Coping Masonry units laid on top of a finished wall. (1) A covering on top of a wall exposed to the weather, usually sloped to carry off water. (2) The materials or masonry units used to form a cap or a finish on top of a wall, pier, chimney, or pilaster to protect the masonry below from water penetration. Commonly extended beyond the wall face and cut with a drip.

Coping Unit Solid masonry unit for use as the top and finished course in wall construction.

Corbel (1) The projection of successive courses of masonry out from the face of the wall to increase the wall thickness or to form a shelf or ledge. (2) A shelf or ledge formed by successive courses of masonry projecting out from the face of a wall, pier, or column.

Core (1) Molded open space in a concrete masonry unit. (2) Hollow space within a concrete masonry unit formed by the face shells and webs. (3) The holes in clay units. Also called cell.

Corner Pole See Story Pole.

Course Horizontal layer of units in masonry other than paving.

Crack Control Methods used to control the extent, size, and location of cracking in masonry, including reinforcing, movement joints, and dimensional stability of masonry materials.

Creep Time-dependent deformation due to sustained load.

Cross-Sectional Area, Gross (1) Area delineated by the out-to-out dimensions of masonry in the plane under consideration. (2) Total cross-sectional area of a specified section.

Cross-Sectional Area, Net (1) Area of masonry units, grout, and mortar crossed by the plane under consideration based on out-to-out dimensions. (2) Gross cross-sectional area minus the area of ungrouted cores, notches, cells, and unbedded areas. Net area is the actual surface area of a cross section of masonry.

Curing Maintenance of proper conditions of moisture and temperature during initial set to develop required strength and reduce shrinkage in concrete products and mortar.

Curtain Wall See Wall, Curtain.

D

Damping Reduction of amplitude of vibrations due to energy loss (as in damping of vibrations from seismic shock).

Damp-proofing Preparation of a wall to prevent moisture from penetrating through it.

Diaphragm Roof or floor system designed to transmit lateral forces to shear walls or other vertical resisting elements.

Differential Movement Movement of two elements relative to one another that differs in rate or direction.

Dimension Stone See Stone, Dimension.

Dimensions of Masonry Units

Actual Measured dimensions of a masonry unit.

Height (1) Vertical dimension of the unit in the face of a wall. (2) Vertical dimension of masonry units or masonry measured parallel to the intended face of the unit or units.

Length (1) Horizontal dimension of the unit in the face of the wall. (2) Horizontal dimension of masonry units or masonry measured parallel to the intended face of the unit or units.

Nominal (1) Dimension greater than the specified (standard) dimension by the thickness of one joint, but not more than 13 mm or ½ in. (2) A dimension that may be greater than the specified masonry dimension by the thickness of a mortar joint.

Specified (1) Nominal dimension less the thickness of a standard mortar joint; that is, net dimension of the masonry unit. (2) Dimensions to which masonry units or constructions are required to conform. (3) Standard dimensions of a masonry unit, plus or minus any allowable size tolerances.

Thickness (1) Dimension designed to lie at right angles to the face of the wall, floor, or other assembly. (2) *Thickness (width)* Horizontal dimension of masonry units or masonry measured perpendicular to the intended face of the masonry unit or units.

Dovetail Anchor Splayed tenon shaped like a dove's tail (i.e., broader at its base) that fits into the recess of a corresponding mortise.

Dowel Straight metal bar used to connect two sections of masonry.

Drip Groove or slot cut beneath and slightly behind the forward edge of a projecting stone member, such as a sill, lintel, or coping, to cause rainwater to drip off and prevent it from penetrating a wall.

Dry-Stack Masonry Masonry work laid without mortar.

Durability Ability of a material to resist weathering action, chemical attack, abrasion, and other conditions of service.

E

Efflorescence, Water-Insoluble Crystalline deposit, usually white, of water-soluble compounds that, on reaching the masonry surface, become water-insoluble, primarily through carbonation (also sometimes called lime run or calcium carbonate stain). Normally requires acid washing for removal.

Efflorescence, Water-Soluble Crystalline deposit, usually white, of water-soluble compounds on the surface of masonry. Normally can be removed with water washing.

Empirical Design Design based on application of physical limitations learned from experience or observations gained through experience, without structural analysis.

Engineered Masonry Masonry that has been analyzed for vertical and lateral load resistance and members proportioned to resist design loads in accordance with working stress design or strength design principles.

Equivalent Thickness (1) Solid thickness to which a hollow unit would be reduced if the material in the unit were recast with the same face dimensions but without voids. (2) Percent solid volume times actual width divided by 100. (3) ET = net volume ÷ (specified unit length × specified unit height). (4) Average thickness of solid material in a unit.

Expansion Joint See **Joint, Expansion.**

F

Face Shell Side wall of a hollow concrete masonry or clay masonry unit.

Face Shell Bedding Mortar is applied only to the horizontal surface of the face shells of hollow masonry units and in the head joints to a depth equal to the thickness of the face shell.

Facing Tile Structural clay tile for exterior and interior masonry with exposed faces (covered by ASTM C212 and ASTM C126).

Flagstone Type of stone that splits easily into flags or slabs; also a term applied to irregular pieces of such stone split into slabs from 1 to 3 in. thick and used for walks, patios, and so forth.

Flashing Impermeable material placed in masonry to provide water drainage or prevent water penetration.

Freeze-Thaw Freezing and thawing of moisture in materials and the resultant effects on these materials and on structures of which they are a part or with which they are in contact.

Frog Indentation in one bed surface of a brick, manufactured by molding or pressing.

Furrowing Striking a V-shaped trough in a mortar bed.

G

Gradation Particle size distribution of aggregate as determined by separation with standard screens. Gradation of aggregate is expressed in terms of the individual percentages passing standard screens. Sieve analysis, screen analysis, and mechanical analysis are terms used synonymously in referring to gradation of aggregate.

Granite Visibly granular, igneous rock generally ranging in color from pink to light or dark gray and consisting mostly of quartz and feldspars, accompanied by one or more dark minerals. The texture is typically homogeneous but may be gneissic or porphyritic. Some dark granular igneous rocks, though not geologically granite, are included in the definition.

Grid Paver Open-type masonry unit that allows grass to grow through openings; used for soil stabilization.

Grout, Masonry Mixture of cementitious materials, aggregates, and water, with or without admixtures, used to fill voids in masonry; initially mixed to a consistency suitable for pouring or pumping without segregation of constituents (covered by ASTM C476).

Grout, Self-Consolidating A highly fluid mixture of cement, water, fine and coarse aggregates, and plasticizing admixtures, which is placed without consolidation.

Grout Lift Grout placed in a single, continuous operation.

Grout Pour Total height of a masonry wall to be grouted prior to erection of additional masonry. A grout pour will consist of one or more grout lifts.

Grout Pumping Method of installing masonry grout.

Grouted Masonry Masonry construction composed of hollow units where designated hollow cells are filled with grout, or multi-wythe construction in which spaces between wythes are filled with grout.

Grouting, High-Lift Technique of grouting in which the masonry is constructed in excess of 5 ft high prior to grouting.

Grouting, Low-Lift Technique of grouting as the wall is constructed, usually to scaffold or bond beam height, but not greater than 4 ft.

H

Head Joint See **Joint, Head.**

Header Masonry unit that overlaps two or more adjacent wythes of masonry to bind or tie them together.

Height of Wall Vertical distance from a foundation wall, or other similar intermediate support, to the top of a wall, or the vertical distance between intermediate supports.

Height-Thickness (*H/T*) Ratio Height of a masonry wall divided by its nominal thickness. The thickness of cavity walls is taken as the overall thickness minus the width of the cavity.

Hot-Weather Construction Procedures used in constructing masonry when ambient air temperature exceeds 100°F or 90°F with a wind velocity greater than 8 mph.

Hybrid Wall See **Wall, Hybrid.**

Hysteresis Irreversible expansion of marble building stone with cycles of heating and cooling.

I

Incidental Water Unplanned water infiltration that penetrates beyond the primary barrier and the flashing or secondary barrier system, of such limited volume that it can escape or evaporate without causing adverse consequences.

Initial Rate of Absorption (IRA) Measure of the capillary suction of water into a dry masonry unit from a bed face during a specified length of time over a specified area.

Initial Set Beginning change from a plastic to a hardened state.

J

Joint In building construction, space or opening between two or more adjoining surfaces.

Bed Joint Horizontal layer of mortar on which a masonry unit is laid.

Collar Joint Vertical, longitudinal joint between wythes of masonry or between masonry wythe and backing.

Control Joint In concrete, concrete masonry, stucco, or coating systems, a formed, sawed, or assembled joint acting to regulate the location of cracking, separation, and distress resulting from dimensional or positional change.

Expansion Joint (1) Continuous joint or plane to accommodate expansion, contraction, and differential movement; does not contain mortar, grout, reinforcement, or other hard material. (2) In building construction, a structural separation between building elements that allows independent movement without damage to the assembly.

Head Joint Vertical transverse mortar joint placed between masonry units within the wythe at the time the masonry units are laid.

Mortar Joint In mortared masonry construction, joints between units that are filled with mortar.

Movement Joint In building construction, a joint designed to accommodate movement of adjacent elements.

Raked Joint Mortar joint in which ¼ to ½ in. of mortar is removed from the outside surface of the joint.

Slushed Joint   Head or collar joint constructed by "throwing" mortar in with the edge of a trowel.

Struck Joint   Joint from which excess mortar has been removed by a stroke of the trowel, leaving an approximately flush joint.

Tooled Joint   Mortar joint between two masonry units shaped manually or compressed with a jointing tool such as a concave or V-notched jointer.

Joint Reinforcement   Metal bar or wire, usually prefabricated, to be placed in mortar bed joints.

K

Keystone   Wedge-shaped stone at the center or summit of an arch or vault, binding the structure actually or symbolically.

L

Lap   (1) The distance that one masonry unit extends over another. (2) The distance that one piece of flashing or reinforcement extends over another.

Lateral Support   Bracing of walls, either vertically or horizontally, by columns, pilasters, cross-walls, beams, floors, roofs, and so forth.

Lightweight Aggregate   Aggregate of low density, such as expanded or sintered clay, shale, slate, diatomaceous shale, perlite, vermiculite, slag, natural pumice, scoria, volcanic cinders, tuff, diatomite, sintered fly ash, or industrial cinders.

Lime   General term for the various chemical and physical forms of quicklime, hydrated lime, and hydraulic hydrated lime.

Limestone   Rock of sedimentary origin composed principally of calcium carbonate.

Lintel   Beam placed or constructed over an opening in a wall to carry the superimposed load and the masonry above the opening.

M

Marble   Carbonate rock that has acquired a distinctive crystalline texture by recrystallization, most commonly by heat and pressure during metamorphism, and is composed principally of the carbonate minerals calcite and dolomite, singly or in combination.

Masonry   Construction, usually set in mortar, of natural building stone or manufactured units such as brick, concrete block, adobe, glass block, tile, manufactured stone, or gypsum block.

Masonry Cement   Hydraulic cement for use in mortars for masonry construction and containing one or more of the following materials: portland cement, portland blast-furnace slag cement, portland-pozzolan cement, natural cement, slag cement, or hydraulic lime; in addition, usually contains one or more materials such as hydrated lime, limestone, chalk, calcareous shell, talc, slag, or clay, as prepared for this purpose.

Masonry Cement Mortar   Mortar produced using masonry cement.

Masonry Unit   Natural or manufactured building unit of clay, concrete, stone, glass, or calcium silicate.

Masonry Unit, Clay   Hollow or solid masonry unit of clay or shale, including clay brick, structural clay tile, and adobe brick.

Masonry Unit, Concrete   Manufactured masonry unit made from portland cement, mineral aggregates, and water, with or without the inclusion of other materials.

Masonry Unit, Hollow   Masonry unit whose net cross-sectional area in any plane parallel to the bearing surface is less than 75% of its gross cross-sectional area measured in the same plane.

Masonry Unit, Solid   Masonry unit whose net cross-sectional area in any plane parallel to the bearing surface is 75% or more of its gross cross-sectional area measured in the same plane.

Mix Design   Proportions of ingredients to produce mortar, grout, or concrete.

Moisture Content   Amount of water contained in a material, expressed as a percentage of total absorption.

Mortar   Mixture of cementitious materials, fine aggregate, and water, with or without admixtures, used to construct masonry.

Fat Mortar   Mortar mixture containing a high ratio of binder to aggregate.

Harsh Mortar   Mortar that is difficult to spread; not workable.

Lean Mortar   Mortar that is deficient in cementitious components, usually harsh and difficult to spread.

Ready-Mixed Mortar   Mortar consisting of cementitious materials, aggregate, water, and set-control admixtures that are measured and mixed at a central location using weight- or volume-control equipment.

Surface Bonding Mortar   Product containing hydraulic cement, glass fiber reinforcement with or without inorganic fillers, or organic modifiers in a prepackaged form, requiring only the addition of water prior to application.

Mortar Bedding   Construction of masonry assemblages with mortar.

Mortar Joint   See **Joint, Mortar**.

Movement Joint   See **Joint, Control; Joint, Expansion;** and **Joint, Movement**.

N

Neat Cement   In masonry, a pure cement undiluted by sand aggregate or admixtures.

Noncombustible   Any material that will neither ignite nor actively support combustion in air at a temperature of 1200°F when exposed to fire.

O

Overhand Work   Masonry laid from the interior side of a wall rather than from the exterior side.

P

Parging   Application of a coat of cement mortar to the back of the facing or the face of the backing in multi-wythe construction.

Perlite (1) Aggregate used in lightweight insulating concrete, concrete masonry units, and plaster. (2) Insulation composed of natural perlite ore expanded to form a cellular structure.

Permeability Property of allowing passage of fluids or gases such as water vapor.

Pier Isolated column of masonry or a bearing wall not bonded at the sides to associated masonry.

Pilaster Portion of a wall serving as an integral vertical column and usually projecting from one or both wall faces. Sometimes called a **pier**.

Pilaster Block Concrete masonry unit designed for use in construction of plain or reinforced concrete masonry pilasters and columns.

Plasticizer Substance incorporated into a material to increase its workability, flexibility, or extensibility.

Pointing Troweling mortar into a joint after the masonry units are laid.

Repointing Filling in cut-out or defective mortar joints in masonry with fresh mortar.

Tuckpointing Decorative method of pointing masonry with a surface mortar that is different from the bedding mortar.

Portland Cement Hydraulic cement produced by pulverizing portland cement clinker, and usually containing calcium sulfate.

Preservation, Building

Conservation Management of a natural resource, structure, or artifact to prevent misuse, destruction, or neglect. It may include detailed characterization and recording (technical or inventory) or provenance and history and application of measures.

Preservation Act or process of applying measures to sustain the existing form, integrity, or materials of a building, structure, or artifact and the existing form or vegetative cover of a site.

Protection Act or process of applying measures designed to affect the physical condition of a building, structure, or artifact by guarding it from deterioration, loss, or attack; or, to cover or shield it from damage.

Rehabilitation, of a Structure Act or process of returning a structure to a state of utility through repair or alteration that makes possible an efficient contemporary use.

Restoration Act or process of reestablishing accurately the form and details of a structure, site, or artifact as it appeared at a particular period in time by means of removal of later work or by the reconstruction of missing earlier work.

Prism Assemblage of masonry units and mortar with or without grout, used as a test specimen for determining properties of the masonry.

Prism Strength Maximum compressive strength (force) resisted per unit of net cross-sectional area of masonry, determined by testing masonry prisms.

Prism Testing Testing an assemblage of masonry units, mortar, or grout to determine the compressive strength of masonry.

R
Racking Stepping back successive courses of masonry.

Rain Penetration See **Water Penetration**.

Reglet (1) Groove or recess to receive and secure metal flashing. (2) Groove in a wall or other surface adjoining a roof surface for the attachment of counterflashing.

Reinforced Masonry Masonry units and reinforcing steel bonded with mortar and/or grout in such a manner that the components act together in resisting forces.

Repointing See **Pointing**.

Retempering To add more water to a hydraulic-setting compound after the initial mixing but before partial set has occurred.

Rowlock Brick laid on its face edge with the end surface visible in the wall face.

Rust Jacking The process of embedded steel rusting, expanding, and cracking the surrounding masonry.

S

Sandstone Sedimentary rock composed mostly of mineral and rock fragments within the sand size range (2 to 0.06 mm) and having a minimum of 60% free silica, cemented or bonded to a greater or lesser degree by various materials including silica, iron oxides, carbonates, or clay, and which fractures around (not through) the constituent grains.

Saturation Coefficient Ratio of the weight of water absorbed by a masonry unit when immersed 24 hours in cold water to the weight of water absorbed after an additional immersion for 5 hours in boiling water. An indication of the probable resistance of brick to freezing and thawing. Also called C/B ratio.

Screen Tile Clay tile manufactured for masonry screen wall construction.

Sedimentary Rock Rock formed from materials deposited as sediments in the sea or freshwater or on land. The materials are transported to their site of deposition by running water, wind, moving ice, marine energy, or gravitational movements, and they may deposit as fragments or by precipitation from solution.

Set Change in consistency from a plastic to a hardened state.

Shelf Angle Metal angle attached to a structural member, used to support masonry.

Shrinkage Volume change due to loss of moisture or decrease in temperature.

Slate Microcrystalline metamorphic rock most commonly derived from shale and composed mostly of micas, chlorite, and quartz. The micaceous minerals have a subparallel orientation and thus impart strong cleavage to the rock, which allows the latter to be split into thin but tough sheets.

Soap Masonry unit of normal face dimension, having a nominal 2-in. thickness.

Soapstone Massive soft rock that contains a high proportion of talc and that is cut into dimension stone.

Soffit The underside of a beam, lintel, or arch.

Soldier Stretcher set on end, with the face showing on the wall surface.

Spall (1) To break away protrusions or edges on stone blocks with a sledge. (2) To flake or split away through frost action or pressure. (3) A chip or flake.

Specified Compressive Strength of Masonry, f'_m   Minimum compressive strength expressed as force per unit of net cross-sectional area required of the masonry used in construction by the project documents and on which the project design is based.

Stone

Building Stone   Natural rock of adequate quality to be quarried and cut as dimension stone, as it exists in nature, and used in the construction industry.

Cast Stone   Architectural precast concrete building unit intended to simulate natural cut stone.

Cut Stone   Stone fabricated to specific dimensions.

Dimension Stone   Natural stone that has been selected, trimmed, or cut to specified or indicated shapes or sizes, with or without one or more mechanically dressed surfaces.

Fieldstone   Natural building stone as found in the field.

Finished Stone   Dimension stone with one or more mechanically dressed surfaces.

Flagstone   Flat stone, thin in relation to its surface area, commonly used as a stepping stone, for a terrace or patio, or for floor paving. Usually either naturally thin or split from rock that cleaves readily.

Stone Masonry   Masonry composed of natural or cast stone.

Ashlar Pattern   Pattern bond of rectangular or square stone units, always of two or more sizes. If the pattern is repeated, it is patterned ashlar. If the pattern is not repeated, it is random ashlar.

Coursed Ashlar   Ashlar masonry laid in courses of stone of equal height for each course, although different courses may be of varying height.

Coursed Rubble   Masonry composed of roughly shaped stones fitting on approximately level beds, well bonded, and brought at vertical intervals to continuous level beds or courses.

Random Ashlar   Stone masonry pattern of rectangular stones set without continuous joints and laid up without drawn patterns. If composed of material cut to modular heights, discontinuous but aligned horizontal joints are discernible.

Random Rubble   Masonry wall built of unsquared or rudely squared stones that are irregular in size and shape.

Squared Rubble   Wall construction in which squared stones of various sizes are combined in patterns that make up courses as high as, or higher than, the tallest stones.

Stone Masonry, Rubble   Stone masonry composed of irregularly shaped units bonded by mortar.

Story Pole   Marked pole for measuring masonry coursing during masonry construction.

Stretcher   Masonry unit laid with its greatest dimension horizontal and its face parallel to the wall face.

Strike   To remove excess mortar from the surface of a joint by cutting it flush with the unit surface using the edge of a trowel. See also **Tooling**.

Structural Clay Tile   Hollow masonry building unit composed of burned clay, shale, fire clay, or combinations of these materials.

Suction   See **Initial Rate of Absorption**.

Surface Bonded Masonry Bonding of masonry units by parging with a thin layer of fiber-reinforced mortar.

Surface Bonding Mortar Product containing hydraulic cement, glass fiber reinforcement with or without inorganic fillers, or organic modifiers in a prepackaged form, requiring only the addition of water prior to application.

T

Terra Cotta A fired clay product used for ornamental work.

Tie See **Connector, Wall Tie.**

Thermal Inertia Lag time required for a mass to heat or cool.

Thermal Mass Dense material capable of absorbing and storing heat.

Thermal Resistance The reciprocal of thermal transmittance (expressed by the notation R).

Tolerance Specified allowance of variation from a size specification.

Tooling Compressing and shaping the surface of a mortar joint. See also Strike.

Toothing Constructing the temporary end of a wall with the end stretcher of every alternate course projecting. Projecting units are toothers.

Travertine Variety of crystalline or microcrystalline limestone distinguished by layered structure. Pores and cavities commonly are concentrated in some of the layers, giving rise to an open texture.

Tuckpointing See **Pointing.**

U

Unit Masonry Construction of brick or block that is set in mortar, dry stacked, or mechanically anchored.

Unreinforced Masonry Masonry constructed without steel reinforcement, except that which may be used for bonding or reducing the effects of dimensional changes due to variations in moisture content or temperature.

V

Veneer A single facing wythe of masonry, anchored or adhered to a structural backing, but not designed to carry axial loads.

Vermiculite Micaceous aggregate used in lightweight insulating concrete, concrete masonry units, and plaster. (2) Insulation composed of natural vermiculite ore expanded to form an exfoliated structure.

W

Wall Vertical element with a horizontal length-to-thickness ratio greater than 3, used to enclose space.

Bearing Wall Wall supporting a vertical load in addition to its own weight.

Bonded Wall Masonry wall in which two or more wythes are bonded to act as a structural unit.

Composite Wall   Multi-wythe wall in which at least one of the wythes is dissimilar to the other wythe or wythes with respect to type of masonry unit.

Curtain Wall   Nonbearing exterior wall, secured to and supported by the structural members of the building.

Hybrid Wall   Masonry integrated structurally with a steel or reinforced concrete structural frame as part of the lateral load resisting system.

Multi-Wythe Wall   Masonry wall composed of two or more wythes.

Panel Wall Exterior   Non-loadbearing wall wholly supported at each story.

Parapet Wall   That part of any wall that is entirely above the roof.

Partition Wall   Interior wall one story or less in height. It is generally non-loadbearing. In Canada, partition means an interior wall of one-story or part-story height that is never loadbearing.

Retaining Wall   Wall that does not enclose a portion of a building; designed to resist the lateral displacement of soil or other material.

Screen Wall   Masonry wall in which an ornamental pierced effect is achieved by alternating rectangular or shaped blocks with open spaces.

Serpentine Wall   Wall that is sine wave in plan.

Single-Wythe Wall   Masonry wall that is only one unit in thickness.

Water Absorption   Process in which water enters a material or system through capillary pores and interstices and is retained without transmission.

Water Infiltration   Process in which water passes through a material or system and reaches an area that is not directly or intentionally exposed to the water source.

Water Leakage   Water infiltration that is unintended or uncontrolled, exceeds the resistance, retention, or discharge capacity of the system, or causes damage or accelerated deterioration.

Water Penetration   Process in which water enters a material or system through an exposed surface, joint, or opening.

Water Permeation   Process in which water enters, flows within, and spreads throughout a material or system.

Water Repellent   Material or treatment for surfaces to provide resistance to penetration by water.

Water Retentivity   That property of a mortar that prevents the rapid loss of water to masonry units of high suction. It prevents bleeding or water gain when mortar is in contact with relatively impervious units.

Water Vapor Permeance   Time rate of water vapor transmission through a unit area of a flat material or construction that is induced by unit vapor pressure difference between two specified surfaces under specified temperature and humidity conditions.

Waterproof   Impervious to water.

Waterproofing   In building construction, treatment of a surface or structure to prevent the passage of liquid water under hydrostatic, dynamic, or static pressure.

Weep Hole   A small hole allowing drainage of fluid.

Wind-Driven Rain Rain driven by the wind.

Workability Ability of mortar to be easily placed and spread.

Workmanship The art or skill of a worker; craftsmanship; the quality imparted to a thing in the process of creating it.

Wythe Each continuous vertical section of a wall, one masonry unit in thickness.

APPENDIX B
ASTM Reference Standards

The following reference standards of the American Society for Testing and Materials (ASTM) are related to masonry products, testing, and construction.

Clay Masonry Units

ASTM C27	*Fire Clay and High Alumina Refractory Brick*
ASTM C32	*Sewer and Manhole Brick*
ASTM C34	*Structural Clay Loadbearing Wall Tile*
ASTM C43	*Terminology Relating to Structural Clay Products*
ASTM C56	*Structural Clay Non-Loadbearing Tile*
ASTM C62	*Building Brick*
ASTM C106	*Fire Brick Flue Lining for Refractories and Incinerators*
ASTM C126	*Ceramic Glazed Structural Clay Facing Tile, Facing Brick, and Solid Masonry Units*
ASTM C155	*Insulating Fire Brick*
ASTM C212	*Structural Clay Facing Tile*
ASTM C216	*Facing Brick*
ASTM C279	*Chemical Resistant Brick*
ASTM C315	*Clay Flue Linings*
ASTM C410	*Industrial Floor Brick*
ASTM C416	*Silica Refractory Brick*
ASTM C530	*Structural Clay Non-Loadbearing Screen Tile*
ASTM C652	*Hollow Brick*
ASTM C902	*Pedestrian and Light Traffic Paving Brick*
ASTM C1088	*Thin Veneer Brick Units Made From Clay or Shale*
ASTM C1261	*Firebox Brick for Residential Fireplaces*
ASTM C1272	*Heavy Vehicular Paving Brick*
ASTM C1405	*Glazed Brick (Single Fired, Solid Brick Units)*

Cementitious Masonry Units

ASTM C55	Concrete Building Brick
ASTM C73	Calcium Silicate Face Brick (Sand-Lime Brick)
ASTM C90	Loadbearing Concrete Masonry Units
ASTM C129	Nonloadbearing Concrete Masonry Units
ASTM C139	Concrete Masonry Units for Construction of Catch Basins and Manholes
ASTM C744	Prefaced Concrete and Calcium Silicate Masonry Units
ASTM C936	Solid Concrete Interlocking Paving Units
ASTM C1319	Concrete Grid Paving Units
ASTM C1364	Architectural Cast Stone
ASTM C1372	Segmental Retaining Wall Units
ASTM C1386	Precast Autoclaved Aerated Concrete (PAAC) Wall Construction Units
ASTM C1452	Reinforced Autoclaved Aerated Concrete Elements
ASTM C1555	Autoclaved Aerated Concrete Masonry

Adhered Manufactured Stone Masonry Veneer

ASTM C482	Test Method for Bond Strength of Ceramic Tile to Portland Cement
ASTM C932	Surface-Applied Bonding Compounds for Exterior Plaster
ASTM C1059	Latex Agents for Bonding Fresh to Hardened Concrete

Natural Stone

ASTM C119	Terminology Relating to Dimension Stone
ASTM C503	Marble Dimension Stone
ASTM C568	Limestone Dimension Stone
ASTM C615	Granite Dimension Stone
ASTM C616	Quartz-Based Dimension Stone
ASTM C629	Slate Dimension Stone

Mortar and Grout

ASTM C5	Quicklime for Structural Purposes
ASTM C33	Aggregates for Concrete
ASTM C91	Masonry Cement
ASTM C144	Aggregate for Masonry Mortar
ASTM C150	Portland Cement
ASTM C199	Pier Test for Refractory Mortar
ASTM C207	Hydrated Lime for Masonry Purposes
ASTM C270	Mortar for Unit Masonry
ASTM C330	Lightweight Aggregates for Structural Concrete

ASTM C331	*Lightweight Aggregates for Concrete Masonry Units*
ASTM C404	*Aggregates for Masonry Grout*
ASTM C476	*Grout for Reinforced and Nonreinforced Masonry*
ASTM C658	*Chemical Resistant Resin Grouts*
ASTM C887	*Packaged, Dry, Combined Materials for Surface Bonding Mortar*
ASTM C932	*Surface Applied Bonding Compounds for Exterior Plastering*
ASTM C1059	*Latex Agents for Bonding Fresh to Hardened Concrete*
ASTM C1142	*Extended Life Mortar for Unit Masonry*
ASTM C1329	*Mortar Cement*
ASTM C1586	*Quality Assurance of Mortars*
ASTM C1660	*Thin Bed Mortar for Autoclaved Aerated Concrete (AAC) Masonry*
ASTM C1713	*Mortars for the Repair of Historic Masonry*
ASTM C1714	*Preblended Dry Mortar Mix for Unit Masonry*
ASTM E2260	*Repointing (Tuckpointing) Historic Masonry*

Reinforcement and Accessories

ASTM A82	*Cold Drawn Steel Wire for Concrete Reinforcement*
ASTM A153	*Zinc Coating (Hot-Dip) on Iron or Steel Hardware*
ASTM A165	*Electro-Deposited Coatings of Cadmium on Steel*
ASTM A167	*Stainless and Heat Resisting Chromium-Nickel Steel Plate, Sheet, and Strip*
ASTM A185	*Welded Steel Wire Fabric for Concrete Reinforcement*
ASTM A496	*Deformed Steel Wire for Concrete Reinforcement*
ASTM A615	*Deformed and Plain Billet-Steel Bars for Concrete Reinforcement*
ASTM A616	*Rail-Steel Deformed and Plain Bars for Concrete Reinforcement*
ASTM A617	*Axle-Steel Deformed and Plain Bars for Concrete Reinforcement*
ASTM A641	*Zinc Coated (Galvanized) Carbon Steel Wire*
ASTM A951	*Masonry Joint Reinforcement for Masonry*
ASTM B227	*Hard-Drawn Copper-Covered Steel Wire, Grade 30HS*
ASTM C1242	*Design, Selection, and Installation of Exterior Dimension Stone Anchors and Anchoring Systems*
ASTM C1472	*Calculating Movement and Other Effects When Establishing Sealant Joint Width*

Sampling and Testing

ASTM C67	*Sampling and Testing Brick and Structural Clay Tile*
ASTM C97	*Absorption and Bulk Specific Gravity of Natural Dimension Stone*
ASTM C109	*Compressive Strength of Hydraulic Cement Mortars*
ASTM C140	*Sampling and Testing Concrete Masonry Units*

Appendix B

ASTM C170	*Compressive Strength of Natural Dimension Stone*
ASTM C241	*Abrasion Resistance of Stone*
ASTM C267	*Chemical Resistance of Mortars, Grouts, and Monolithic Surfacings*
ASTM C426	*Drying Shrinkage of Concrete Block*
ASTM C780	*Preconstruction and Construction Evaluation of Mortars for Plain and Reinforced Unit Masonry*
ASTM C856	*Petrographic Examination of Hardened Concrete*
ASTM C880	*Flexural Strength of Natural Dimension Stone*
ASTM C952	*Bond Strength of Mortar to Masonry Units*
ASTM C1006	*Splitting Tensile Strength of Masonry Units*
ASTM C1019	*Sampling and Testing Grout*
ASTM C1028	*Determining the Static Coefficient of Friction of Ceramic Tile and Other Like Surfaces by the Horizontal Dynamometer Pull-Meter Method* [used to measure slip-resistance of stone flooring and paving]
ASTM C1072	*Measurement of Masonry Flexural Bond Strength*
ASTM C1093	*Accreditation of Testing Agencies for Unit Masonry*
ASTM C1148	*Measuring the Drying Shrinkage of Masonry Mortar*
ASTM C1194	*Compressive Strength of Architectural Cast Stone*
ASTM C1195	*Absorption of Architectural Cast Stone*
ASTM C1196	*In Situ Compressive Stress within Solid Unit Masonry Estimated Using Flatjack Method*
ASTM C1197	*In Situ Measurement of Masonry Deformability Using the Flatjack Method*
ASTM C1262	*Evaluating the Freeze-Thaw Durability of Manufactured Concrete Masonry and Related Concrete Units*
ASTM C1314	*Constructing and Testing Masonry Prisms Used to Determine Compliance with Specified Compressive Strength of Masonry*
ASTM C1324	*Examination and Analysis of Hardened Masonry Mortar*
ASTM C1357	*Evaluating Masonry Bond Strength*
ASTM C1403	*Rate of Water Absorption of Masonry Mortars*
ASTM C1601	*Field Determination of Water Penetration of Masonry Wall Surfaces*
ASTM C1611	*Slump Flow of Self-Consolidating Concrete*
ASTM C1715	*Evaluation of Water Leakage Performance of Masonry Drainage Systems*
ASTM D75	*Sampling Aggregates*
ASTM E72	*Conducting Strength Tests of Panels for Building Construction*
ASTM E119	*Fire Tests of Building Construction and Materials*
ASTM E447	*Compressive Strength of Masonry Prisms*
ASTM E488	*Strength of Anchors in Concrete and Masonry Elements*
ASTM E514	*Water Permeance of Masonry*

ASTM E518	*Flexural Bond Strength of Masonry*
ASTM E519	*Diagonal Tension in Masonry Assemblages*
ASTM E754	*Pullout Resistance of Ties and Anchors Embedded in Masonry Mortar Joints*

Assemblages

ASTM C901	*Prefabricated Masonry Panels*
ASTM C946	*Construction of Dry Stacked, Surface Bonded Walls*
ASTM E835	*Dimensional Coordination of Structural Clay Units, Concrete Masonry Units, and Clay Flue Linings*
ASTM C1283	*Installing Clay Flue Lining*
ASTM C1400	*Reduction of Efflorescence Potential in New Masonry Walls*
ASTM E1602	*Construction of Solid Fuel-Burning Masonry Heaters*
ASTM E2112	*Practice for Installation of Exterior Windows, Doors, and Skylights*
ASTM E2128	*Evaluating Water Leakage of Building Walls*
ASTM E2266	*Design and Construction of Low-Rise Frame Building Wall Systems to Resist Water Intrusion*

Bibliography

ALLEN, EDWARD. *Fundamentals of Building Construction.* New York: John Wiley & Sons, 1999.

AMRHEIN, JAMES E. *Steel in Masonry.* Los Angeles: Masonry Institute of America, 1977.

AMRHEIN, JAMES E. *Grout . . . The Third Ingredient.* Los Angeles: Masonry Institute of America, 1980.

AMRHEIN, JAMES E. *Reinforced Masonry Engineering Handbook*, 5th ed. Los Angeles: Masonry Institute of America, 1992.

AMRHEIN, JAMES E., AND MICHAEL W. MERRIGAN. *Marble and Stone Slab Veneer,* 2nd ed. Los Angeles: The Masonry Institute of America, 1989.

AMRHEIN, JAMES E., AND MICHAEL W. MERRIGAN. *Reinforced Concrete Masonry Construction Inspector's Handbook,* 2nd ed. Los Angeles: The Masonry Institute of America and the International Conference of Building Officials, 1989.

AMRHEIN, JAMES E., ET AL. *Masonry Design Manual,* 3rd ed. Los Angeles: Masonry Industry Advancement Committee, 1979.

BALLAST, DAVID KENT. *Handbook of Construction Tolerances.* New York: McGraw-Hill, 1994.

BALLAST, DAVID KENT. *Architect's Handbook of Construction Detailing, 2nd edition.* New York: McGraw-Hill, 2009.

BEALL, CHRISTINE. *Thermal and Moisture Protection Manual.* New York: McGraw-Hill, 1999.

BEALL, CHRISTINE. *Masonry and Concrete for Residential Construction.* New York: McGraw-Hill Complete Construction Series, 2001.

BEALL, CHRISTINE. *Masonry Mortar.* The Line, Texas Masonry Council, 2003.

BEALL, CHRISTINE, AND ROCHELLE JAFFE. *Concrete and Masonry Databook.* New York: McGraw-Hill, 2003.

BORCHELT, J. GREGG, ED. *Masonry: Materials, Properties, and Performance,* ASTM STP 778. Philadelphia: American Society for Testing and Materials, 1982.

BRICK INDUSTRY ASSOCIATION. *Technical Notes on Brick Construction,* Nos. 1–45. Reston, VA: BIA.

CANADIAN HOMEBUILDERS ASSOCIATION. *Builder's Manual.* Ottawa, ON, Canada: CHBA, 1994.

CAST STONE INSTITUTE. *Technical Manual with Case Histories.* Tucker, GA: Cast Stone Institute.

CHRYSLER, JOHN, ET AL. *Masonry Design Manual,* 4th ed. Los Angeles: Masonry Institute of America and International Code Council, 2007.

CONSTRUCTION SPECIFICATIONS INSTITUTE. *Manual of Practice.* Alexandria, VA: CSI, 1992/1994.

DRYSDALE, ROBERT, AHMAD HAMID, AND LAWRIE BAKER. *Masonry Structures Behavior and Design,* 2nd ed. Boulder, CO: The Masonry Society, 1999.

FARNY, J. A., J. M. MELANDER, AND WILLIAM C. PANARESE. *Concrete Masonry Handbook,* 6th ed. Skokie, IL: Portland Cement Association, 2008.

GRIMM, CLAYFORD T. *Cleaning Masonry—A Review of the Literature.* Arlington, TX: Construction Research Center of the University of Texas at Arlington, 1988.

GRIMM, CLAYFORD T. *Conventional Masonry Mortar—A Review of the Literature.* Arlington, TX: Construction Research Center of the University of Texas at Arlington, 1994.

GROGAN, J. C., AND J. T. CONWAY, EDS. *Masonry: Research, Application, and Problems,* ASTM STP 871. Philadelphia: American Society for Testing and Materials, 1985.

GUNN, RICK. *Listen to the Mason: Portland Cement-Lime Type N Mortar (1:1:6) Provides the Necessary Workability and Strength.* International Building Lime Symposium, 2005.

HARRIS, HARRY A., ED. *Masonry: Materials, Design, Construction, and Maintenance,* ASTM STP 992. Philadelphia: American Society for Testing and Materials, 1988.

HEYMAN, JACQUES. *The Masonry Arch.* New York: John Wiley & Sons, 1982.

HENSHELL, JUSTIN. *Moisture Related Problems in Masonry Parapets.* Fifth North American Masonry Conference, 1990.

HUCKABEE, CHRIS. *Are You Building A School or a Liability? A Guide to Using Total Masonry Construction in Public Schools.* Austin, TX: Texas Masonry Council, 2004.

INDIANA LIMESTONE INSTITUTE. *Indiana Limestone Handbook,* 17th ed. Bedford, IN: ILI.

INTERNATIONAL MASONRY INSTITUTE. *Masonry Grout.* 2009.

JAFFE, ROCHELLE C. *Masonry Basics.* Boulder, CO: The Masonry Society, 2002.

JAFFE, ROCHELLE C. *Masonry Instant Answers.* New York: McGraw-Hill, 2004.

JOHNSON, FRANKLIN B., ED. *Designing, Engineering and Constructing with Masonry Products.* In *Proceedings of the International Conference on Masonry Structural Systems,* The University of Texas at Austin, 1967. Houston, TX: Gulf Publishing, 1969.

KLINGNER, RICHARD. *Masonry Structural Design.* New York: McGraw-Hill, 2010.

KPFF CONSULTING ENGINEERS. *Notes on the Selection, Design and Construction of Reinforced Hollow Clay Masonry.* Los Angeles: Western States Clay Products Association, 1995.

KREH, R. T., SR. *Masonry Skills.* New York: Van Nostrand Reinhold, 1976.

KUBBA, SAM. *Architectural Forensics.* New York: McGraw-Hill, 2008.

LIEFF, M., AND H. R. TRECHSEL, EDS. *Moisture Migration in Buildings,* ASTM STP 779. Philadelphia: American Society for Testing and Materials, 1982.

MASONRY INDUSTRY COUNCIL. *Hot and Cold Weather Masonry Construction Manual.* Lombard, IL: Masonry Industry Council, 1999.

MASONRY INSTITUTE OF AMERICA. *Masonry Veneer.* Los Angeles: MIA, 1974.

MASONRY INSTITUTE OF AMERICA. *Reinforcing Steel in Masonry.* Los Angeles: MIA.

MASONRY INSTITUTE OF AMERICA. *Grout . . . The Third Ingredient.*

MASONRY INSTITUTE OF BRITISH COLUMBIA. *A Guide to Rain Resistant Masonry Construction for the British Columbia Coastal Climate.* Burnaby, BC, Canada: Masonry Institute of British Columbia, 1985.

MASONRY STANDARDS JOINT COMMITTEE. *Building Code Requirements for Masonry Structures* (ACI 530/ASCE 5/TMS 402). Detroit, MI: American Concrete Institute, 2008.

MASONRY STANDARDS JOINT COMMITTEE. *Specification for Masonry Structures* (ACI 530.1/ASCE 6/TMS 602). Detroit, MI: American Concrete Institute, 2008.

MASONRY VENEER MANUFACTURER'S ASSOCIATION. *Installation Guide for Adhered Concrete Masonry Veneer.* Washington, DC: MVMA, 2010.

MATTHYS, JOHN H., ED. *Masonry: Components to Assemblages,* ASTM STP 1063. Philadelphia: American Society for Testing and Materials, 1990.

MATTHYS, JOHN H., ED. *Masonry Designers' Guide, 6th ed., Based on Building Code Requirements for Masonry Structures (ACI 530-99/ASCE 5-99/TMS 402-08 and Specification for Masonry Structures (ACI 530.1-99/ASCE 699/TMS 602).* Boulder, Colorado: The Masonry Society, 2010.

MELANDER, JOHN. "Choosing the Right Mortar." *Masonry,* September/October 2000.

NASHED, FRED. *Time Saver Details for Exterior Wall Design.* New York: McGraw-Hill, 1996.

NATIONAL CONCRETE MASONRY ASSOCIATION. *TEK Bulletins,* Nos. 1-01 through 19-04. Herndon, VA: NCMA.

NATIONAL LIME ASSOCIATION. *Masonry Mortar Technical Notes Series.* Arlington, VA: NLA.

NATIONAL PARK SERVICE, U.S. Department of the Interior. *Preservation Briefs,* Nos. 1–23. Washington, DC: U.S. Government Printing Office.

NEWMAN, MORTON. *Structural Details for Masonry Construction.* New York: McGraw-Hill, 1986.

O'BRIEN, JAMES J. *Construction Inspection Handbook,* 3rd ed. New York: John Wiley & Sons, 1991.

O'CONNOR, THOMAS F., ED. *Building Sealants: Materials, Properties and Performance,* ASTM STP 1069. Philadelphia: American Society for Testing and Materials, 1990.

PATTERSON, STEPHEN AND MEDAN MEHTA. *Roofing Design and Practice.* Upper Saddle River, NJ: Prentice-Hall, 2001.

PATTERSON, TERRY L. *Illustrated 2006 Building Code Handbook.* New York: McGraw-Hill, 2006.

QUIROUETTE, R. L. *The Difference Between a Vapor Barrier and an Air Barrier. Ottawa, Ontario:* National Research Council of Canada, 1985.

RANDALL, FRANK, AND WILLIAM PANARESE. *Concrete Masonry Handbook, 5th ed. Skokie, IL: Portland Cement Association, 1991.*

RITCHIE, THOMAS, AND J. IVAN DAVISON. *Cement-Lime Mortars.* Ottawa, ON, Canada: National Research Council, Division of Building Research, 1964.

ROCKY MOUNTAIN MASONRY INSTITUTE. *Adhered Natural Stone Veneer Installation Guide.* Denver, CO: RMMI, 2010.

SCHEFFLER, MICHAEL J., IAN R. CHIN, AND DEBORAH SLATON. "Moisture Expansion of Fired Bricks." In *Proceedings 5NAMC,* June 1990, Vol. II, pp. 549–562.

SCHNEIDER, ROBERT, AND WALTER DICKEY. *Reinforced Masonry Design, 2d ed. Englewood Cliffs, N.J.: Prentice Hall, 1989*

SCHWARTZ, THOMAS A., ED. *Water in Exterior Building Walls: Problems and Solutions,* ASTM STP 1107. Philadelphia: American Society for Testing and Materials, 1991.

TMS ARCHITECTURAL PRACTICES COMMITTEE. *TMS Annotated Guide to Masonry Specifications, 2nd Edition.* Boulder, CO: TMS, 2007.

TOBIASSON, WAYNE. "Vapor Retarders to Control Summer Condensation." In *Proceedings of the ASHRAE/DOE/BTECC Conference on Thermal Performance of the Exterior Envelope of Buildings IV,* Orlando, FL, 1989.

TOBIASSON, WAYNE, AND MARCUS HARRINGTON. "Vapor Drive Maps of the U.S." In *Proceedings of the ASHRAE/DOE/BTECC Conference on Thermal Performance of the Exterior Envelope of Buildings III,* Clearwater Beach, FL, 1985.

TRECHSEL, HEINZ R., AND MARK BOMBERG, EDS. *Water Vapor Transmission Through Building Materials and Systems: Mechanisms and Measurement,* ASTM STP 1039. Philadelphia: American Society for Testing and Materials, 1989.

Index

A

Acoustical characteristics, 87–90
    sound absorption, 22, 88–89, 323
    sound transmission, 88–93, 323
    sound transmission class (STC), 88–93
Actual dimensions, 32
Adhered veneer, 6, 31, 44, 205–213
Admixtures, 15, 22, 27, 37–38, 51, 54
Aggregates, 13–14, 21–22, 40, 51, 78–79, 100, 312, 324–325
Air barrier, 86–87, 124, 126–127, 154, 156–158, 178, 186
Analytical design, 191, 261–274
Anchor bolts, 60, 256, 262
Anchored veneer, 123, 152, 157, 191–203
Anchors, 55–57, 58, 60, 64–67, 191–194, 202, 232, 296–297, 300, 312, 317, 325, 326, 330, 332, 336
Arches, 136, 237, 245–247
Arching action, 237–239
ASTM A82, 64, 312, 325, 369
ASTM A153, 57, 59, 64, 67, 194, 232, 312, 326, 369
ASTM A167, 57–59, 63, 312, 326, 369
ASTM A185, 64, 312, 369
ASTM A615, 56, 312, 369
ASTM A616, 56, 312, 369
ASTM A617, 56, 312, 369
ASTM A641, 57, 59, 369
ASTM A951, 59, 312, 325, 369
ASTM C5, 325, 367
ASTM C32, 367
ASTM C33, 13, 39, 324, 368
ASTM C34, 367
ASTM C55, 14, 21, 313, 324, 368
ASTM C56, 367
ASTM C62, 4, 9, 23, 313, 322, 323, 367
ASTM C67, 75, 318, 322, 369
ASTM C73, 14, 21, 368
ASTM C90, 14, 18, 21, 242, 304, 313, 317, 323, 324, 367
ASTM C91, 36–37, 39, 43, 312, 325, 368
ASTM C97, 23, 38, 369
ASTM C126, 4–5, 356, 367
ASTM C129, 14, 21, 217, 304, 313, 324, 368
ASTM C140, 4–5, 142, 312, 323–324, 369
ASTM C144, 37, 39, 312, 325, 368
ASTM C150, 35, 39, 43, 312, 324, 368
ASTM C170, 23, 370
ASTM C199, 368
ASTM C207, 36, 39, 43, 312, 325, 368
ASTM C212, 356, 367
ASTM C216, 3, 4, 5, 9, 75, 127, 217, 304, 312, 320, 322–323, 367
ASTM C241, 23, 370
ASTM C270, 12, 35, 36, 41, 44–50, 185, 242, 264, 312–313, 324, 325–327, 330, 368
ASTM C279, 367
ASTM C315, 367
ASTM C331, 13, 324, 369
ASTM C404, 39, 51, 52, 54, 312, 325, 369
ASTM C410, 367
ASTM C476, 50, 52–54, 242, 264, 271, 312, 325, 356, 369
ASTM C503, 23, 24, 368
ASTM C530, 217, 367
ASTM C568, 23, 368
ASTM C615, 23, 368
ASTM C616, 23, 25, 368
ASTM C629, 23, 368
ASTM C652, 4, 9–11, 313, 323, 367
ASTM C744, 17, 368

ASTM C780, 49–50, 266, 314, 325–326, 370
ASTM C880, 378
ASTM C901, 317, 371
ASTM C902, 367
ASTM C936, 368
ASTM C946, 371
ASTM C952, 370
ASTM C1019, 52–53, 266, 314, 326, 370
ASTM C1072, 370
ASTM C1093, 370
ASTM C1142, 278, 326, 369
ASTM C1242, 369
ASTM C1261, 367
ASTM C1272, 367
ASTM C1283, 371
ASTM C1314, 265–266, 272, 314, 327, 370
ASTM C1319, 368
ASTM C1324, 50, 370
ASTM C1329, 37, 39, 43, 44, 312, 324, 369
ASTM C1357, 370
ASTM C1364, 29, 31, 368
ASTM C1372, 368
ASTM C1386, 15, 368
ASTM C1400, 142, 371
ASTM C1405, 4–5, 312, 323, 367
ASTM C1452, 15, 368
ASTM C1472, 109–110, 369
ASTM C1601, 342, 370
ASTM C1602, 371
ASTM E72, 370
ASTM E835, 371
ASTM E488, 370
ASTM E514, 342, 370
ASTM E518, 371
ASTM E519, 371
ASTM E754, 371
ASTM E2112, 148, 201, 371
ASTM E2128 121, 339, 340, 371
ASTM E2260, 369
ASTM E2266, 142, 371
Autoclaved aerated concrete (AAC), 14–15, 94, 350, 352, 368
Autogenous healing, 36
Axial load distribution, 253

B

Barrier wall, 123–125, 138, 175
Beams, 189, 216, 237, 239, 255–256, 260–261, 264,
Bond:
    beams, 107, 132, 159, 253, 257, 270, 274

Bond (*Cont.*):
    mortar, 8, 38, 40–42, 72, 122, 124–125, 128, 138, 157, 163, 179, 208, 269, 270, 280, 336, 350
    patterns, 4, 96, 218, 260, 267, 305, 350
    stack, 4, 96, 218, 222, 231, 253, 273, 274, 286, 304, 305, 322, 351
    strength, 12, 36–38, 40–45, 128, 139, 269, 278, 280, 351, 368, 370–371
Bracing, 303, 313, 358
Brick:
    calcium silicate, 14, 17, 19, 21
    fire, 352, 367
    glazed, 4–6, 8, 75, 312, 367
    grade, 3–6, 10–11, 14, 21, 75–76, 127, 205, 217, 312–313, 317, 320, 322–323, 352
    hollow, 4, 9–11, 77, 78, 90, 159, 215, 283, 313, 323, 351, 367
    initial rate of absorption, 11, 12, 27, 40, 52, 128, 130, 280, 313, 318, 335, 349, 357
    modular, 7–8, 11, 19, 151, 206, 283, 311
    paving, 36, 43, 264, 272, 352, 364, 367–368, 372
    refractory, 352, 367
    thin, 6–7, 205–206, 367
    type, 3–5, 10, 304, 305, 312, 322–323, 352

C

Calcium carbonate stains, 355
Calcium chloride, 37, 51, 55, 312, 330
Calcium silicate brick, 14, 17, 19, 21, 351, 368
Carbonation, 55, 355
Cast stone, 27–28, 30–31, 76, 108, 111, 130, 132–133, 167, 169, 180, 182, 186, 242, 362, 368
Cavity drainage, 1, 122–127, 139, 146, 148–149, 151–152, 155–156, 162–163, 165, 167, 176–177, 179–180, 186–188, 191, 194–197, 200–203, 206, 241, 252, 275, 284, 286–287, 290, 293–295, 312, 328, 333, 335–337, 370
Cavity walls, 58, 62, 70, 82, 84, 86, 90, 101, 123, 125–126, 152, 155
Cellular concrete (*See* Autoclaved aerated concrete)
Cleaning, 318, 328, 330–331, 337
Coatings, 14, 22, 75, 122, 136–139, 155, 158, 168–169, 337
Cold weather construction, 35, 37, 51, 69, 280, 306–307, 309, 312–313, 328, 331, 337, 353
Composite walls, 175–176, 179

Concrete masonry units (CMU)
    calcium silicate brick, 14, 17, 19, 21, 351, 352
    concrete block, 12–22, 52, 70–71, 73, 84, 89, 125, 137–139, 150, 159–173, 175–184, 186–189, 216, 221, 224, 227, 237, 239, 282–283, 285, 304, 323, 353
    concrete brick, 12, 14, 17, 21, 59, 313, 324, 351, 353
    paving, 43, 368
    prefaced, 17, 368
    sand–lime brick, 12, 14, 17, 352
    screen block, 15, 218, 353
    segmental retaining walls, 368
    slump block, 14
Condensation, 55, 82, 84, 86–87, 122, 152, 154–157, 162, 177, 179, 196
Connectors, 60–67, 191, 279, 313–314, 321, 328, 333, 350
    anchors, 55–58, 60–61, 64–67, 80, 99, 101–105, 107–110, 127, 162, 179–180, 184–185, 187, 191–196, 199, 202, 223, 232, 266, 286, 296–297, 300, 305, 312, 317, 321, 325–326, 332–333, 336, 349, 353
    fasteners, 353
    ties, 55–57, 59–60, 62–64, 84, 126, 178, 260, 268–269, 296–297, 300, 312, 325–326, 332, 336, 350, 353, 363, 371
Construction tolerances, 5, 17, 29, 53, 109–110, 128, 241, 293, 296–297, 299, 313–314, 318, 322–325, 332, 335, 352, 355, 363
Control joints, 18–20, 62, 99–109, 113, 115–117, 119, 220–221, 296, 312–314, 318
Coping, 6, 20, 28, 30, 108, 111, 130–134, 140–141, 165–168, 180–182, 184, 189, 195, 197–198, 208–209, 227–228, 230, 336, 354
Corbelling, 237–238, 261, 270, 354
Corrosion, 29, 31, 37, 51, 55–60, 67, 177, 191–194, 242, 294, 296, 312, 326
Coursing, 7–8, 10, 169, 171–172, 218, 282–283, 329, 336, 362
Curing, 14, 21, 36, 38, 42, 54, 56, 100–101, 130, 269, 278, 280, 296, 309, 313, 337, 350, 354
Curtain walls, 83, 96, 123–124, 143, 145, 159–160, 162, 165, 169, 178, 180, 185, 199–200, 265, 290–291, 354, 364

D

Dampproofing, 148–149, 155, 180, 184, 186–187 (*See also* Waterproofing)

Deflection, 103, 108, 110, 122–123, 152, 192–193, 195, 206, 231, 240, 253, 262, 346–347
Design:
    analytical, 96, 191, 251, 261–274
    empirical, 46, 159, 191, 215–217, 224, 251, 257–258, 264, 270, 272, 320, 327, 355
    strength, 251, 261–274
Differential movement, 58, 64, 99, 101–103, 108, 111, 117, 130, 165, 175, 178, 185, 192, 208, 241–242, 253, 296, 354, 357
Dimensions:
    actual, 32
    nominal, 7, 19, 32, 206, 244, 311, 355
Drainage accessories, 72–73, 294–295, 312, 328, 335
Drainage wall, 1, 122–126

E

Efflorescence, 37, 51, 136, 142, 152, 177, 179, 228, 276, 294, 322–323, 355, 371
Empirical design, 46, 159, 191, 215–217, 224, 251, 257–258, 264, 270, 272, 320, 327, 355
Environmental impact, 90–94
Equivalent thickness, 77–79, 355
Expansion joints, 99–100, 102–103, 105–107, 109–119, 127, 143, 193, 199, 208, 224, 227, 296, 312–313, 332, 336, 355, 357

F

Fasteners, 353
FBA brick, 4, 10, 304, 305, 312, 322–323, 352
FBS brick, 3–5, 10, 304, 305, 312, 322–323, 352
FBX brick, 3–5, 10, 304, 305, 312, 322–323, 352
Fences, 215–231
Field observation, 96, 334–338 (*See also* Inspection)
Fire resistance, 17, 19, 76–81, 89, 96, 323
Flashing, 1–2, 20, 44, 55–56, 67–70, 72–73, 76, 85, 94, 97, 101–102, 108–109, 116, 118, 121–128, 131–133, 135–136, 138–151, 162–172, 177–189, 192, 195–203, 207–210, 211, 213, 228, 230, 232, 241, 255, 275–276, 280, 288, 290–294, 305–306, 311–314, 316–317, 326, 328–329, 332–333, 335–336, 338, 356–358, 361
Flexible anchorage, 99, 101, 103–105, 159, 165, 169
Foundation walls, 15, 46, 219, 255

G

Glass block, 32–33, 38, 45, 80, 90, 215, 231–235, 358
Glazed brick, 3–6, 8, 75, 312, 323, 352, 367
Granite, 22–24, 356, 368
Grout, 22, 39–40, 49–58, 67, 90–93, 106–107, 139, 149, 159, 161, 164, 192, 230, 242–245, 251, 255, 259, 264, 266, 270–273, 276–280, 284–287, 296, 299–303, 306–309, 312–314, 317, 321, 324–326, 328, 330–331, 334–337, 344, 349–352, 354, 356–357, 359–361, 368–370
Grout slump, 51–52, 54, 278, 370
Grout strength, 317
Grouted/Ungrouted, 18, 50–53, 55–56, 62–63, 68, 77, 79, 109, 125, 140, 158–162, 165, 169–170, 173, 175–176, 186, 188, 203, 208, 216, 219, 230, 240, 242, 252, 257–258, 270, 272–274, 285, 297–298, 318, 329, 336, 344–345, 351, 354, 356
Grouting, 20, 51–53, 67, 159, 188, 203, 228, 251, 299, 301–303, 306, 313, 321, 325, 328, 331, 334, 336, 356

H

h/t ratios, 215–217, 264, 357
Headers, 222, 258, 260, 267, 350–351
Hot weather construction, 38, 51, 307–308
Hysteresis, 27, 357

I

Initial rate of absorption, 11, 12, 27, 40, 52, 128, 130, 280, 313, 318, 335, 349, 357
Inspection, 20, 96, 262, 280–281, 303, 311, 315, 317, 320–321, 328, 331–336, 339–340 (*See also* Field observation)
Insulation, 22, 58, 60, 63, 72–73, 80–86, 92, 101–102, 110, 125–126, 155, 161–163, 178–180, 184, 186–187, 195–196, 205, 213, 217, 266, 280, 286, 360, 363

J

Joint reinforcement, 56–61, 63–64, 100, 106–107, 115–117, 159, 161, 169, 175, 193–194, 233, 260, 268–269, 273–274, 296–297, 312–313, 325–326, 328, 336, 344–345, 358, 369
Joints:
  concave tooled, 128, 162, 228, 288, 336, 358
  control, 18–20, 62, 99–109, 113, 115–117, 119, 220–221, 296, 312–314, 318

Joints (*Cont.*):
  expansion, 99–100, 102–103, 105–107, 109–119, 127, 143, 193, 199, 208, 224, 227, 296, 312–313, 332, 336, 355, 357
  mortar, 1, 36, 38, 42, 128–129, 132–133, 162, 182, 200, 228, 245, 275, 284, 288, 297, 304–305, 314, 321, 335–336, 357, 360, 363
  sealant, 62, 70, 106, 108–111, 118–119, 123–124, 130, 131–134, 167–169, 182, 200, 209, 369
  tooled, 128, 162, 228, 288, 336, 358

L

Lateral support, 103, 159, 215–217, 219, 224, 258, 262, 264–266, 272, 296, 358
Lime, 12, 22, 36–47, 50, 53, 128, 231, 277–278, 308, 312, 325, 328, 352, 358, 368
Limestone, 23–24, 26, 78, 358, 368
Limit states design (*See* Design, strength)
Lintels, 192, 237–245
Lipped brick (also called lintel brick), 9–10, 147, 184–185, 286
Loadbearing masonry, 14, 21, 42, 46, 50, 77, 92, 96, 125, 159, 171, 175, 217, 237, 251–274, 317, 322–323, 326, 367–368

M

Marble, 22, 24, 26–27, 357–358, 368
Masonry cement, 36–39, 43, 47, 128, 277, 312, 325, 328, 330, 352, 358, 368
Minimum wall thickness, 77, 258
Mock–ups, 313–315, 322–324, 330–333
Moist curing (*See* Curing)
Moisture:
  expansion, 99–101, 103, 108, 175–176, 178, 227, 253, 296
  shrinkage, 22, 99–100, 104, 108, 175–176, 178, 296
  vapor (*See* Condensation)
Mortar:
  bond, 8, 40–42, 72, 122, 124–125, 128, 138, 157, 163, 179, 208, 269–270, 280, 336, 350
  bond strength, 12, 36–38, 40–45, 128, 139, 269, 278, 280, 351, 368, 370–371
  cement, 37, 39, 43–44, 47, 312, 324, 328, 330, 369
  joints, 1, 36, 38, 42, 128–129, 132–133, 162, 182, 200, 228, 245, 275, 284, 288, 297, 304–305, 314, 321, 335–336, 357–358, 360, 363

Mortar (*Cont.*):
  mixes, 12, 35, 40, 46, 277, 313, 328, 337, 369
  placement, 128, 162, 228, 269, 275, 284, 286, 288, 313, 318, 336, 356, 358, 360, 362
  repointing, 360, 369
  retempering, 36, 40–42, 279, 308, 313, 335, 361
  slump, 50, 279
  testing, 50, 266, 279, 370
  tuckpointing, 45, 360, 369
  type, 12, 41, 44–47, 76, 128, 159, 178, 195, 205, 217, 231, 242, 263–264, 271, 311–312, 325, 327
Multi–wythe walls, 58, 62, 70, 82, 84, 86, 90, 101, 123, 125–126, 152, 155, 175–176, 179

N

Noise reduction coefficient (NRC), 89
Nominal dimensions, 7, 19, 32, 206, 244, 311, 355

P

Parapets, 46, 102, 108, 122, 130, 134–136, 140–141, 146, 165, 168, 180, 182–183, 185, 189, 197–199, 209, 333, 364
Partitions, 10, 14–15, 43, 46, 87–88, 100, 108, 215–217, 258, 264, 272, 352, 364, 367–368, 372
Paving, 36, 43, 352, 367–368, 372
Pilasters, 19, 52, 80, 111, 115–116, 159, 215–216 219–223, 225–226, 258, 318, 358–360

Q

Quality assurance, 48, 50, 317, 319–321, 328, 330–331, 340, 369
Quality control, 50, 126, 276, 313, 319–320, 328, 330–331

R

Retempering, 36, 40–42, 279, 308, 313, 335, 361
Rowlok, 130, 218, 286, 361
Rubble stone, 26, 31, 72, 231

S

Sample panel, 38, 131, 313–314, 322–324, 330–332, 334
Sandstone, 23, 25–26, 361
Screen walls, 215, 219, 225–256, 361, 364
Seismic requirements, 58, 60, 66, 185, 191, 193, 200, 205, 242, 251, 253–254, 257, 272–274, 297, 354, 364

Serpentine walls, 224–225, 227
Shelf angles, 102–103, 108, 116, 126–127, 354, 364
Single–wythe walls, 58, 73, 83, 86, 96, 125, 140, 159, 161–173, 176, 288
Slump, 14, 50–52, 54, 278–279
Soldier, 287, 361
Spalling, 17, 75–76, 102, 106, 136, 179, 240, 284, 296, 361
Stack bond, 4, 96, 126, 185, 200, 218, 231, 253, 273, 286, 304–305, 322, 351

T

Thermal expansion and contraction, 13, 23, 27, 58, 99–100, 103–104, 109, 143, 205, 227–228, 253, 257, 262, 288, 296
Ties, 55–57, 59–60, 62–64, 84, 126, 178, 260, 268–269, 296–297, 300, 312, 325–326, 332, 336, 350, 353, 363, 371
Tolerances, 5, 17, 29, 53, 109–110, 128, 241, 293, 296–297, 299, 313–314, 318, 322–325, 332, 335, 352, 355, 363
Tuckpointing, 45, 360, 369

V

Vapor (*See* Moisture, vapor)
Vapor retarder,
Veneer:
  adhered, 6, 31, 44, 205–213, 349
  anchored, 123, 152, 157, 163, 191–193

W

Walls:
  barrier, 123–125, 138, 175
  cavity, 58, 62, 70, 82, 84, 86, 90, 101, 123, 125–126, 152, 155
  composite, 175–176, 179
  curtain, 83, 96, 123–124, 143, 145, 159–160, 162, 165, 169, 178, 180, 185, 199–200, 265, 290–291, 354, 364
  drainage, 1, 122–126
  loadbearing, 14, 21, 42, 46, 50, 77, 92, 96, 125, 159, 171, 175, 217, 237, 251–274, 317, 322–323, 326, 367–368
  multi–wythe, 58, 62, 70, 82, 84, 86, 90, 101, 123, 125–126, 152, 155, 175–176, 179
  screen, 215, 219, 225–256, 361, 364
  serpentine, 224–225, 227

Walls (*Cont.*):
 single–wythe, 58, 73, 83, 86, 96, 125, 140, 159, 161–173, 176, 288
 veneer
  adhered, 6, 31, 44, 205–213, 349
  anchored, 123, 152, 157, 163, 191–193
Water repellents, 15, 21–22, 24, 38, 51–52, 73, 84, 125, 137–139, 151, 162–163, 172, 178, 208, 313, 331, 337, 364

Water vapor (*see* Condensation)
Waterproofing, 149, 155, 364 (*See also* Dampproofing)
Weep holes, 1–2, 44, 70–73, 76, 84, 108, 122–128, 131, 134, 138–141, 146, 149–154, 156, 161–162, 166, 169, 177–179, 184, 188, 195, 197, 199–201, 203, 207–208, 212, 275, 286, 288, 293–295, 312–314, 316–317, 328, 330, 332, 335–338, 364